Chemistry in the Oil Industry VII
Performance in a Challenging Environment

Chemistry in the Oil Industry VII
Performance in a Challenging Environment

Edited by

T. Balson
Consultant

H. A. Craddock
TR Oil Services Ltd, Dyce, Aberdeen, UK

J. Dunlop
JD Horizons Ltd, Macclesfield, Cheshire, UK

H. Frampton
BP Exploration Operating Co Ltd, Sunbury on Thames, Middlesex, UK

G. Payne
Briar Technical Services Ltd, Cults, Aberdeen, UK

P. Reid
Schlumberger Cambridge Research, Cambridge, UK

RS•C
ROYAL SOCIETY OF CHEMISTRY

The proceedings of the Chemistry in the Oil Industry VII meeting organised jointly by the RSC and EOSCA held at Manchester Conference Centre, Manchester on 13–14 November 2001.

Special Publication No. 280

ISBN 0-85404-861-8

A catalogue record for this book is available from the British Library

Published by The Royal Society of Chemistry,
Thomas Graham House, Science Park, Milton Road,
Cambridge CB4 0WF, UK
Registered Charity No. 207890

For further information see our web site at www.rsc.org

Printed by Athenaeum Press Ltd, Gateshead, Tyne and Wear, UK

Preface

The North West region of the Royal Society of Chemistry, Industrial Division held it's first Chemicals in the Oil Industry Symposium at Manchester University in 1983. This conference, reconvened once again in Manchester, represents the seventh in a series of highly successful industry events. After much debate the timing of this event, organised jointly by the RSC and EOSCA (European Oilfield Speciality Chemicals Association), was changed to November to better fit within a busy oil industry event calendar. It is anticipated that future Chemistry in the Oil Industry conferences will be held biannually in the Manchester area.

On this occasion the conference was organised in four main sessions – Environmental Issues, New Technology, Applications and Flow Assurance – reflecting the increasingly important role for additive technologies in offshore, deepwater and challenging environments allied to developments of low environmental impact chemistry. Keynote papers were presented by The Ministry of Economic Affairs, Netherlands and BP Exploration. In addition to the primary technical programme, the conference also hosted a poster session and an exhibition event supported by eleven oil industry technology companies.

Key themes for oilfield chemistry emerging from the conference and exhibition included:

- Increased cooperation between oil producers and additive technology suppliers
- Industry commitment toward development of low environmental impact chemistry
- New application developments, testing methodology and field deployment techniques
- Ongoing investments in chemistry R&D programmes for specialised oilfield applications, particularly related to offshore field developments

The RSC and EOSCA would like to express their thanks to the organising committee – Terry Balson (Consultant), Henry Craddock (TR Oil Services), Jack Dunlop (JD Horizons), Harry Frampton (BP), David Karsa (Akzo Nobel), Graham Payne (Briar Technical Services & EOSCA), Paul Reid (Schlumberger Cambridge Research) and Ruth Lane, Conference Organising Secretary – for their efforts and enthusiasm in reviving this industry conference. Additionally, the organisers would like to acknowledge the contributions of the conference sponsors – Baker Hughes, BP, Clariant, Drilling Specialities, Kernow Analytical Technology, Ondeo Nalco Energy Services, Schlumberger and TR Oil Services.

Dr Jack Dunlop
March 2002

Contents

Environmental Issues

New Technology

Applications

Flow Assurance

Environmental Issues

AN OVERVIEW OF THE HARMONISED MANDATORY CONTROL SYSTEM

L.R. Henriquez

State Supervision of Mines, Ministry of Economic Affairs, P.O. Box 8, 2270 AA Voorburg, The Netherlands

1 INTRODUCTION

Since the 1970s there has been a major concern by the public in general regarding the potential pollution of the North Sea marine environment by discharges of chemicals used in the offshore Oil Exploration and Production Industry (E&P).

In 1974 most of the North Sea countries experiencing these offshore activities signed the Convention for the Prevention of Pollution from land based sources (the so-called Paris Convention) which came into force in 1978. However the policy applied by the countries party to this convention about the prevention of pollution by the use and discharge of offshore chemicals until recently did suffer from harmonisation. For the discharge will take place in the same North Sea and consequently the potential pollution due to these discharges does not have boundaries.

First steps towards harmonisation started in 1985 with discussions about protocols how to carry out toxicity testing, for there was a lack in seawater tests. Ring tests carried on toxicity and biodegradability of offshore chemicals resulted in harmonised test protocols accepted by all countries party to the 1978 Paris Convention. This resulted in 1995 in the acceptance by all parties of the Harmonised Offshore Chemical Notification Format (HOCNF 1995). This format contains all necessary information for the assessment and evaluation of offshore chemicals prior their use and discharge.

Meanwhile a risk based approach for the assessment and evaluation of the use and discharge of offshore chemicals became a more important instrument. At the 4th International Conference on the Protection of the North Sea, the Ministers agreed to invite Paris Commission to adopt a Harmonised Mandatory Control System (HMCS). This should be adopted if possible at the Paris Commission Meeting in 1996, taking into account of the Chemical Hazard Assessment and Risk Management (CHARM).

This resulted in the PARCOM Decision 96/3 on a Harmonised Mandatory Control System for the Use and Reduction on the Discharge of Offshore Chemicals having the HOCNF as an Annex to the decision.

In 1998, a new convention, called the Convention for the protection of the marine environment of the North - East Atlantic, succeeded the 1978 Paris Convention. All decisions taken by Contracting Parties to this so-called 1998 OSPAR Convention shall have binding force. Consequently, the PARCOM Decision 96/3 had to be adapted to the new binding condition, which resulted in 2000 in a new OSPAR Decision 2000/2 on

HMCS. This paper gives an overview of this new decision and the latest developments with regard its implementation within the framework of the OSPAR Convention.

2 THE OSPAR CONVENTION, STRATEGIES AND DECISIONS

The OSPAR Convention entered into force on 25 March 1998. It replaces the OSLO and PARIS Conventions of 1978. The Convention has been signed and ratified by all contracting parties (15 countries including the EU) to the last mentioned conventions and by Luxembourg and Switzerland.

The long-term objective of the OSPAR Convention is to prevent and eliminate pollution and to protect the maritime area against the adverse effects of human activities so as to safeguard human health and to conserve marine ecosystems, where practicable, restore marine areas which have been adversely affected. For that, the OSPAR Commission will take all necessary measures to realise that objective. Contracting Parties to the Convention shall adopt programmes and measures which contain, where appropriate, time – limits for their completion. They should also take full account of the use of the latest technological developments (Best Available Techniques or BAT) and practices (Best Environmental Practice or BEP) designed to prevent and eliminate pollution fully.

In 1998, the OSPAR Commission adopted the OSPAR Strategy with regard to Hazardous Substances (Reference number: 1998 – 16) and in 1999 the OSPAR Strategy on Environmental Goals and Management Mechanisms for Offshore Activities (Reference number: 1999 – 12; the Offshore Strategy).

These strategies led to the new OSPAR Decision 2000/2 on a Harmonised Mandatory Control System for the Use and Reduction of the Discharge of Offshore Chemicals.

2.1 Framework of the OSPAR Convention

The OSPAR Convention also establishes the OSPAR Commission to administer the Convention and to develop policy and international agreements in this field. The Commission is supported by an International secretariat based in London. Information about the OSPAR Convention, the organisation and the OSPAR Secretariat can be found on the following website: http://www.ospar.org.

All official documents of the OSPAR Commission as the OSPAR Strategies, decisions, recommendations and agreements, can also be downloaded from the fore mentioned website.

2.2 Organisation – Committees and Working Groups of the OSPAR Commission

In 2000 OSPAR examined proposals for a new working structure and working procedures and agreed to retain the second tier Environmental Assessment and Monitoring Committee (ASMO), and to establish second tier Committees for each of the five OSPAR Strategies. OSPAR 2000 adopted net Terms of Reference for ASMO and Terms of Reference for the five strategy Committees, i.e. the Offshore Industry Committee or OIC.

The function of OIC is to facilitate the implementation of the OSPAR Strategy on Environmental Goals and Management Mechanisms for Offshore Activities (Reference number 1999-12) by the OSPAR Commission. In accordance with the OSPAR Action Plan, OIC shall:

• Identify the environmental pressures and their impact on the marine environment.

- Assess the effectiveness of programmes and measures and the need for and scope of further action.
- Develop the basis for programmes and measures.
- Develop programmes and measures.
- Assess the implementation of programmes and measures by Contracting Parties.

2.3 OSPAR Convention Mechanism

In Appendix 1 of this paper, a schematic view of the mechanism established by the OSPAR Convention is shown. The long-term objective and guiding principles are the basis for the strategies, i.e. the Offshore Strategy. This strategy, explained later in this paper, contains a management mechanism for setting goals and establishing programmes and measures to ensure the achievement of these goals within a specific timeframe. Goals should comply with the SMART (Specific, Measurable, Achievable, Realistic and Time limited) principles. The OSPAR Commission adopt programmes (plans) and measures, i.e. decisions or recommendations, to realise the goals. Contracting Parties should implement the programmes and measures in the national laws and regulations to ensure that the goals are met. Contracting Parties, on a yearly basis, should also report the progress of implementing these programmes and measures. If necessary, the OSPAR Commission decides on the actions to ensure a continuous improvement of the performance about the achievement of its overall long-term objective.

3 THE OSPAR STRATEGY WITH REGARD TO HAZARDOUS SUBSTANCES

3.1 Objective

The objective of the OSPAR Strategy with regard to hazardous substances is to prevent pollution of the maritime area by continuously reducing discharges, emissions and losses of hazardous substances as defined in Annex 1 of the Strategy. The ultimate aim is achieving concentrations in the marine environment near background values for naturally occurring substances and close to zero for man-made synthetic substances. At the Ministerial Meeting of the OSPAR Commission at Sintra in 1998 it was also agreed to make every endeavour to move towards the target of cessation of discharges, emissions and losses of hazardous substances by the year 2020 and adoption of this strategy in order to make this agreement operational.

3.2 Definitions

For the purpose of this Strategy, hazardous substances have been defined to be substances or groups of substances that are toxic, persistent and liable to bioaccumulate. The OSPAR Commission may also categorise other substances or group of substances as hazardous substances. Even if these substances do not meet all the criteria for toxicity, persistence and bioaccumulation, but which give rise to an equivalent level of concern.

The Strategy also defines substances and group of substances and toxicity. Toxicity is defined as the capacity of a substance to cause toxic effects, to organisms or their progeny such as:

a.   a reduction in survival, growth and reproduction;
b.   carcinogenicity, mutagenicity or teratogenicity;
c.   adverse effects as result of endocrine disruption.

Description of other definitions like persistent, bioaccumulation, bioconcentration, risk assessment, exposure assessment, hazard identification, dose – response assessment, risk characterisation and endocrine disruptor are also presented in the glossary of the Strategy (Annex 5).

3.3 Guiding principles

The OSPAR Hazardous Substance Strategy will use principles like the precautionary principle and the polluter pays principle as a guide. The application of Best Available Techniques (BAT) and Best Environmental Practice (BEP) should also be promoted when dealing with hazardous substances. In addition, the principle of substitution, i.e. the substitution of hazardous substances by less hazardous substances or preferably non-hazardous substances, where such alternatives are available, is a mean to reach this objective.

3.4 Strategy of OSPAR with regard to Hazardous Substances

Based on the strategy programmes and measures will be developed to identify, prioritise, monitor and control (i.e., to prevent and/or reduce and/or eliminate) the emissions, discharges and losses of hazardous substances which reach, or could reach, the marine environment. To this end, the OSPAR Commission will complete the development of a dynamic selection and prioritisation mechanism. In Annex 2 of the OSPAR Hazardous Substances Strategy, a list of chemicals for priority action has been agreed upon initially at Sintra in 1998. Meanwhile this list has been up-dated at the last OSPAR Commission Meeting in Valencia in 2001 (Reference number 2001-2). The OSPAR Commission also discussed cut-off Values for the selection criteria used in the initial selection procedure of the OSPAR Dynamic Selection and Prioritisation Mechanism for Hazardous Substances.

3.5 Cut off values

At its Commission Meeting in Valencia (25–29 June 2001) OSPAR agreed on cut-off values for the intrinsic properties of individual substance (Reference Number: 2001-1). These are specifically whether the substances are persistent (P), toxic (T) or liable to bioaccumulate (B), which determine whether these substances fall within the definition of hazardous substances given in the OSPAR Hazardous Substances Strategy. These PTB criteria are used for selecting substances in the initial selection procedure of the dynamic selection and prioritisation mechanism. The cut-off values are as follows:

Persistent (P):   Half-life ($T_{\frac{1}{2}}$) of 50 days **and**
Liability to Bioaccumulate (B): log P_{ow}>=4 or BCF>=500 **and**
Toxicity (T) T_{aq}: acute $L(E)C_{50}$=<1 mg/l, long-term NOEC=<0,1 mg/l
        or $T_{mammalian}$: CMR or chronic

By applying these values OSPAR will continue to select substances for priority action in coming years in order to meet its objective by 2020.

4 OSPAR STRATEGY ON ENVIRONMENTAL GOALS AND MANAGEMENT MECHANISMS FOR OFFSHORE ACTIVITIES (OFFSHORE STRATEGY)

4.1 Objective

To achieve the general objective of the OSPAR Convention, the aim of the Offshore Strategy is to set environmental goals for the offshore oil and gas industry and to establish of improved management mechanisms to prevent and eliminate pollution. If necessary to take measures to protect the maritime area against adverse effects of offshore activities so as to safeguard human health and to conserve marine ecosystems and, when practicable, restore marine areas which have been adversely affected.

Other OSPAR Strategies, like the Hazardous Substances Strategy, apply in so far as they relate to offshore activities.

4.2 Guiding principles

Besides the other mentioned guiding principles the Offshore Strategy is also referring to the application of the principle of sustainable development and principles agreed in the new Annex V of the OSPAR Convention on Biological Diversity (Biodiversity). Waste management should be based on the application of the hierarchy of avoidance, reduction, re-use, recycling, recovery, and residue disposal (the 5 R's hierarchy).

4.3 General process of establishing goals and measures

In addition to work in hand, the OSPAR Commission will establish and periodically review environmental goals and timeframes for achieving the objective of this strategy. These goals should be in measurable terms, wherever practicable, in order to facilitate monitoring. To this end, the Commission by its Ministerial Meeting in 2003 will take the following intermediate steps:

(i)   establish environmental goals, and, where appropriate, intermediate goals, in respect of prevention and elimination of pollution from offshore sources;

(ii)   provide for the machinery required for implementing and enforcing any programme or measure adopted under this strategy.

The Commission with the support of the Contracting Parties concerned will promote the development and implementation by the offshore industry of environmental management mechanisms. These mechanisms should include elements for auditing and reporting, which are designed to achieve both continuous improvements in environmental performance and the environmental goals referred to here above.

4.4 Implementation of the Offshore Strategy

The strategy will be implemented and developed under the OSPAR Commission's Action Plan, which will establish priorities, assign tasks, and set deadlines to make the best use of resources. The Action Plan will concentrate on those offshore activities identified as being of greatest concern to the marine environment like:

- the use and discharge of hazardous substances, consistent with the OSPAR Hazardous Substances Strategy;
- discharges of oil and other chemicals in water and from well operations.

The implementation of the Offshore Strategy will be through the developing of programmes and measures by the Commission. The following programmes and measures have already been adopted for the above mentioned priority issues.

4.5 An overview of OSPAR Measures in place

The OSPAR Commission adopted the following measures during the last 2 years:

4.5.1 OSPAR Decision 2000/3 on the Use of Organic-Phase Drilling Fluids (OPF) and the Discharge of OPF-Contaminated Cuttings. The objective of this decision is to regulate the use and discharge of drilling fluids based on mineral oils and synthetic fluids. The main guidance principle applied here are the BAT and BEP principles based on the application of the 5 R hierarchy principles for waste management. The use of diesel oil is prohibited. The discharge into the sea of cuttings contaminated with OPF at a concentration greater than 1% by weight on dry cuttings is prohibited. In exceptional circumstances, discharge of contaminated cuttings with synthetic fluids may be authorised by competent authorities. This is only allowed on a case-by-case basis and the based of criteria which take into account the toxicity, biodegradability and liability to bioaccumulate of drilling fluid concerned and of the hydrography of the receiving environment.

4.5.2 OSPAR Recommendation 2001/1 for the Management of Produced Water from Offshore Installations. The objective of this recommendation is to reduce the input of oil and other substances into the sea resulting from produced water discharges from offshore installations. For the first time in the OSPAR history, the Commission adopted a goal, which is to reduce the total quantity of oil by 2006 with 15% compared with 2000. This goal can be achieved either by reduction of the volume of produced water discharged or by lowering the concentration of oil in the discharge water. Besides that, OSPAR agreed to adopt a new performance standard of 30 mg/l for 2006 to replace the 1978 standard of 40 mg/l. This recommendation also addresses a programme to reduce the input of aromatic hydrocarbons and other substances (like heavy metals) into the sea.

4.5.3 OSPAR Decision 2000/2 on a Harmonised Mandatory Control System for the Use and Reduction of the Discharge of Offshore Chemicals (HMCS). The objective of this decision is that authorities shall ensure and actively promote the continued shift towards the use of less hazardous substances by application of the main guiding principle of substitution. Preferably, this should result in the use of non-hazardous substances or a reduction of the overall environmental impact from the use and discharge of offshore chemicals. A more detailed overview about this decision will be presented in the following chapters of this paper.

5   THE HARMONISED MANDATORY CONTROL SYSTEM (HMCS)

5.1 Definitions

The HMCS decision defines the following descriptions like CHARM, Generic PEC / PNEC ratio, hazardous substances, substance and preparation, offshore chemicals, P_{ow}, use and discharge. It also refers to the OSPAR Recommendation on Harmonised Offshore Chemical Notification Format or HOCNF. The OSPAR List of Substances / Preparations

Used and Discharged Offshore, which are considered to Pose Little or No Risk to the Environment or PLONOR, is also part of these definitions.

5.2 Guiding principles

Besides the principle of substitution, this HCMS Decision is also prescribing those authorities to follow the following guiding principles:

- To avoid emissions, discharge and losses of new hazardous substances, or preparations containing hazardous substances, except where the use of these substances / preparations is justified by the application of substitution;
- To encourage the development of less hazardous substances and preparations, and techniques for minimising the discharge of hazardous substances;
- To encourage the reduction of the uses and discharge of substances and preparations from offshore installations that might otherwise be harmful to the marine environment, such as substances causing taint or oxygen depletion.

In doing so authorities shall take health, safety and economic factors and technical performance into account, as appropriate. Processes, methods and equipment, which might lead to, lowered use and discharge of chemicals or the use and discharge of less hazardous chemicals shall be taken into account when assessing substitutes.

5.3 Data requirements

Any application to an authority for the use and discharge of offshore chemicals shall include information as mentioned in the OSPAR Recommendation 2000/5 on a Harmonised Offshore Chemical Notification Format (HOCNF). OSPAR also has also issued guidelines for completing the HOCNF format, and for toxicity testing of substances and preparations used and discharged offshore.

The HOCNF format contains information about the application, the amount of use and discharge and the necessary PTB data. It also gives information on tainting and other properties like carcinogenicity, mutagenicity, teratogenicity or endocrine disruption. The EU SDS data are also mentioned in the HOCNF format. Based on the HOCNF information competent authorities have to carry out an assessment and evaluation of the offshore chemical before use and discharge into the sea by applying a pre-screening and ranking mechanism. Before explaining this mechanism an overview of the management decisions, which should be taken by the competent authorities based on the HMCS decision, is presented here.

5.4 Management decisions based on the HMCS and Pre-screening

Authorities shall take the following management decisions after assessment based on the pre-screening and ranking mechanism:

5.4.1 Permission. Permits or approval for the use and discharge of offshore chemicals may contain conditions e.g. regarding the amount to be discharged, period of validity etc.

5.4.2 Substitution. Taking into account the outcome of the pre-screening and ranking mechanism, competent authorities may request operators to apply a substitute for the offshore chemical. Alternatively, if deemed necessary, the operator may be requested to provide additional data.

In case of substitution for economic or performance reasons the generic PEC / PNEC ratio of the substitute and the overall environmental impact associated with its use and discharge shall be lower than, or equal to, that of original offshore chemical.

5.4.3 Temporary Permission. Authorities shall grant a temporary permission for a maximum period of three years, whilst a less hazardous (or preferably non-hazardous) substitute is sought. In case of substitution for non-environmental reasons (e.g. for reasons of safety, health, or technical performance) and if the generic PEC / PNEC ratio of the substitute and the overall impact associated with its use and discharge is higher then that of the original chemical authorities may issue a special temporary permission for a maximum of three years.

5.4.4 Refusal of permission. Authorities refuse permission for those offshore chemicals, which they consider unsuitable for use and discharge offshore.

5.4.5 Pre-screening. All offshore chemicals shall be subject to a harmonised pre-screening (on a substance by substance basis, where possible) in accordance with the pre-screening criteria adopted in the OSPAR Recommendation 2000/4 on a Harmonised Pre-screening Scheme for Offshore Chemicals. The Pre-screening scheme, as shown in Appendix 2 of this paper, is not only presenting the criteria but also gives the necessary input for the management decision process. Based on this input the competent authorities regulate the use and discharge of offshore chemicals by giving (temporary) permission or refusal or act with regard to substitution. The Pre-screening scheme contains boxes to guide the assessment of the offshore chemical before the evaluation for use and discharge. The information on use and discharge is given in § 1.3 and § 1.4 of the HOCNF. This assessment is dependable on this information, e.g. the application function (i.e. a drilling or production chemical) or whether the chemical is applied in an open or closed system etc. In the next chapters, these guiding boxes will be explained.

Start

Before starting the assessment the authority may already decide, based on the preliminary information received about the offshore chemical, that further actions are not necessary because of the hazardous properties of the substances or preparations concerned.

Is the substance on the PLONOR list?

As mentioned earlier this list contains substances or preparations, which are considered to pose little or no risk to the environment. An important issue here is that the substances or preparations listed do not need to be strongly regulated as, from experience of their discharge. This list includes natural constituents of seawater, natural products e.g. nutshells, and other substances / preparations where some relevant toxicity data is available. In § 1.6 of the HOCNF format this information is given. In case the offshore chemical is listed on the PLONOR list then the decision whether to discharge or not is dependable on the receiving environment. In that case, the following box refers to an expert judgement by the competent authorities for deciding that.

Expert Judgement
Authorities may use expert judgement to regulate the discharge of PLONOR listed substances. In accordance with the precautionary principle competent authorities should take into account sensitive areas and the discharge amounts of chemicals which may have unacceptable effects on the receiving environment. These criteria are the input to determine whether to give permission or to refuse permission. Examples are e.g. the fact that the location where the discharge will take place may be an area of special environmental concern, like coastal or fisheries zones on the Continental Shelf of the North Sea.

Is the substance a Hazardous substance or of equivalent concern?

Information on this subject is given in § 1.7 of the HOCNF format. Substances listed on Annex 2 of the Hazardous Substances Strategy should be substituted. Authorities may also consider substances, i.e. heavy metals, organohalogen compounds, as mentioned on Appendix 2 of the OSPAR Convention of equal concern and should therefore be substituted. If there are no alternatives competent authorities may grant temporary permission for the use and discharge of these substances. CHARM can be used as a decision-supporting tool together with expert judgement by the competent authority. The outcome of CHARM may determine the time limit of the temporary permission and other permit conditions, i.e. the concentration or amount discharged. Other permit conditions may also refer to the encouragement of the development of less hazardous or preferably non-hazardous substances or preparations for that specific application.

Is the substance inorganic?

If the substance is inorganic then the toxicity level determines whether the substance should be substituted or not (information in § 1.6 of the HOCNF format). Again, in this case permission is given after an expert judgement by the competent authority.
In case the substance is organic, all mandatory information required by the HOCNF should be completed for further assessment and evaluation by the competent authority.

Is biodegradation of substance < 20% in 28 days?

Ecotoxicological information on biodegradation, bioaccumulation and toxicity are given in Part 2 of the HOCNF format.
Aerobic biodegradability tests are mandatory for all organic substances. If this readily biodegradation test results in a biodegradation lower then 20% then the substance is assumed to be persistent. In such a case, the substance should be substituted, unless an aerobic inherent biodegradation test shows the opposite. Most OSPAR countries also consider the outcome of an inherent biodegradation test of less than 20% to be persistent (e.g. polymers). Substances ended up in the sediment are considered persistent unless an anaerobic biodegradation test shows the opposite.

Does the substance meet 2 of the 3 criteria?

The eco-toxicological characteristics of the substance should not meet 2 of the 3 following criteria; otherwise, it is liable for substitution:

- Ready biodegradation of 70% or 60% in 28 days dependable on the test protocol
- Bioaccumulation potential (or log Pow) greater or equal to 3 or BCF greater then 100 taking into account the molecular weight
- Toxicity level LC_{50} or EC_{50} lower then 10 mg/l.

Toxicity testing

Full OSPAR data sets on the following marine species tests are required:

a. *Skeletonema costatum* (algae)
b. *Acartia tonsa* (crustacean)
c. *Scophthalmus maximus* (fish) and
d. If the substance or preparation ends up in the sediment then instead of the *Acartia tonsa* test a sediment reworker test with the *Corophium volutator* should be carried out.

continued . . .

continuation ...

The OSPAR Commission adopted the Protocols on Methods for the Testing of Chemicals Used in the Offshore Industry, which can be found on the OSPAR website.
Within the OSPAR framework, it is agreed to carry out toxicity testing on preparation level. Toxicity testing on substances level is preferred and may be requested by the competent authorities.

Substances versus preparations

The OSPAR HMCS Decision 2000/2 requires that all offshore chemicals shall, where possible, is subject to a harmonised pre-screening on a substance by substance basis. Tests of Biodegradation and bioaccumulation tests are carried out on substance level. However, toxicity tests are carried out on preparation level and not on a substance level. Since the pre-screening mechanism is subject to a substance by substance approach OSPAR should agreed on a harmonised way on how to estimate the toxicity on a substance level. This issue also plays an important role when applying the ranking mechanism, which is explained in the following chapter of this paper.

Scientifically impossible it is impossible to calculate the toxicity of the substance from the toxicity result on a preparation level. The latest proposal within the OSPAR framework is to use for the time being the following formula:

$$LC_{50} \text{ of substance } X = C_x \cdot LC_{50} \text{ of the preparation, where}$$

C_x = weight percentage of the substance X present in the preparation.

This formula represents a conservative approach, suggesting that any one of the substances solely is responsible for the measured toxicity. Other approaches have been discussed but they gave less conservative results however, there is no agreement yet within the OSPAR framework how to deal with this issue.

Ranking mechanism by applying CHARM

Ranking of the offshore chemicals according to the generic PEC / PNEC ratio gives an indication of the relative risks of these offshore chemicals. The PEC / PNEC ratio, referred to as the 'hazard quotient' in the CHARM model, shall be calculated by using the standardised reference oil / gas platforms and dilution factors as defined in that model. The CHARM 'hazard assessment' module shall be used as a primary tool for ranking. Other suitable assessment methods may be used additionally for comparative evaluation of the ranking. Generic PEC / PNEC ratios shall be used for ranking purposes only, and not as the sole factor to control the use and discharge of offshore chemicals.

continued . . . .

continuation ...

The results of these calculations, together with the uncertainty factors identified by CHARM, shall be taken into account by authorities when establishing:

a. a ranking list of offshore chemicals;
b. the appropriate regulatory actions.

According to the EU, ranking based on PEC / PNEC ratios should be done on substance by substance level. However, within the OSPAR framework toxicity testing is carried out on preparation level. For that it is agreed within the CHARM model to use a worst case approach which is given in formulas (30) and (31) of the model:

1) If both data for PEC and PNEC are available on substance level (*formula 30*):

$$HQ_{preparation} = Maximum \left[\frac{PEC_{substance\ i}}{PNEC_{substance\ i}}_{\ substance\ i\ to\ n} \right]$$

2) If data for PEC is available on substance level and data for PNEC is available on preparation level (*formula 31*):

$$HQ_{preparation} = Maximum \left[\frac{PEC_{substance\ i}}{PNEC_{preparation}}_{\ substance\ i\ to\ n} \right]$$

6 CONCLUSION

6.1 Level of harmonisation within the OSPAR Framework

Within the OSPAR framework, so far most of the requirements with regard to the OSPAR HMCS Decision are harmonised. The OSPAR competent authorities require from operators of oil and gas installations or suppliers of chemicals the same information for their assessment of offshore chemicals prior their use and discharge. Protocols on testing of eco-toxicological properties of those chemicals are harmonised. Even there is an agreement about the pre-screening and ranking mechanism to support decision making by the competent authorities in relation to the management of the use and discharge of offshore chemicals.

However, there are still issues unresolved about harmonisation among the OSPAR Countries. In the following paragraphs, some of these issues are addressed here.

6.2 Substances versus preparations

As mentioned earlier within the framework of OSPAR toxicity testing is carried out on preparation level and not on substance by substance level. According to the EU and to eco-toxicity scientists in general, the assessment of risks based on the PEC / PNEC ratio is only valid when all eco-toxicological data on substance level are available. The reason is that up to now it is impossible to assess the interactions within the formulation of the preparation, which is a mixture of substances, in relation to its fate and effects on the marine environment when discharged. Therefore all eco-toxicological and risk models derived to date are based on substance level.

Consequently, when applying the pre-screening and ranking mechanism on preparation level as proposed within the OSPAR framework, may lead to false management decisions

6.3 Surfactants

Bioaccumulation potential using the protocols for determining log P_{ow} is not applicable for organic substances having surface-active properties. Therefore when applying box 6 of the pre-screening scheme these substances can only be assessed on the basis of their biodegradation rate and toxicity and not on their bioaccumulation potential.

According to § 2.5 of the HOCNF format, measured adsorbability (K_{oc}) data for surface-active substances are mandatory. However, up to now no standard method for determining the adsorbability is agreed but within the OSPAR framework some proposals have been made for determining the K_{oc}.

6.4 Polymers

Polymers, most of the time, have a biodegradation rate lower than the specified 70% in 28 days (or 60% in 28 days dependable on the test method). Some OSPAR countries do require an inherent biodegradation test. In that case the inherent biodegradation rate should also exceed 20% in 28 days or otherwise be liable for substitution. However, there is no agreement yet within the OSPAR framework on this issue.

6.5 Goal setting mechanism for substances / preparations

As described in chapter 4 of this paper a general process for establishing goals and measures have been agreed within the Offshore Strategy.

At the Meeting of the Offshore Industry Committee (OIC) in Oslo (13 – 16February 2001), reference was made to the evaluation process of environmental performance used in the ISO 14031. Schematically this evaluation process as described by ISO 14031 is shown in Appendix 3 of this paper. A similar process will be followed for evaluating progress on any goal set by OSPAR. Consequently, indicators must be specified for achieving those goals and the supporting data to be collected and analysed.

Thus, the first step for any goal set is to select indicators for environmental performance evaluation. The second step is to collect data relevant to the selected indicators. Currently there is lack of data on offshore chemicals collected and reported to OSPAR to continue with the following steps. So far, analysis and conversion of the data into information for describing the environmental performance about the use and discharge of substances are not possible yet within the OSPAR framework.

6.6 Selecting indicators for goal setting

The ISO 14031 gives general guidance for selecting indicators as a mean of presenting quantitative or qualitative data or information in a useful form. The following characteristics may be required for selecting the indicators:

- Basic data can be directly measurable or based on calculations, i.e. such as in tonnes of hazardous substances used and discharged

- Data could also be relative to another parameter, e.g. such as in tonnes hazardous substances used or discharged per unit of production for a specific year.

The Netherlands so far have worked out some examples for such an approach. By using the pre-screening scheme and categorisation of those candidate substances liable for substitution or ranking the number of substances used offshore the Netherlands have been identified and selected as indicators for future monitoring. Appendices 4 and 5 show the results for this exercise. In Appendix 4 the apple pie represents a break down of total number of substances, as reported in 1998 to the competent authorities in the Netherlands. In due time if there is a progress based on the continuous improvement mechanism the apple pie should show bigger green or blue coloured pieces. These pieces represent substances ranked in acceptable risk categories or substances categorised as PLONOR's. Consequently, a trend like that will prove that the OSPAR Programmes and Measures, i.e. the OSPAR Decision 2000/2 on HMCS, are working. Hence, the objective of the HMCS is to ensure and actively promote the continued shift towards the use of less hazardous substances by application of main principle of substitution. Preferably, this should result in the use of non-hazardous substances or a reduction of the overall environmental impact from the use and discharge of offshore chemicals. Appendix 5 is an example for the corrosion inhibitors in use offshore the Netherlands for 1998.

6.7 Collecting data

For the time being collecting of data may be the main objective to provide input for calculating values for the selection of indicators. For that, the Offshore Industry Committee agreed in 2001 to come up with formats for reporting by the OSPAR Contracting Parties dealing with the use and discharge of offshore chemicals. The aim of OIC in 2002 is to come with a mechanism for setting goals based on the data collected. A proposal will be forwarded to the Ministerial Meeting in 2003 for establishment of goals or where appropriate intermediate goals on this issue.

Appendix 1

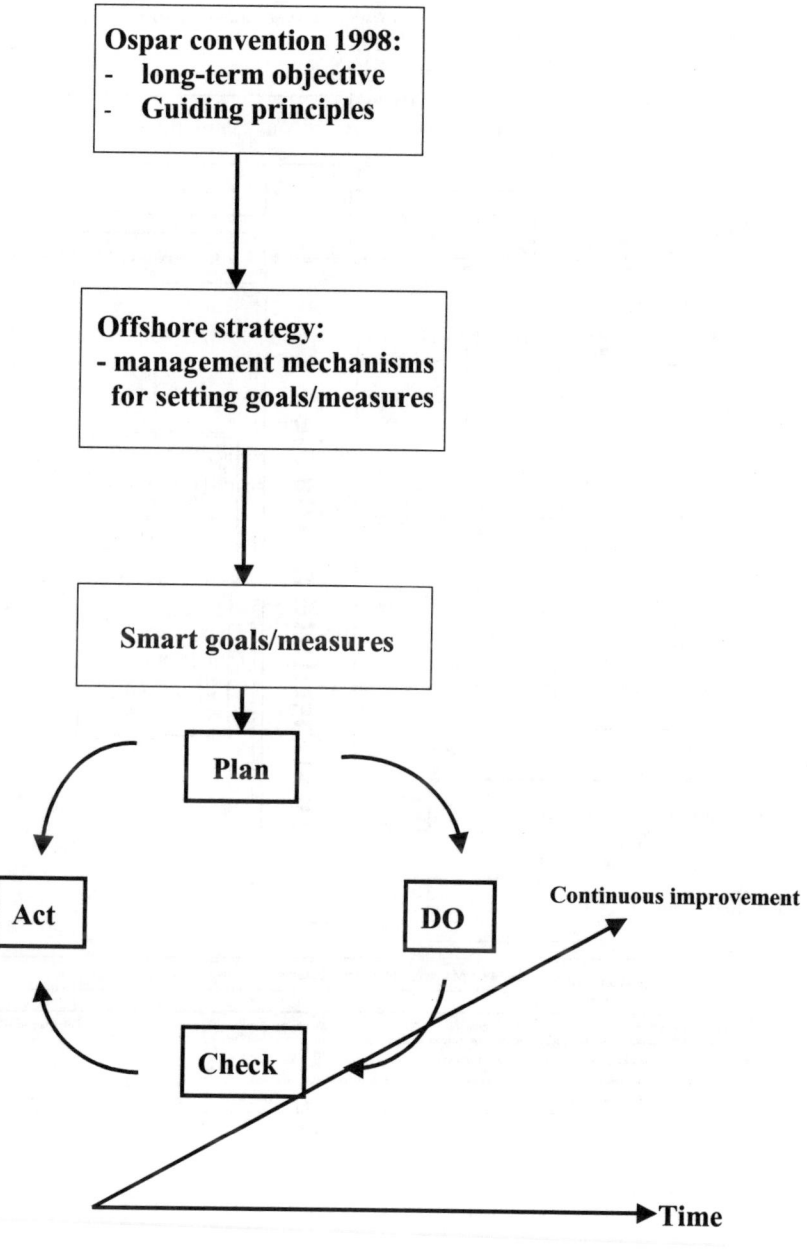

Appendix 2

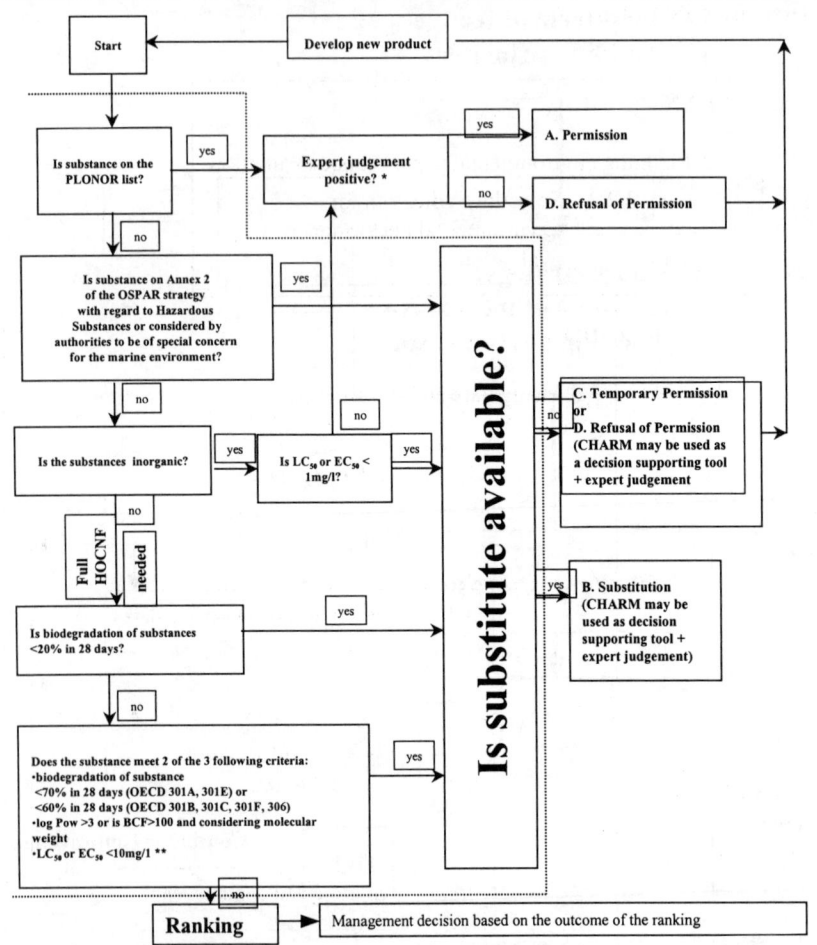

The Harmonised Pre-Screening Scheme (dotted part) as Part of the Whole Harmonised Mandatory Control System for Offshore Substances set out in the applicable OSPAR Decision (Appendix 1)

* In accordance with the precautionary principle, expert judgement on a PLONOR substance should take into account sensitive areas, where the discharge of certain amounts of such a PLONOR substance may have unacceptable effects on the receiving environment.

** if toxicity data are available only for a preparation, the authority should, on a case by case basis:

a. seek further information from the supplier to identify that substance, which is the major contributor to the overall toxicity of the preparation; or

b. use the toxicity data of the preparation to estimate the toxicity of a substance contained in it, taking into account the concentration of the substance in the preparation.

Appendix 3

Environmental Performance Evaluation

Extract from ISO 14031

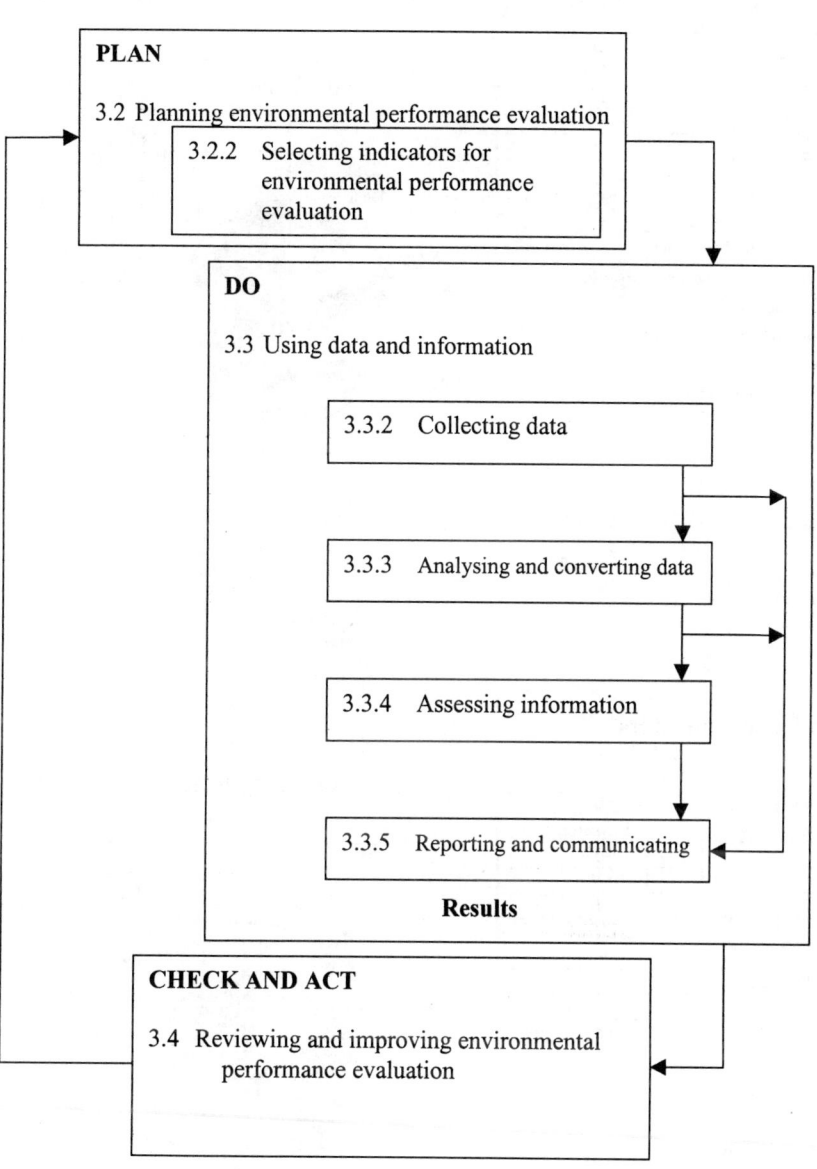

Appendix 4

Result of Table X 1998 for The Netherlands

Pre-screening category	Number of substances
A	1
B	0
C	11
D	27
R	14
PLONOR	3
NSD	13

Total 35 products Total 70 substances

A = HAZ Subst Strategy list

B = Inorganic < 1 mg/l tox

C = Organic < 20% biod

D = substitution 2 of 3 criteria

R = Ranking

NSD = No suff. data

ALL SUBSTANCES 1998

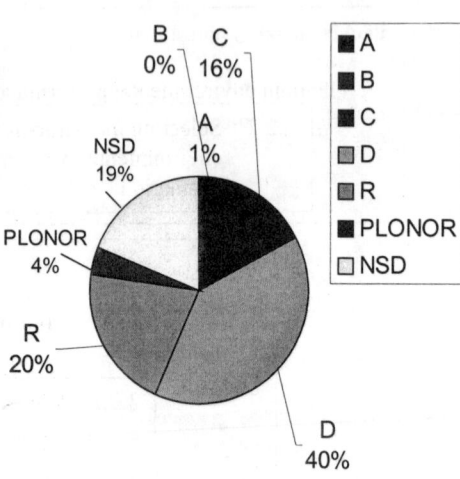

Appendix 5

Pie-chart Corr. Inhibitors

Result of Table X 1998 for The Netherlands

Pre-screening category	Number of substances
A	0
B	0
C	1
D	10
R	4
PLONOR	1
NSD	3

Total 9 products Total 19 substances

A = HAZ Subst Strategy list

B = Inorganic < 1 mg/l tox

C = Organic < 20% biod

D = substitution 2 of 3 criteria

R = Ranking

NSD = No suff. data

All COR. INHIBITORS 1998

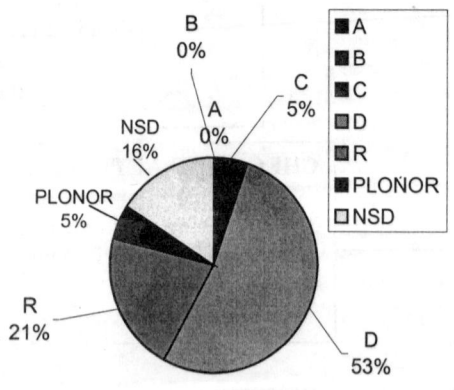

IMPACT OF THE OSPAR DECISION ON THE HARMONISED MANDATORY
CONTROL SYSTEM ON THE OFFSHORE CHEMICAL SUPPLY INDUSTRY

Melanie Thatcher[1] and Graham Payne[2]

[1]EOSCA, c/o Baker Hughes INTEQ, Barclayhill Place, Portlethen, Aberdeen AB12 4PF
[2]EOSCA, c/o Briar Technical Services Ltd, 501 North Deeside Road, Cults, Aberdeen
 AB15 9ES, UK

1 ABSTRACT

By November 2001, the OSPAR Decision on the Harmonised Mandatory Control System
(HMCS) should have been in force and operating for nearly a year. The objective of the
HMCS is to protect the marine environment by identifying those chemicals used in
offshore oil and gas operations with the potential for causing an adverse environmental
impact and restricting their use and discharge to the sea. Accordingly, this legislation will
drive the development and selection of offshore chemicals that have the lowest impact on
the marine environment. A series of associated Recommendations provide guidance on
how to compare the potential environmental impact of different chemicals. This involves
the generation of an environmental data set (i.e. toxicity, persistence and bioaccumulation
potential) and its evaluation using pre-screening criteria and a decision-support tool called
CHARM (Chemical Hazard Assessment and Risk Management) Model.
    While the Decision provides a standardised framework, those countries having oil and
gas operations in the Northeast Atlantic have implemented the requirements in different
ways. An account is given of the national schemes operating in the UK, Denmark, Norway
and the Netherlands that focuses upon the involvement of the chemical supplier. The
paper draws conclusions about the current and future impacts of the HMCS on the offshore
chemical supply industry.

2 INTRODUCTION

Contracting Parties to OSPAR, i.e. government agencies representing those countries
bordering the Northeast Atlantic are charged with protecting the marine environment of the
North Sea. In June 2000, the OSPAR Commission adopted Decision 2000/2 on a
Harmonised Mandatory Control System for the Use and Reduction of the Discharge of
Offshore Chemicals[1]. The aim of this legislation is to establish a consistent framework
within which the amount and harmfulness of chemicals used and discharged in the course
of offshore oil and gas exploration and production processes can be reduced. These
chemicals include those used for drilling, production, cementing, completions and
workovers.
    The common framework outlined in OSPAR Decision 2000/2 will be incorporated into
the National legislation of the contracting parties to OSPAR. The Decision is supported by

a number of Recommendations that describe how the Mandatory Control Scheme will work in practice and this is summarised in Figure 1. The responsibilities of the chemical supplier, operating company and regulatory agency differ according to the national sector in which the chemical is to be used and will be clarified later when the different national schemes are described.

Under the HMCS, a chemical developed for use on an offshore installation will not be permitted without authorisation from the authorities of the intended sector of the North Sea. If a new product is to be considered for use, the first step is to complete a standard form known as the Harmonised Offshore Chemical Notification Format or HOCNF which is described in Recommendation 2000/5[2]. The HOCNF requires details of the chemical composition, information on the quantities to be used and discharged, how the chemical will be applied, and the environmental properties of the products including toxicity to aquatic organisms and the fate and effects of component substances.

3 ENVIRONMENTAL TESTING

The toxicity tests to be conducted are specified in the guidelines accompanying Recommendation 2000/5. Those marine species selected for the scheme not only represent

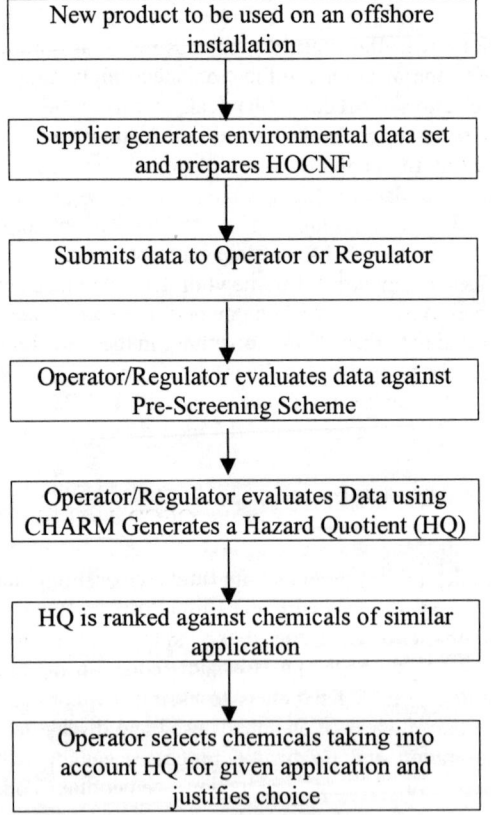

Figure 1   *Outline of the Harmonised Mandatory Control Scheme*

different physical positions within the marine environment (i.e. water surface, water column and seabed), but also represent links in the food chain i.e. fish feed on crustacea which feed on algae.

The usual toxicity tests conducted for the registration process and their typical costs are given in Table 1.

Table 1　　*Environmental Tests required under the HMCS and Typical Cost*

Test Required	Test protocol	Typical Cost (£)
Algae	72hr EC_{50}: *Skeletonema costatum* ISO/DIS 10253	950
Crustacean	48 hr LC_{50}: *Acartia tonsa* ISO TC147/SC5/WG2	850
Fish	96hr LC_{50}: *Schophthalamus maximus*, juvenile OECD 203 modified for marine species	960
Crustacean – sediment reworker	10 day LC_{50}: *Corophium volutator* PARCOM	900
Biodegradation – Water soluble substances	28 day aerobic, marine OECD 306	660
Biodegradation – Water insoluble substances	28 day aerobic, marine BODIS (BOD for Insoluble Substances)	660
Bioaccumulation Potential	Octanol/water partition co-efficient (log P_{ow}) OECD 117 or 107	400 or 900

The tests on the water-dwelling species (*Skeletonema, Acartia* and *Schophthalamus*) are mandatory whereas the sediment reworker test is conditional upon the possibility that the chemical will reach the seabed. Other test species are permitted and these are outlined in the Draft OSPAR Guidelines for Toxicity Testing of Substances and Preparations Used and Discharged Offshore[3].

Biodegradation data on each deliberately-added organic substance is required in addition to the toxicity tests. Two 28-day aerobic marine protocols are preferred: OECD 306[4] and the BODIS test[5].

Bioaccumulation potential data on each deliberately-added organic substance is also required. Most commonly, the test conducted is the OECD 117[6] HPLC test although OECD 107[7] is also accepted for pure substances, and the blue mussel bioconcentration factor test OECD 305[8] was required for synthetic base fluids for drilling muds.

It is only substances which appear on the PLONOR list[9] (formerly the PARCOM A list) that are not required to be tested as described above. PLONOR substances are those considered to Pose Little Or NO Risk to the environment and their environmental effects are considered to be well known. Over 100 substances appear on this list.

4 PRE-SCREENING SCHEME

Once the HOCNF is complete, it is passed to the Operator or Regulator for appraisal of the environmental profile of the product. The first phase of the assessment will be to evaluate the data against the Pre-Screening Scheme. This is a flow-chart outlined in OSPAR Recommendation 2000/4[10]. There are a number of possible outcomes from the flow-chart. A PLONOR substance will generally receive immediate approval although special features of the receiving environment e.g. fish spawning season, may dictate conditions for use. Conversely, a few substances e.g. those appearing on Annex 2 to OSPAR Strategy with regard to Hazardous Substances[11] may be prohibited from use.

The remaining offshore chemicals will go to one of two other outcomes. Those substances having a low rate of biodegradation, or a combination of this with low toxicity or high potential for bioaccumulation will go to the "Substitute" box. The Operating Company would be expected to try to find an alternative product for the same application, but which has a better environmental profile. If an alternative can not be found, temporary permission for use of the product will be granted. The duration of the temporary permission will range between 6 months and 3 years depending upon the level of concern about the potential environmental effects of the substance.

Those substances which pass through the scheme to the "Ranking" box of the flow-chart and those given temporary permission go to the second stage of the assessment. This involves evaluation by CHARM (Chemical Hazard Assessment and Risk Management) model.

In considering the impact of the pre-screening scheme on the Chemical Supply Industry, we can look to the evaluation that CEFAS (Centre for Environment, Fisheries and Aquaculture Science) performed about a year ago. They reviewed 1990 oilfield chemicals in their database of registered products to determine the proportion of chemicals arriving at each outcome from the flow-chart. This breakdown, which is based upon the current environmental data available for the products is given in Table 2.

The table indicates that a proportion of chemicals will go to the "Substitute" box. These are predominantly products containing substances having a low rate of biodegradation and are mostly of a polymeric nature. It will be very difficult to find alternatives to these in the short term, but this is the future challenge for the industry.

Table 2   *Proportion of oilfield chemicals arriving at different outcomes of the pre-screening scheme*

Rebrand of Substances or Products containing Substances...	*Number of Chemicals*	*Percentage (%)*	*Pre-Screening Outcome*
PLONOR chemicals	604	30	Permitted for use
Listed on Annex 2 to OSPAR Strategy on Hazardous Substances	43	2	Prohibited for use
Rebrand of inorganic substances (if LC/EC_{50} >1 mg/l)	119	6	Expert Judgement
Products containing inorganic substances (if LC/EC_{50} >1 mg/l)	398	20	Expert Judgement
Biodegrade <20% in 28 days	615	31	Substitute
Meets 2 of the 3 criteria	193	10	Substitute
Go to Ranking	377	19	CHARM Assessment

5 CHARM

The CHARM model comprises a set of calculation rules from which the outcome is a single number that represents the likelihood that a chemical will cause harm when used and ultimately discharged into the marine environment. The outcome is called a Hazard Quotient (HQ) and this represents the ratio of the Predicted Environmental Concentration (PEC): Predicted No Effect Concentration (PNEC).

There are different sets of calculation rules for production chemicals, surfactants, water-based drilling muds, cementing, completion and workover chemicals that reflect the different ways that they are applied on the offshore installation. These are described in the CHARM User Guide[12]. Where a chemical is made up of a number of component substances, a CHARM assessment is run on each substance.

The input information needed to calculate the HQ for each substance comprises only the environmental data, the dose rate and the percentage of the substance in the preparation or mixture. Parameters representing a standard gas or oil platform or drilling rig are kept the same for all assessments of the HQ. The dose given must represent that dose which would provide optimal technical performance under the conditions of the standard platform/rig and so might not reflect actual dosages being used in the field.

Since these "standard installations" do not exist, dose rates for them must be of a somewhat arbitrary nature especially where new and possibly untrialed products are concerned. Where an actual dose rate can be shown to be significantly different from that for the standard installation, then RQs should be generated and compared rather than HQs.

The CHARM model also does not fully cover all offshore operations in which chemicals are used. Activities like downhole scale inhibitor squeeze treatments do not fall easily into the current production chemical usage proforma within CHARM. The CHARM model will need to develop further to encompass such everyday operations as this and others not presently catered for. Involvement of the chemical supply companies in this further development of CHARM is essential if these are to be accurately reflected in the model.

5.1 Hazard Quotient Ranking

The generation of a single number (HQ) for each substance in principle, means that the environmental properties of two substances can be directly compared and gives an Operator visibility to select the chemical having the better environmental performance.

The significance of HQs and the inherent uncertainties in the numbers generated must be fully understood. There will be some who will take the HQs calculated as definitive and will differentiate between products having HQs of, say, 1.3 (being "bad" as it is greater than 1) and 0.7 (being "good" as it is less than 1). Uncertainty analysis for production chemicals has shown that the 90% confidence interval for each HQ can be set at HQ/3 and HQ*3 for the lower and upper limits[13]. For a product with an HQ of 1 these become a range between 0.33 and 3.0. Therefore, to differentiate between products with HQs in this range can not really be justified. Similar uncertainty analysis is being considered for drilling chemicals.

5.2 Justification for Use/Risk Assessment

The operating company must justify the selection of the suite of chemicals to be used on a drilling rig or production platform to the authorities. Of course, the environmental effects of the chemical in the marine environment are only one parameter in a number of considerations that must be given to the selection. Most importantly, the chemical must perform effectively. Factors such as human health effects and cost should also come into the equation.

The CHARM model will permit a site-specific assessment of risk by allowing the user to enter actual rig or platform-specific data. The Risk Quotient (RQ) resulting from this set of calculations can be used to assist the selection process. However, only the UK authorities have stated that the use of the risk assessment module in CHARM will be acceptable as part of the justification process.

6 IMPLEMENTATION OF THE HMCS INTO NATIONAL LEGISLATION

The four countries that have major offshore activities in the North Sea are Denmark, The Netherlands, Norway and the UK. Each is in the process of incorporating the HMCS into their national legislation. Slight differences exist in the way that each country is operating the scheme and this is commented on below. Comments are made on how these differences affect the offshore chemical supply industry.

6.1   Denmark

In Denmark, the Danish Environmental Protection Agency (EPA) and operators are working closely on implementation of the HMCS. This will be initially by administrative action and then via an amended marine law (the present marine law does not allow regulation of chemical use, only discharge). The new law will be laid before the Parliament before the end of this year and the administrative process for all offshore chemicals should be in place by 2002.

The registration process is described in a document entitled "New rules and requirements concerning offshore chemicals used in the Danish Sector of the North Sea"[14]. It involves the submission of the completed HOCNF with full composition to the Danish Product Register who assigns a Pr-number if acceptable. The data will be entered into a database which the Danish EPA have access to. Products currently in use must be re-registered over the next three years on a prioritised basis.

For the environmental testing, Denmark is the most strict on requiring toxicity data at the substance level. This has huge cost implications. The testing cost for a demulsifier comprising four component substances would be nearly £20,000.

The chemical supplier will also give an HOCNF with generic composition to the Operating Company. This provides the information the Operator needs to perform the pre-screening and CHARM assessments. Health and Safety criteria will also be integrated into the decision-making process. Permits will be granted to operators for up to three years, depending upon the outcome of these evaluations. These permits would apply across all installations operated and would not include site-specific evaluations. The authorities, however, could still impose site-specific conditions, regulating the use and discharge of chemicals based on their intrinsic properties rather than an assessment of risk to the marine environment.

CHARM is only to be used for generic ranking purposes. Only the "hazard module" is to be used. The site-specific risk assessment module is not permitted for use in justifying the selection of particular chemicals.

6.2   Netherlands

In the Netherlands, implementation of HMCS will be through a new mining law.   It is expected to come into force on 1 January 2002 and until then, the HMCS will be implemented by "administrative action".   The inspector general (IGM) of the State Supervision of Mines can write a so-called Order in Council that effectively means that use and discharge will be controlled via the HMCS. In 2002, the Inspector General of Mines for the Netherlands will issue a letter stating how the OSPAR Decision is to be implemented.   The appropriate OSPAR Decisions and Recommendations will be attached to this letter.

The HMCS will work under the framework of the Environmental Covenant within a broader goal of phasing out harmful substances by 2010.   The objectives of the Dutch approach will be to reduce progressively the use and discharge of all chemicals.

Operators in the Netherlands have begun to develop an inventory of chemicals.   At this early stage, it is clear that of 35 preparations in use on the Netherlands Continental Shelf, almost 50% of the individual substances are candidates for substitution.

6.3   Norway

The State Pollution Control Authority (SFT) who regulate the Norwegian sector of the North Sea have issued a draft of the new Norwegian regulations incorporating the HMCS within a broader HSE regulatory framework[15].   These new regulations are likely to be implemented at the end of 2001/early 2002.

In Norway, the chemical supply company prepares an HOCNF with the generic composition.   On behalf of an operating company, a copy is sent to Novatech who run the KPD Centre.   They quality-check the data and enter it into the Chems database that is available to operating companies and the SFT.

In Norway, a full HOCNF is required for each chemical additive even for closed system chemicals i.e. those that will not be discharged such as organic phase drilling fluids and pipeline chemicals.   In a drilling fluid, this could amount to more than 50 data points and a cost of over £50,000.   The value of all this data has to be questioned.   The UK require a reduced data set for organic phase fluids (Toxicity of the whole mud to *Corophium* and aerobic biodegradation on each organic substance) on the whole mud which is considered sufficient to give an indication of the environmental impact in the event of an accidental spillage.

The SFT also require substance level testing for organic substances having a biodegradability of <20% in 28 days.

The operating companies are granted Frame Permits by SFT and within these permits can select chemicals given consideration of their environmental profile.   The evaluation will include assessment according to the pre-screening scheme for products containing persistent and bioaccumulative substances in particular, a phase-out plan is agreed between the operating company and the chemical supplier.

CHARM is not as central to the process as in other countries.   The operators have developed a more advanced model for performing environmental impact assessments and this is increasingly used in Norway.

6.4   UK

In the UK, the new Offshore Chemical Regulations, 2001 will be issued by the Department of Trade and Industry (DTI) under the Integrated Pollution Prevention and Control Act.

Draft Regulations, a Regulatory Impact Assessment and a set of Guidance Notes have been reviewed in a public consultation process. Latest drafts of the Regulations and Guidelines are posted on the DTI website[16]. At present, the expected date for introducing the Regulations is the end of October.

Operators will be required to have a permit for use and discharge of offshore chemicals for each installation. A "grace period" of 3 months starting on the day the Regulations come into force will apply. The new permitting regime will be phased in. All existing installations will be granted 'deemed permits' (for up to two years) which will be called in for determination according to the size of the operation and the sensitivity of the area in which the operation is taking place. All new activities will need a new permit. Applications will include a 28-day public notice period when applicants have to signal their intention to seek a discharge through an appropriate medium.

Under the UK Offshore Chemical Notification Scheme, the HOCNF with full compositional information is sent to CEFAS. The guidelines for registering products for use in the UK sector are given on their website[17].

In terms of testing requirements, CEFAS are the most strict on having toxicity tests performed at the preparation level, arguing that this takes into account synergistic or antagonistic effects of combining chemicals. Unlike the other authorities, they are more flexible in permitting the submission of freshwater biodegradation data although the result is penalised in the CHARM assessment.

CEFAS will evaluate the data according to the pre-screening scheme and CHARM Hazard Assessment module. The product will be assigned to a colour band depending upon the CHARM HQ. It will be ranked against products having a similar application. The importance of colour banding ensures that, except at boundaries to the bands, small, insignificant differences in HQ are masked.

A "certificate" will be issued to the chemical supplier by CEFAS after completing the quality-check and the environmental assessment. This certificate will contain all the information required to for the operating company to use for in the risk assessment module of CHARM. CEFAS and the Fisheries Research Services in Aberdeen will evaluate the risk assessments.

Fees for data registration would be rolled into those for permit applications and be charged back to Operators applying for permits. For this reason, checks will be made to ensure that products registered under the new scheme are in use in the UK sector as it has been reported that products may be registered in the UK for use elsewhere in the world.

6.5   Other OSPAR Countries

Countries like France, Germany, Ireland and Spain who have limited exploration or production activity at present must still implement the HMCS and this may be problematic. For example, French law covers not only 'European France' but also a range of dependent territories where OSPAR regulations were 'inappropriate'. The authorities are asking companies to apply the OSPAR controls on its Atlantic coast. The new French administration will need to find a generic solution to this problem.

7   IMPACT ON THE OFFSHORE CHEMICAL SUPPLY INDUSTRY

The introduction of HMCS will impact upon the chemical supply industry in a number of ways.

The degree of harmonisation achieved by the framework of the HMCS is very positive for the chemical supply industry. Standardisation of the reporting formats (HOCNF), environmental test protocols, and the use of the pre-screening scheme and CHARM helps chemical suppliers to source the required data more efficiently. The transparency of the system enables suppliers to invest resources into products that will be more successful under the scheme i.e. those with good environmental performance.

On the other hand, the differences encountered in the National schemes lead to confusion for companies which register products for use in more than one country. Frequently, companies (particularly those handling registrations from the USA) believe that if they have registered a product in the UK, they can also sell it in the Netherlands or Norway and this is not the case.

Indeed, it may be the case, that the environmental data generated for registration in one county is not valid for registration in the others. For example, if toxicity data is generated on a preparation for the UK, it may not be accepted in Denmark. Additional testing to satisfy these differences adds to the compliance costs, not to mention the extra weeks needed to generate the data.

The increased cost of environmental testing (for example with the introduction of the mandatory fish test) could result in companies shortening their list of available products for use in the OSPAR area. As seen with the introduction of the Biocide Directive, the HMCS is likely to hinder the development of new products by reducing the level of research and development that companies are willing to invest in. This is contrary to the objectives of continual improvement through the HMCS. However, the cost of regulatory requirements must be out-weighed several fold by the return on sales otherwise companies will not invest.

While greener chemistries do exist, they are much more expensive than traditional products. For example, a "green" scale inhibitor has been developed, but is five times the price of phosphonates or polymers. Unsurprisingly, the industry is still to realise a market for these products.

Some countries e.g. UK are considering publishing the ranking list on the Internet. This is a concern for the chemical supply industry as it could have a significant negative commercial impact if misinterpreted or misused. First, the HQ is generated from a set of calculations using parameters of "standard installations" (water depth, tidal flow rate etc..). The conditions applied to this assessment may, therefore not reflect the actual conditions in which the chemical will be used in the field. The outcome of the standardised assessment may incorrectly suggest the chemical is a bad actor.

Second, the dosage parameter has a very strong influence in the calculation of HQ. Since the dose used is that for the "standard platform" rather than actual dose rate used, it is somewhat arbitrary. Visibility of the HQs of competitive products may enable the manipulation of the estimated dose to give a better HQ for the same product supplied by a different vendor. The dose rate for new products will need to be carefully scrutinised to ensure that it is not out of line with those of a similar composition and application.

8 FUTURE FOR THE OFFSHORE CHEMICAL SUPPLY INDUSTRY

Despite the concerns over testing costs, recent history has frequently shown that the chemical supply industry is developing increasingly environmentally acceptable chemicals. This is sure to continue as the relative positions on the hazard-ranking list will stimulate competition among companies for the best position.

The major challenge for the chemical supply industry is to develop products with high technical performance and good environmental performance. This is particularly difficult for corrosion inhibitors (traditionally comprising fairly toxic chemistries such as imidazolines and quaternary ammonium compounds) and demulsifiers that comprise persistent polymeric chemistries in organic solvents. Given time, alternatives will be found for these oilfield chemicals and others.

While the HMCS provides a common framework for OSPAR countries, it is clear that there are many differences in the way that the national schemes work in practice. In the future, we may see increasing harmonisation at the national level. For example, a common database of environmental data, a common OSPAR ranking list and a single registration authority for the OSPAR area have already been discussed.

References

1. OSPAR 00/20/1-E Summary Record, OSPAR Commission Meeting 26-30 JUNE 2000. http://www.ospar.org/eng/html/welcome.html
2. OSPAR Recommendation 2000/5 on a Harmonised Offshore Chemical Notification Format (HOCNF)
3. OIC 01/6/2-E Draft OSPAR Guidelines for Toxicity Testing of Substances and Preparations Used and Discharged Offshore presented by Denmark at the Oil Industries Committee meeting, February 2001.
4. OECD (1992): OECD Guideline 306 for Testing of Chemicals: "Biodegradability in Seawater – Closed Bottle Test", 17 July 1992.
5. ISO (1991): "BOD Test for Insoluble Substances – Two Phase Closed Bottle Test", TC/147.SC5/WG4 N141.
6. OECD (1989): "Partition Coefficients (n-octanol/water). High Performance Liquid Chromatography (HPLC) Method". OECD Guideline 117 for Testing of Chemicals, 30 March 1989.
7. OECD 107
8. OECD (1984): Guidelines for Bioaccumulation Testing 305A-E.
9. PLONOR list (1999-9) OSPAR List of Substances/Preparations Used and Discharged Offshore, which Pose Little or No Risk to the Environment (PLONOR)
10. OSPAR Recommendation 2000/4 on a Harmonised Pre-screening Scheme for Offshore Chemicals
11. OSPAR Strategy with regard to Hazardous Substances (1998-16). Annex 2: OSPAR List of Chemicals for Priority Action
12. Thatcher, M, L Henriquez and M Robson (2001): CHARM: User Guide. Ref http://www.ogp.org.uk/publications/index.html choose Non-OGP publications.
13. Karman, C.C and H.PM. Schobben (1996). CHARM Technical Note 42. "Uncertainty Analysis of the CHARM model".
14. New rules and requirements concerning offshore chemicals used in the Danish sector of the North Sea. Letter to all suppliers of offshore chemicals in Denmark, dated 15 January 2001.
15. Draft Norwegian Regulations http://www.sft.no/english/
16. Draft UK Regs and Guidelines http://www.og.dti.gov.uk
17. Guidelines for UK Offshore Chemical Notification Scheme http://www.cefas.co.uk/ocns/index.htm

THE DEVELOPMENT AND INTRODUCTION OF CHEMICAL HAZARD ASSESSMENT AND RISK MANAGEMENT (CHARM) INTO THE REGULATION OF OFFSHORE CHEMICALS IN THE OSPAR CONVENTION AREA; A GOOD EXAMPLE OF GOVERNMENT/INDUSTRY CO-OPERATION OR A WARNING TO INDUSTRY FOR THE FUTURE?

I. Still

M-I Drilling Fluids L.L.C., Dubai United Arab Emirates

1 INTRODUCTION

The OSPAR Convention area has recently implemented a Harmonised Mandatory Control Scheme (HMCS) for the approval for use/discharge of chemicals offshore. The main intention of the scheme is to enable the end-user to justify his selection of chemicals on the basis of their environmental performance, hence to actively promote the continued shift toward the use of less hazardous products. The scheme requires the Harmonised Offshore Chemicals Notification Format (HOCNF) data-set, and, following pre-screening, appropriate chemicals are evaluated using Chemical Hazard Assessment and Risk Management (CHARM) an empirical model. This is intended to allow the final selection for use of those chemicals with the lowest hazard quotient to be made.

The development of the CHARM model has required a considerable input from the offshore industry, especially EOSCA, and has taken nearly 10 years of continual effort to produce a workable regulatory tool. During this time the original concept for the use of the model has changed from a risk-based assessment using any suitable available environmental data to a hazard ranking involving a specific environmental dataset. This has been further compromised by the model being best able to handle data on a single substance basis, when the majority of chemicals used/discharged offshore are preparations composed of two or more substances. This issue has been further complicated with the general difficulty of adequately interpreting the bioaccumulation potential of surfactants, specifically for preparations containing two or more organic components that give a range of bioaccumulation values.

As a consequence, some chemical supply companies are becoming concerned that replacing the original risk based approach with hazard interpretation will result in unnecessary restrictions on the discharge and use of chemicals offshore where risk assessment could demonstrate that there is no environmental cost benefit to do so. This approach may meet the HMCS requirements but does not lead to any measurable reduction in the overall environmental impact resulting from these industrial activities.

2 THE CHARM MODEL

2.1 CHARM Concept

The CHARM concept was initiated in 1992 and proposed in a Memorandum of Understanding (MOU) between the Dutch and Norwegian offshore regulatory authorities. It was then discussed at the Seventeenth Meeting of the PARCOM Working Group on Oil Pollution in 1993, where the terms of reference for an ad hoc working group to elaborate a harmonised mandatory control system for the use and reduction of the discharge of chemicals were formerly set[1]. The offshore supply industry then became involved through EOSCA and latterly the UK offshore Regulatory authorities joined the development group. The budget for development of the model was 250,000 Euros, with EOSCA having to finance 20,000 Euros to take part in the development (approx 8.0% of costs). This sum amounted to an extra 50 % on the annual membership subscription for the EOSCA members for that year.

It is not the intention of this paper to chronicle all the meetings and changes to the original CHARM Model proposal. However, it should be mentioned that the development timetable was revised several times due to problems with the original software development, and changed following some irresolvable issues associated with the use and interpretation of the available environmental data. Additional costs for development were also required as the programme progressed, and EOSCA subsequently paid an additional 20,000 and 10,000 Euros for Phases II and III respectively.

2.2 CHARM Objectives

The requirements for CHARM were set out in the 1993 MOU Documents as follows:
The objective is to develop an environmental risk evaluation model for the use of offshore E & P production chemicals (including hazard assessment, risk analysis and risk management procedures), in order to guide the harmonisation of governmental regulations with respect to discharges of E & P chemicals. The system will be named CHARM (Chemical Hazard Assessment and Risk Management), as it will include both a hazard assessment and a risk analysis and management part.

The CHARM model is a tool to support the environmental evaluation of the use of E& P production chemicals on the basis of available data on these production chemicals and platform related conditions. CHARM will not cover the potential harm during the production and transport of chemicals or the handling of unused remainders. CHARM will only produce information on the potential harm to occur in the marine environment, i.e. the North and Norwegian Seas. Potential air pollution problems and human health problems are not within the scope of CHARM. In principle CHARM can handle more or less complete datasets. The user has to define the criteria for using CHARM, i.e. with respect to specific data requirements, such as PARCOM test results and whether or not CHARM has to be used if the chemical will not release into the environment when carefully used, e.g. in a closed system. The user has also to define the basis for decisions to be made on the results of CHARM, as it is a decision support tool and not a decision imposing method. The CHARM method enables a stepwise evaluation of E & P production chemicals by means of a successive pre-screening – hazard assessment –risk analysis – risk management.   The model will be derived from currently available models for the environmental evaluation of substances and will be based on internationally accepted, principles, e.g. sustainable development, within the framework of international conventions (PARCOM, OECD, EG). The model will guide the harmonization of governmental regulations for pre- and disqualification of chemicals with respect to discharges, e.g. permits. The model will be worked out and described in detail and a prototype; running with existing software, e.g. EXCEL, will be delivered[2].

The initially proposed development timetable was as follows:

- Initiated during December 1992
- First phase 1993 – Framework and hazard assessment, production chemicals.
- Second phase 1994 - Risk analysis, drilling chemicals
- Third phase 1995 – Implementation

In Phase II different hazard/risk assessment models used in the Netherlands and in Norway, as well as models described in the literature will be evaluated. These models will include the comparison of exposure and effects (toxicity), like the PEC/NEC ratio (potential environmental concentration/no effect concentration = margin of safety). Jointly with Norwegian and Dutch users, one model will be selected for further adaptation to offshore chemicals. The adaptation to the specific offshore conditions means that the following factors will be taken into consideration: 1) the oil/water distribution factor for offshore chemicals (OWDF), 2) the usage and water treatment systems, 3) long term exposure (biodegradation and bioaccumulation), and 4) effects on marine environment. After the adaptation the model will be prepared for harmonisation within PARCOM. A standardised procedure will also be proposed to the OECD[2].

It was decided to include drilling chemicals in the Phase II proposal so as to broaden the applicability of the model. This development was perceived positively by the offshore supply industry in general and EOSCA in particular. It offered a standard approach based on assessment of environmental risk that would benefit all stakeholders involved (regulators, chemical suppliers and end users). It also became apparent at the start of the development that the project would require industry participation, as they were the only source for the required discharge and environmental data for the chemicals used.

However, it soon became obvious that the original concept of the model should now be restricted to using only the HOCNF dataset agreed by OSPAR in 1994, and to only assess production chemicals, and water based drilling and cementing chemicals[3]. The control of synthetic fluids was already being discussed within OSPAR, and was eventually controlled by other Decision 2000/3 so their control using CHARM was increasingly being seen as unnecessary. This effectively imposed the same discharge restrictions for synthetic fluids as had existed for oil based discharges since 1992, setting a discharge limit of 1% oil on cuttings, or where the discharge into the sea of cuttings contaminated with synthetic fluids shall only be authorised in exceptional circumstances[4].

3 REGULATORY DEVELOPMENTS

3.1 Development of HMCS

The ad hoc Government/Industry working group set up following the GOP meeting in 1993 resulted in a proposal for a Harmonised Mandatory Control System (HMCS) for the Use and Reduction of the Discharge of Offshore Chemicals before the year 2000. This was discussed and finally agreed at the PARCOM meeting in 1996 as Decision 96/3. The main points of this Decision are as follows:

3.1.1 In all applications concerning the use and discharge of chemicals used offshore, the names of all deliberately added substances shall be reported for all preparations. Applications made after June 1995 shall only be considered by competent national authorities if a fully completed HOCNF is supplied. In the case of currently allowed chemicals permitted as a result of older and more limited tests, the full HOCNF dataset shall be requested on a prioritised basis, i.e. the highest hazard chemicals shall be tested earliest. The highest hazard chemicals (as determined by current national schemes) shall comply with the full HOCNF dataset within 12 months of the introduction of the scheme (1997). The remaining existing chemicals shall comply with the full HOCNF dataset within the subsequent 3 years, i.e. by 2000.

3.1.2 All offshore chemicals shall be run through a pre-screening scheme. This shall, where possible, be done on a substance-by-substance basis.

3.1.3 Such a pre-screening scheme should enable the competent national authorities to assess whether the offshore chemical concerned should be subject to:

A. permission
B. substitution
C. Temporary permission with the aim of seeking less hazardous alternatives
D. Ranking, e.g. according to CHARM, and subjected to continuing evaluation

When the pre-screening scheme arrives at one of these four outcomes, an action is required as specified below.

A. Permission: The competent national authorities may take regulatory action e.g., permit or approve chemicals without further evaluation. This regulatory action may control the amount to be discharged.

B. Substitution: The competent national authorities shall request the operator to apply a substitute for the offshore chemical or, if deemed necessary, request the operator to provide additional data.

C. Temporary permission with the aim of seeking less hazardous alternatives: Where no substitute for the chemical is available a temporary permission for a maximum period of 3 years is applicable whilst a substitute is sought. This temporary permission may be renewed after expiry, if the operator can demonstrate to the satisfaction of the competent national authorities that, despite considerable efforts, no alternative is yet available. Alternatively, additional data may be requested to allow reassessment of hazard.

D. Ranking, e.g. according to CHARM, and subjected to continuing evaluation: The relative PEC/NEC ratio will be calculated using CHARM. Regulatory action will be taken by the competent national authorities, e.g. permits or approvals. If the operator wishes to substitute the offshore chemical for economic reasons or for reasons of performance, the relative PEC/NEC ratio calculated by CHARM should be lower than, or equal to, that of the original chemical (in the application of this paragraph national authorities can take into account uncertainty factors identified by CHARM).

For all offshore chemicals a ranking list shall be drawn up by the competent national authorities based on the relative PEC/NEC ratios calculated by CHARM. In this ranking list the national authorities shall take into account the uncertainties in the CHARM calculations, e.g. in hazard intervals. This ranking list shall be grouped in function categories according to the categorisation in the annual reporting system for the use and discharge of chemicals from offshore installations. The competent national authorities shall, when taking regulatory action, ensure that over time a shift is realized towards lower relative PEC/NEC ratios. If the operator intends to replace a chemical with an alternative chemical with a higher relative PEC/NEC ratio, special permission must be issued. In most cases special permission will only be issued on a temporary basis in cases of health, safety, technical or economic difficulties[5]. Note: NEC was later changed to Potential No Effect Concentration (PNEC).

The proposal was to have a workable CHARM model and the HMCS implemented by 1998. However, this timetable slipped, and a two-year trial implementation period using CHARM was begun in 1997. The findings were reviewed at SEBA 1999, but no agreement to proceed was reached so a final decision was deferred to SEBA 2000. During 1999 a joint Government/Industry Group revised the CHARM model equations and developed a user guide. A software copy of the model was then developed and issued by EOSCA. These developments enabled the final format of the scheme to be finally agreed at the SEBA 2000 meeting, with implementation during 2001. [Refer to Appendix 1 for further details on CHARM].

4 IMPLEMENTATION

4.1 Practical Application of CHARM

The original intention was for the model to be available as a software package (to be supplied by TNO one of the consultants specified in the MOU) where input of HOCNF data would allow hazard quotients to be easily calculated. However, there were problems with development of, both, the model and the software, the most fundamental being:
1. Instability of the original software to run the model in an acceptable "user friendly" way.
2. Conflict between chemical suppliers, end users and regulatory authorities regarding confidentiality of product formulations and environmental data.
3. Difficulties of resolving data-set requirements for pure substances and preparations (a preparation being a combination of two or more substances).
4. Restricting development of CHARM to hazard assessment, when the original concept was to include risk assessment also.

The software development was abandoned and was replaced by an EOSCA funded initiative to develop a simple EXCEL based spreadsheet. The confidentiality issues are still of concern, especially regarding the publishing of the calculated Hazard Quotient (HQ) values.

The dataset requirements for assessments using substances/preparations has been harder to resolve. Historically, toxicity testing has been done on **preparations,** whereas biodegradation and bioaccumulation data on **components** is scientifically more relevant. The majority of oil field chemicals discharged are preparations comprising of two or more substances. To use component data in combination to represent or indicate a result for a preparation assumes component data is readily available, and that component toxicity effects are additive, which is not necessarily the case[6].

A compromise was eventually reached whereby the model should now accept data on components or preparations as required. The PEC: PNEC approach, which is the basis of the CHARM-model, is a methodology in which single PEC and PNEC values need to be available for comparison. This is not a problem in those cases in which the chemical is a single substance. The physico-chemical parameters of the substance can be used to calculate the PEC, while the PNEC can be derived from toxicity tests performed with the substance. However, the majority of the chemicals used, as offshore E&P chemicals are preparations composed of a number of substances.

Although a PNEC for preparations can be derived in the same way as for substances, difficulties arise when trying to interpret the practical meaning of this information. The toxicity test is performed on the preparation before it is discharged, while when using it in the relevant process the preparation may change. Individual substances may partition according to their physico-chemical properties, react with other chemicals, etc. Furthermore, after discharge other (biochemical) processes, such as biodegradation, may also influence the fate and effect of the individual substances of the preparation. In order to make a valid PEC: PNEC analysis, preparations should also be assessed on a substance basis.

While data for calculating a PEC is required be made available on a substance level, this is not the case for the toxicity data. Although several options have been studied to work around this problem there is not yet enough scientific support for determining a PNEC for the individual substances on the basis of the toxicity data for the preparation. It was decided to use a simple approach to determine the HQ of the preparation (see below), until a better and scientifically sound method is available. Further research in this area is therefore encouraged.

One of the following approaches (depending on the type of data available) should be used to calculate a Hazard Quotient for a preparation:

1) If both data for PEC and PNEC are available on substance level:

$$HQ_{preparation} = Maximum\left[\frac{PEC_i}{PNEC_i}\right]^{i=substance\,1\,to\,n}$$

2) If data for PEC is available on substance level and data for PNEC is available on preparation level:

$$HQ_{preparation} = Maximum\left[\frac{PEC_i}{PNEC_{preparation}}\right]^{i=substance\,1\,to\,n}$$

5   CONCLUSION

The introduction of a Harmonised Mandatory Control Scheme will require all new and existing classified chemicals to undergo pre-screening and CHARM hazard evaluation for approval for use/discharge offshore. It has only been with the intervention and support from the industry group EOSCA that CHARM development has been revived and a general consensus of agreement reached to enable its general acceptance by the industry, with the added proviso that expert judgement is required on the part of the user. This industry participation may be very commendable, but therein lies the issue. Does the industry stand back and let the regulatory authorities develop regulations in isolation, or do we participate to jointly develop a Best Practicable Environmental Option (BPEO), which the regulators will be prepared to adopt?

If so, the mechanism must be transparent and both parties must be accountable for their actions. The only way to demonstrate responsible environmental care is to quantify the potential risk to the environment from (in this case) discharges of chemicals to the aquatic environment. For CHARM to do this, the risk assessment part of the model must be developed, and field validation of actual discharges, and measurement of environmental effects carried out. This would at least quantify the uncertainties associated with the assumptions made in the CHARM model and allow a more informed decision on whether a hazard based approach was justified.

This is both costly and time consuming, but from an industry perspective important. Otherwise CHARM in hazard ranking mode coupled with the precautionary principle, may eliminate the use of some of the most effective chemicals, or at best require their replacement with others of lower hazard ranking, but with no demonstrable benefit to the aquatic environment.

In the UK the Government's interpretation, [of risk management and the precautionary principle is] based on the Rio definition. It states that precautionary action requires assessment of the costs and benefits of action and transparency in decision-making. The precautionary principle means that it is not acceptable just to say 'we can't be sure that serious damage will happen, so we'll do nothing to prevent it' Precaution is not just relevant to environmental damage-for example chemicals which may effect wildlife may also effect human health.

At the same time, precautionary action must be based on objective assessments of the costs and benefits of action. The principle does not mean that we only permit activities if we are sure that serious harm will not arise, or there is proof that the benefits outweigh all possible risks. That would severely hinder progress towards improvements in the quality of life. The extent to which precautionary action is necessary should be given careful thought for three reasons. First, action that is taken to protect one aspect of the environment can sometimes cause damage elsewhere (unintended consequences). Second, it may be better in certain circumstances not to take action if the consequences of doing so are irreversible (reversibility). Third, a decision on whether to take precautionary action should take account of the potential benefits forgone as a result of such action.

Because of the general lack of consensus over practical application of the precautionary approach, the use of risk assessments to inform decisions about environmental protection has sometimes been presented as being in conflict with the precautionary principle. In reality, risk assessment is often employed where issues are not clear and can be used to identify effects considered serious enough to warrant precautionary action. Risk assessments can sometimes point to the possibility of significant environmental damage, albeit in the presence of large uncertainties, and it is in such cases that precautionary action is valid[8].

The only way to fully address the principles of sustainability is to have a transparent approach to risk management, employing the precautionary principle only where the uncertainty is unknown or unacceptably high. Therefore the best way forward for the industry is to emphasise risk assessment and attempt field validation of the model to quantify the uncertainties and assess the environmental validity of CHARM within the HMCS (which must also include the pre-screening requirement) overall.

The issue is further complicated by the environmental regulatory burden shifting by OSPAR, as their remit is only the aquatic environment. A more holistic regulatory approach would be needed to give a proper interpretation of the principles of sustainability, but both Government and Industry should take note that this and future generations will judge our actions using these same principles of sustainability.

References

1 *Proposed Terms Of Reference For An Ad Hoc Working Group To Elaborate A Harmonised Mandatory Control System For The Use And Reduction Of The Discharge Of Chemicals:* PARCOM GOP Meeting, GHENT 9-12 February 1993.
2 *MOU Project: An environmental risk evaluation model for offshore E & P Chemicals.* Prepared by TNO, Aquateam, 3 September 1993.
3 I.. Still, *Development of New Drilling Fluids, BICS Conference: Offshore Emissions & Discharges-The Next Ten Years*, Aberdeen, July 5-6 1994.
4 *OSPAR 2000/3.*
5 *PARCOM 96/3.*
6 D. Calamari and M. Vighi, Institute of Agricultural Entomology University of Milan *Scientific Basis For the Assessment of Toxic potential of Several Chemcial Substances in Combination at Low Level,* July 1991.
7 M. Thatcher (Baker Hughes INTEQ/EOSCA), M. Robson (MAERSK Olie OG GAS AS/NSOC/D), L.R. Henriquez (State Supervision of Mines, the Netherlands) and C. Karman (TNO, the Netherlands). *User Guide for the Evaluation of Chemicals used and Discharged Offshore,* Version 1.0, 1999.
8 *Guidelines For Environmental Risk Assessment And Management,* DETR, Environment Agency, Institute for Environment And Health, **13-14**, July 2000.

Glossary

BPEO	Best Possible Environmental Option
CHARM	Chemical Hazard Assessment and Risk Management
EOSCA	European Oilfield Speciality Chemicals Association
HMCS	Harmonised Mandatory Control Scheme
HOCNF	Harmonised Offshore Chemical Notification Format
MOU	Memorandum Of Understanding
OSPAR	Oslo and Paris Commissions
PARCOM	Paris Commissions
PEC	Predicted Environmental Concentration
NEC	No Effect Concentration
PNEC	Predicted No Effect Concentration
TNO	Netherlandse organisatie voor toegepast natuurwenschappelijk onderzoek

Appendix 1

THE CHARM MODEL

A schematic of the representation of the CHARM model is given in Figure 1.

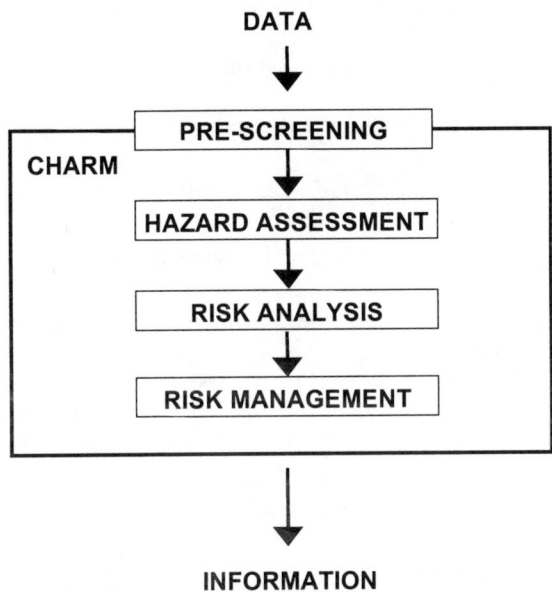

Figure 1 *A schematic representation of the CHARM model*

Pre-screening *identifies chemicals that might lead to specific long term 'chronic' effects since these cannot be assessed using a PEC:PNEC comparison. Those chemicals are characterised by long term persistency and a high potential for bioaccumulation. Pre-screening is, therefore, used to screen substances prior to the use of the CHARM model.*
Hazard Assessment *provides a general environmental evaluation of a chemical based on its intrinsic properties under "realistic worst case" conditions. The Hazard Assessment is primarily intended for relative comparison of single production chemicals, i.e. for ranking or classifying production chemicals.*
Risk Analysis *covers a more specific evaluation of the environmental impact from the use of a production chemical or a combination of production chemicals under actual conditions. Such a specific analysis enables risk management on the basis of various scenarios for environmental care options and input of various cost options.*
Risk Management *is used to compare various risk reducing measures based on cost/benefit (benefit = risk reduction) analyses for a combination of chemicals*

Calculation of Hazard and Risk Assessment, PEC:PNEC Approach

The hazard evaluation and risk assessment is made by comparison of the Predicted Environmental Concentration of the chemicals in the environment (PEC) with the Predicted No Effect Concentration of the chemicals (PNEC) expressed as a ratio i.e. PEC: PNEC.

Where the PEC is less than or equal to the PNEC we can say that the discharge is acceptable (lower hazard). Where the PEC is greater than the PNEC potentially damaging environmental effects may occur.

Three PEC: PNEC quotients have been identified:

$Q_{Pelagic} = PEC_{water}$: PNEC aquatic biota

$Q_{Benthic} = PEC_{Sediment}$:PNEC aquatic biota

$Q_{Food\ chain} = PEC_{Biota}$: PNEC birds, mammals

The PNEC values are determined from marine eco-toxicity and mammalian testing using existing OECD methods.

Within CHARM, environmental Hazard Assessment, Risk Analysis and Risk Management are all based on Hazard and Risk Quotients (HQ and RQ), which are calculated using the internationally accepted PEC:PNEC method. The traditional method of comparing single PEC and PNEC values by calculating the ratio of PEC and PNEC is illustrated in Figure 2.

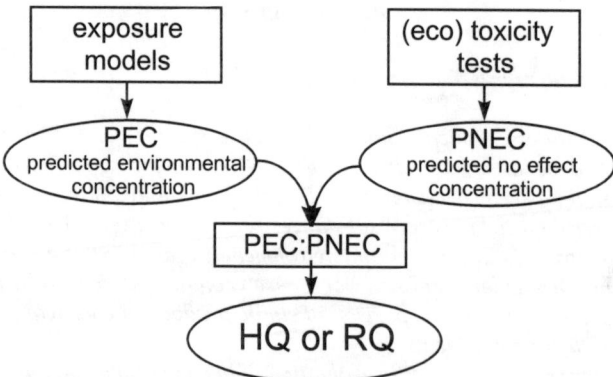

Figure 2   *The traditional method of comparing PEC and PNEC in order to calculate a Hazard or Risk Quotient.*

The Predicted Environmental Concentration (PEC) is an estimate of the expected concentration of a chemical to which the environment will be exposed during and after the discharge of that chemical. The actual exposure depends upon the intrinsic properties of the chemical (such as its partition coefficient, degradation and bioconcentration factor), the concentration in the waste stream, and the dilution in the receiving environmental compartment.

Most of the calculations within CHARM are concerned with the estimation of the concentration of a chemical in the waste stream. This is dependent upon the process in which it is used, the dosage of the chemical, its partitioning characteristics, the oil (or condensate) and water production at the platform, the in-process degradation mechanisms and the residence time before release.

As the name suggests, the Predicted No Effect Concentration (PNEC) is an estimate of the highest concentration of a chemical in a particular environmental compartment at which no adverse effects are expected. It is, thus, an estimate of the sensitivity of the ecosystem to a certain chemical. In general the PNEC represents a toxicity threshold, derived from standard toxicity data (NOECs, LC_{50}s, EC_{50}s)[1].

Within the CHARM model, a $PNEC_{water}$ is extrapolated from toxicity data using the OECD method, which is accepted by most OSPAR Countries. In this method, the PNEC for a certain ecosystem is determined by applying an empirical extrapolation factor to the lowest available toxicity value. The magnitude of the extrapolation factor depends upon the suitability of the available ecotoxicological data.

By calculating a PEC:PNEC ratio for a certain chemical, the CHARM model compares the expected environmental exposure to a chemical (quantified as the PEC) with the sensitivity of the environment to that chemical (quantified as the PNEC). If the PEC:PNEC ratio (an indication of the likelihood that adverse effects will occur) is larger than 1, an environmental effect may be expected. It must be noted, however, that these results should be interpreted with care, and only used as a means to estimate potential adverse environmental effects of chemicals. Furthermore, in order to acknowledge uncertainty in the results of the model, the raw data should be considered as well when comparing chemicals

Within CHARM the offshore environment is divided into two compartments: water and sediment. This is done in order to acknowledge the fact that a chemical present in the environment will partition between the water and organic matrix in the sediment. This is illustrated in Figure 3. The concentration of a chemical may, therefore, vary greatly from one compartment to another. Consequently, two PEC values are calculated: PEC_{water} and $PEC_{sediment}$.

Chemicals dissolved in water may have adverse effects on the pelagic biota, i.e. plankton and most fish species. Those which accumulate in the sediment may affect the benthic biota, i.e. worms, echinoderms, crabs and bivalves. For this reason, two PNEC values are calculated: $PNEC_{pelagic}$ and $PNEC_{benthic}$.

In order to estimate a chemical's potential to cause environmental impacts, a PEC:PNEC ratio is calculated for each compartment ($PEC:PNEC_{water}$ and $PEC:PNEC_{sediment}$). The higher of the two ratios is used to characterise the maximum environmental hazard or risk associated with the discharge of a product. This approach avoids arbitrary weighing of the compartments and yet ensures protection of the other compartment by measures to minimise or reduce risks.

---

[1] NOEC, LC_{50}, and EC_{50} are parameters derived from ecotoxicity tests.

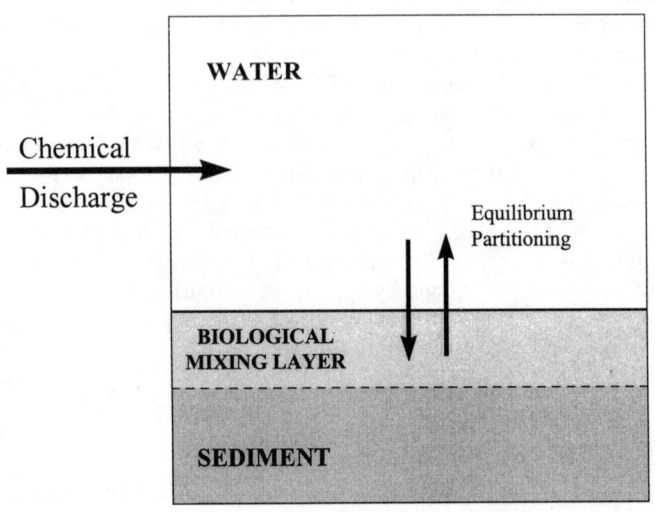

Figure 3 *Schematic representation of the environmental compartments considered within the CHARM model.*

Table 1 *An overview of the names used to indicate the compartment to which the PEC, PNEC and PEC:PNEC ratio is referring*

PEC	PNEC	PEC:PNEC-ratio
Water	Pelagic	Water
Sediment	Benthic	Sediment

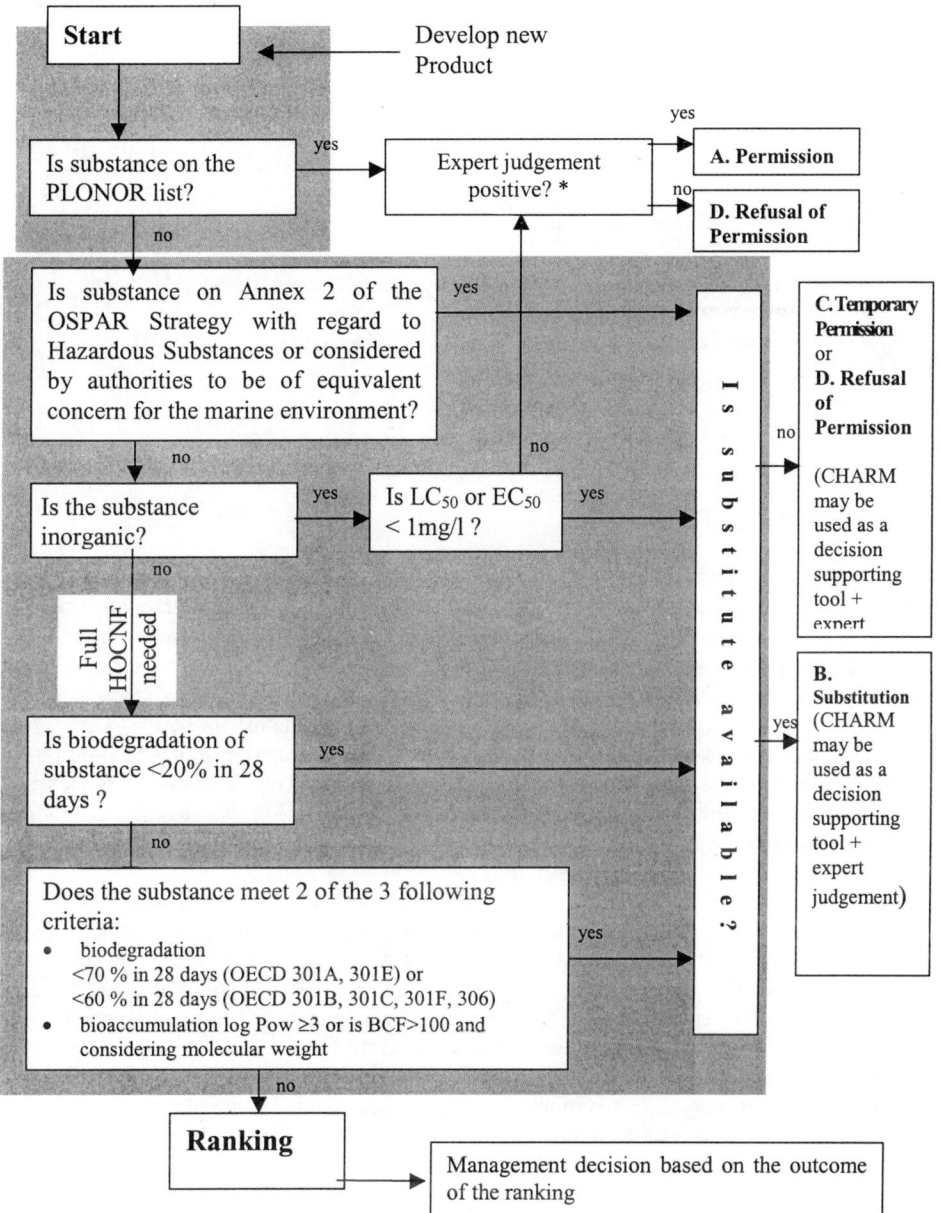

Explanatory notes:

* In accordance with the precautionary principle, expert judgement on a PLONOR substance should take into account sensitive areas, where the discharge of certain amounts of such a PLONOR substance may have unacceptable effects on the receiving environment.

** If toxicity data are available only for a preparation, the authority should, on a case by case basis:

a. seek further information from the supplier to identify that substance, which is the major contributor to the overall toxicity of the preparation; or

b. use the toxicity data of the preparation to estimate the toxicity of a substance contained in it, taking into account the concentration of the substance in the preparation.

BIOACCUMULATION POTENTIAL OF SURFACTANTS: A REVIEW

Phil McWilliams[1] and Graham Payne[2]

[1]EOSCA, c/o ILAB, Bergen High-Technology Center, PO Box 4300 Nygardstangen, N-5837 Bergen, Norway.
[2]EOSCA, c/o Briar Technical Services Ltd, 501 North Deeside Road, Cults, Aberdeen AB15 9ES, UK

1 INTRODUCTION

Surfactants are a chemical group for which it is difficult to obtain reliable partitioning (log P_{ow}) or bioconcentration factor (BCF) data for inclusion in current models used in performing environmental risk assessments. The difficulties revolve largely around the intrinsic property of surface-active substances to adsorb to surfaces and to accumulate at phase interfaces. Despite the apparent limitations of surrogate analytical approaches to estimation of bioaccumulation potential for a surfactant, regulatory authorities have, with few exceptions, insisted on the submission of log P_{ow} data for surfactants for the purposes of environmental risk assessments (OSPAR HOCNF 1995). The alternative approaches – experimental determination of a BCF, or derivation of a log P_{ow} using quantitative structure-activity relationships (QSARs) – would appear to be equally unreliable for surfactants.

A wide range of surfactants is used offshore, for a number of different purposes, although the quantities of each class of surfactant used are difficult to estimate. It is considered that the most important environmental issues in relation to surfactant use/discharge offshore are whether the surfactants pose a risk as a result of direct toxicity in the aqueous environment, or whether biodegradation, bioaccumulation and biomagnification of surfactants poses a greater risk to the marine environment.

The European Oilfield Speciality Chemicals Association (EOSCA) commissioned a review to collate and assess currently available data on bioaccumulation potential of surfactants (log P_{ow} and BCF) in order to address a number of issues. This paper gives a brief summary of the salient points of this review.

2 ENVIRONMENTAL ISSUES

All major surfactant groups (anionic, cationic, nonionic and amphoteric) are currently used to some extent by the offshore oil industry Table 1.

Nonionic surfactants are the most widely used, with perhaps the greatest concern focusing on bioaccumulation potential of alkylphenolethoxylates, for some of which there is tentative evidence of weak endocrine disruption activity. Interest in the bioaccumulation of surfactants has increased over recent years due to the large quantities of these materials

Table 1   *Summary classification of currently used/discharged oilfield surfactants and their general applications in the North Sea.*

Surfactant category	Type	Used in products of type*	Currently in use in North Sea
Alkyl aryl sulfonates	Anionic	EB, CI	Yes
Alkyl sulfates	Anionic	AF	Yes
Alkyl ethoxylate sulfates	Anionic	AF	Yes
Phosphate esters	Anionic	CI	Yes
Quaternary ammonium compounds	Cationic	CI, BC	Yes
Fatty amine salts	Cationic	CI	Yes
Fatty acid amides	Cationic	EB	Yes
Imidazolines	Cationic	CI	Yes
Alkyl phenol ethoxylates	Non-ionic	CI, BC, EB	No
Alkyl poly glycosides	Non-ionic	CI	Yes
Ethoxylate-Propoxylate polymers	Non-ionic	EB	Yes
Fatty alcohol ethoxylates	Non-ionic	BC, CI, EB	Yes
Betaines	Amphoteric	CI	Yes

*Key: **AF**, antifoam; **BC**, biocide; **CI**, corrosion inhibitor; **EB**, emulsion breaker

manufactured and the relatively high proportion discharged to the environment. In environmental terms, surfactants possess properties that mean that their fate and behaviour in an aqueous environment will differ from that predicted for non-surface-active chemicals. In particular, they all have a combined lipophilic/hydrophilic structure which gives them a tendency to collect at aqueous/organic-phase boundaries, and they will form micelles in water when present above critical levels (CMC). Most surfactants are susceptible to biodegradation, metabolism and other breakdown reactions that may lead to metabolites with significantly different chemical properties.

The quantities of each class of surfactant used are difficult to estimate. As an approximation, anionic surfactants are the most important, representing 60-70% of surfactants currently in use. Non-ionic compounds constitute around 30% but their use is increasing, while cationic and amphoteric products make up the smallest proportion. Currently adopted approaches to hazard assessment and risk management of chemicals, including surfactants, used and discharged offshore in the North Sea are based on a harmonised scheme of testing and evaluation (Harmonised Offshore Chemical Notification Format, OSPAR HOCNF 1995; and CHARM). The octanol-water partition coefficient (log P_{ow}) has been defined as a central parameter in the risk assessment of offshore chemicals, being used to estimate predicted environmental concentration (PEC) through its use in partitioning calculations (CHARM), but the evidence is not convincing enough to support the view that it *is* a key parameter, especially for surfactants or complex mixtures. There are a lot of experimental data which indicate that it is often useless in this respect for oilfield chemicals, and that indeed there is no single partition coefficient for many chemicals (i.e., their partitioning behaviour depends on various factors such as salinity, pH and temperature). Log P_{ow} demonstrably does *not* determine environmental fate, although it is used for this purpose.

If log P_{ow} is not considered to be satisfactory for bioaccumulation prediction, then it is not only unsatisfactory for sediment partitioning estimation, but it is *a priori* unsatisfactory for estimating the amount released in produced water. Surfactants are important, and often significant (in terms of quantity) components of production chemicals, and using current approaches, log P_{ow} is clearly an unsatisfactory parameter as a basis for hazard and risk assessment of surfactants and/or highly hydrophobic chemicals. The current (mandatory) test methods adopted in the HOCNF (OECD 117 HPLC method or OECD 107 Shake Flask method) are inherently unsuitable for the determination of a log P_{ow} for surface-active chemicals, not least because of the tendency for surfactant molecules to accumulate at phase interfaces or form emulsions, thereby giving spurious and unreliable results. Despite these obvious limitations, regulatory authorities have based environmental hazard and risk assessment of surfactants on log P_{ow} data obtained from these tests (HOCNF). In reality the existing OECD 117 HPLC method is being misused by being applied to formulations of "unknown" content, and in particular the estimation of a weighted-average log P_{ow} for anything other than a group of homologues cannot be construed as scientifically valid. Intended changes to the present requirements of the HOCNF (Summary Record SEBA 2000) propose that log P_{ow} determinations for surfactants should be abandoned in favour of a sediment-water partitioning coefficient (K_{oc}), and that default values should be used for fraction released to water and for bioconcentration factor (BCF). This should be regarded as only a temporary measure, until industry and regulatory authorities have explored other approaches or looked at ways of improving current methodology, particularly focusing on some of the large-volume surfactants currently in use.

A factor that has been largely overlooked in the environmental assessment of surfactants, apart from the intrinsic toxicity of the surfactant, is that of the potential synergistic effects on migration, dispersion, bioavailability, etc. of otherwise low-toxicity chemical compounds in a formulation. The current HOCNF guidelines accept that surfactants may increase the bioavailability of other substances in preparations, and suggest that a bioconcentration test may be required in such cases. However, it is difficult to justify a "black box" regulatory approach that relies on a single and often arbitrary measurement. Any assessment of bioaccumulation potential should, realistically, take into account as much information as possible on the chemistry, metabolism, degradability and potential breakdown products of the chemical. With oilfield chemicals, this can be difficult, since they are often quite complex mixtures and their chemistry is often very poorly described.

Default fraction released values estimated from available log P_{ow} data and adopted in CHARM evaluations are viewed as extremely conservative, as exemplified by the often significant disagreement (up to an order of magnitude or more) between adopted values and those determined by field validation studies on various surfactants Table 2.

The list of default fraction released values, i.e. chemical discharge factors established in CHARM table some surfactant categories should be expanded to include all the relevant surfactant categories/classes included in this review. There are doubts that it is practical to relate such default values to the water-cut. Measured values are "real", but can only be related to the particular operation at the time of the measurement, since the process is unlikely ever to be in equilibrium. Factors determined this way may thus be a valid tool for documentation, but the results may be inappropriate for modelling over the lifetime of a field. Site-specific environmental risk assessment should preferably be based on experimentally determined discharge factors obtained from mass-balance studies (e.g. Sæten et al. 1999; Bakke et al. 2000). If the circumstances upon which the site-specific discharge factors have been determined are studied in detail, it could be judged whether the same figures could be applied under other conditions (expert judgement).

Table 2 *Default values (from Thatcher et al. 1999) and results from field validation studies for the fraction released of surface-active production chemicals*

Type of surfactant	Default fraction released	Fraction released in field validation studies
Primary amines (cationic type C>12)	0.1 (10%)	0.038 (3.8 %)[1]
Quaternary amines	1.0 (100%)	
Ethoxylate-Propoxylate (Eo-Po) Block polymer demulsifier	0.4 (40%)	
Imidazolines	0.1 (10%)	0.01 (1.0 %)[2]
Amines	0.1 (10%)	
Phosphate esters (anionic type C>13)	0.1 (10%)	0.002 (0.2 %)[1]
Other	1.0 (100%)	
		[1] TNO: Fokema et al. (1998)
		[2] Statoil: Sæten et al. (1999)

3 RELIABILITY OF EXISTING DATA

Physico-chemical properties of a substance, such as solubility, Pow and sorption properties, are parameters that can be used early in an evaluation process to assess its likely fate and to determine the environmental compartments into which it will partition. An octanol-water partition coefficient can be used to predict BCF, and in many cases molecular structure has been used to estimate Pow, using so-called 'fragment contribution' methods. These fragment methods do not, however, take into account the branching positions on the molecule, and may therefore not give a true representation of bioaccumulation potential. For some molecules there are significant differences between the results obtained using different calculation methods, and as the complexity of the surfactant molecule increases the reliability of the methods decreases. The development of QSARs to predict partition coefficients has been a useful approach to reducing the need for extensive live animal or surrogate testing, but such approaches require extensive validation before they can be adopted and used with any degree of confidence. The available data indicate that the use of QSARs to estimate log Pow for some classes of surfactant are not reliable. Not least, the development of QSARs depends on valid data on which to develop the relationship. For surfactants, the reliability of existing Pow data is questionable. The OECD 117 HPLC method, for example, adopts a QSAR approach to the estimation of a log Pow for a substance, but for surfactants there are insufficient established log Pow values for specific surfactant molecules to enable a valid calibration of the system.

Experimentally derived log P_{ow} values were found for a small number of surfactants (Tolls et al. 1995). However, the formation of emulsions must be regarded as a serious problem when determining octanol-water partition coefficients for surfactants, and for ionic surfactants the use of current techniques will most likely yield distribution ratios rather than partition coefficients. For this reason, P_{ow} cannot be regarded as characterizing the partitioning of ionic surfactants, and current data obtained using OECD 107 or 117 tests cannot be viewed as valid. The majority of surfactant log P_{ow} data have been derived by calculation, many using equations based on the fragment contribution methods of Leo and Hansch (1979). Calculation methods are based on the theoretical fragmentation of the molecule into suitable substructures for which reliable log P_{ow} values are known. The log P_{ow} is obtained by summing these fragment values and applying correction factors for bonding, branching etc. However, the validity of calculated values must be questioned since the reliability of the various calculation methods decreases as the complexity of the molecule increases, and interpretations may often be subjective.

The existing BCF data set for surfactants is relatively small, with the majority of data relating to anionic surfactants, particularly LAS. Some data is available for cationic and nonionic surfactants, but no data were found for amphoteric surfactants. The usefulness of the data is limited by the lack of a unified approach to experimental determination of a BCF. Measurement of a BCF for a surfactant is an alternative to estimation of P_{ow}, but this approach can also be problematic. There is often significant variability in BCFs determined for the same surfactant with different species, and also for the same surfactant tested on the same species (e.g. Tolls et al. 1994). In addition, the vast majority of studies have been carried out on freshwater species. As indicated by Tolls et. al. (1995), much of the available data can only be used tentatively since it has been derived from experiments using radiolabelled compounds. Very few such studies can differentiate between parent compounds and metabolites or other breakdown products. Because of this limitation, many reported BCFs are probably significant overestimates. In general, BCFs for surfactants are reported as being comparatively low, and are generally below the conventional criteria for concern (i.e. log P_{ow} value of 3 - 4).

4 RELEVANCE OF log P_{ow} /BCF TO SURFACTANTS

In principle, partition coefficients are not relevant to surfactants since they do not partition between immiscible solvents such as octanol and water, but will tend to accumulate at the phase interface or form emulsions at high concentrations. The question should really be 'how relevant are existing (or potentially new) techniques to assessing the passage of surfactants across a biological membrane?', or 'how likely is it that a surfactant molecule will cross a biological membrane?'. In view of the surface-active properties of this class of chemicals, this consideration naturally leads on to the question of whether discharge of surfactants poses a risk as a result of *direct toxicity* in the marine environment, or whether *biotransformation, bioaccumulation and/or biomagnification* of surfactants constitute a greater risk.

In the longer term, the exposure of organisms to surfactants in the marine environment will be dependent on the fate and behaviour of this class of chemicals when discharged. In general terms, surfactants may be removed from the marine environment by mechanisms such as volatilisation, abiotic degradation, adsorption to particles, microbial degradation or uptake by marine organisms, factors that are applicable for any type of chemical. Volatilisation is not likely to be a significant factor because of the relatively high aqueous solubility and low/negligible vapour pressures of most surfactants. Surfactants are likely to adsorb to sediments, although sorption of surfactants on marine sediments has received little attention. Generally speaking, sorption behaviour of surfactants on marine sediments is consistent with observed characteristics in freshwater sediments, although other factors such as salinity, organic carbon content, temperature and pH may be important.

The studies and data reviewed in this report indicate that the majority of surfactants are susceptible to biodegradation, both aerobic and anaerobic. Compared to freshwater studies, there is a limited data set of biodegradation values for surfactants in the marine environment. The majority of studies on the environmental fate and behaviour of surfactants in the marine environment have been carried out on LAS and other anionic surfactants. The general conclusion must be that surfactants are not likely to be persistent in the marine environment, although there is an observed trend of slower rates of biodegradation in marine compared to freshwater environments. For this reason a safety factor is applied in CHARM when only freshwater data are available. Therefore, while sediment sorption processes are undoubtedly of significance in reducing water column exposure concentrations of surfactants in aqueous environments, the most important process controlling the environmental fate of surfactants in the marine environment is undoubtedly biodegradation. Sorption will result in a redistribution of surfactants from water to sediments, while biodegradation results in a net loss of chemical from environmental compartments. However, with regard to environmental exposures, the primary consideration when reviewing biodegradation characteristics of surfactants, or any chemical for that matter, is that it is not the extent of biodegradation over an arbitrary time period that is important, but rather the rate of biodegradation compared to residence time in an environmental compartment that will ultimately determine exposure. Environmental exposure will vary, depending on solution strength, application method and rate, the degree of dilution and dispersion, and meteorological conditions. Subsequent biodegradation of surfactants will affect exposure concentration and duration, although the toxicity of surfactant metabolites is an issue on which no studies were found. Lewis (1991) notes that although comprehensive data on effect and exposure exists for LAS, comparable information is not available for other surfactants, especially in the marine environment. Consequently, existing risk assessments should be considered to be of limited validity since they are based on extrapolated data and may be inapplicable to all marine species and all surfactant classes without extensive validation

Current scientific understanding of the toxic effects of surfactants is based mainly on laboratory experiments for a few freshwater species. As a result, extrapolation of existing laboratory data to the marine environment is difficult. As a general observation, most surfactants appear to be less toxic in the environment than would be inferred from laboratory tests (Lewis 1990). Current awareness of surfactant toxicity to aquatic organisms, and apparent trends in toxicity in relation to different surfactant classes should be viewed with caution and broad generalisations avoided as the range of species tested and the number of different surfactants involved is limited. A taxonomic cross-comparison of the surfactant toxicity data in this review highlights the difficulties in identifying trends in surfactant toxicity. For acute toxicity studies with anionic surfactants the algae and fish species tested appear to be most sensitive, with the molluscs showing an intermediate sensitivity and crustaceans being the least sensitive. However, larval stages of crustacean species appear to show significantly higher sensitivity to this class of surfactant than adults.

Surfactants generally seem to impact on higher aquatic organisms via their respiratory structures. In invertebrates such as crustaceans these may be simple external gills or areas of specialised cells on the body surface. In higher organisms such as fish the respiratory structures (gills) consist of epithelial membranes that may be extensively folded to provide large surface areas for gaseous exchanges. Destabilisation of these epithelial membranes, as may occur when exposed to surfactants, results in changes in membrane permeability, cellular lysis, and impairment of cellular respiration. In lower organisms, in which exchange of respiratory gases is via mechanisms of simple diffusion across membrane surfaces, surfactant toxicity appears to result from an initial disruption of normal membrane function followed by physical disruption of the cellular membrane. As might be expected, charged surfactants (anionic and cationic) appear to have a greater denaturing effect than neutral surfactants. Cationic surfactants also appear to be the most toxic to both freshwater and marine species of algae, invertebrates and fish.

Although only a limited range of surfactants has been investigated for aquatic toxicity, a few studies have illustrated a difference in toxicity between surfactant classes. Lewis (1990) noted that the toxicity of different surfactants on the same algal test species might vary over four orders of magnitude. Charged surfactants (anionic and cationic) have been reported to have a greater denaturing effect than neutral chemicals, and cationic surfactants are generally considered to be most toxic to both freshwater and marine algae, invertebrates and fish (Ukeles 1965; Lewis 1991). It is possible that existing HOCNF data includes reference to toxicity of various oilfield surfactants to marine organisms, and if made available, these could usefully supplement the comparatively limited marine data available in the public domain. However, the current emphasis on toxicity testing of complete preparations will mean that few such studies will be relevant.

Surfactant toxicity has also been found to vary between homologues within a given surfactant type and may also depend on chemical structure. Increasing the length of the alkyl chain can modify toxicity of LAS, and toxicity of nonionic ethoxylated surfactants depends on the length of the ethoxylate chain (Lewis 1991 and references therein). In some cases, toxicity may be predicted from the ethylene oxide molar ratio, with a ratio of 15 or less being associated with the most toxic surfactants and ratios of 30-50 being consistent with observations of low toxicity (Scott Hall et al. 1989). This observation applied both for a given series of homologues and across various surfactant types.

In reviewing the potential of surfactants to bioaccumulate, a general association of increasing alkyl chain length (i.e., increasing hydrophobicity) with an increase in BCF was noted (Tolls et al. 1997, 2000) for LAS compounds and isomers, and alcohol ethoxylate components. Conversely, increasing the length of the hydrophilic section of a surfactant molecule (i.e., decreasing overall hydrophobicity) results in a reduction in BCF (reviewed in Staples et al. 1998). Tolls et al. (2000) also found increased uptake rates and BCFs for alcohol ethoxylate surfactants when hydrophobicity was increased. These apparent steric influences on surfactant toxicity and BCF appear to be consistent, and may offer a means of predicting likely toxic effects of surfactants on marine organisms through a consideration of steric factors. A more thorough evaluation of existing data may be useful, particularly if combined with further investigative studies, to establish and validate some general principles describing the relationship between surfactant chemistry (molecular/steric factors) and toxicity/BCF. If modifications to the molecular structure of surfactants can result in predictable influences on bioaccumulation and toxicity to aqueous organisms, then environmental effects of new formulations could be predicted at an early stage in product development.

A tendency for surfactant molecules to be retained on epithelial surfaces, rather than to cross cellular/epithelial membranes (uptake) and hence bioaccumulate, may be a possible explanation for the longer-chain/lower-toxicity observations. Surfactant molecules residing (bound) on an epithelial membrane surface may be expected to disrupt membrane integrity (permeability/fluidity), and interact with mucus (a charged, fibrous glycoprotein-carbohydrate matrix). Studies of the effects of sodium lauryl sulphate (SLS) and LAS at concentrations of 100 mg l^{-1} showed that the integrity of the upper layers of the epithelium of fish gills was severely disrupted, resulting in severe water imbalance. However, the test concentrations used are several orders of magnitude greater than would be expected in the environment. At low concentrations (e.g. 6 μg l^{-1} of SLS) some effects are reversible, indicating temporary binding to specific sites (Stagg et al. 1981). The number of binding sites on epithelial or cellular membranes is usually limited, resulting, for example, in transmembrane transport mechanisms that display saturation kinetics. If a critical number of (surfactant) molecules must occupy the available binding (transport) sites in order for lethal poisoning to occur, then surfactants that can more easily cross the membrane and bioaccumulate (as indicated by a higher BCF) are less likely to exhibit acute toxic effects. In general, BCFs for surfactants are reported as being comparatively low, and are generally below the conventional level for concern (i.e. log P_{ow} value of 3 - 4). Although considerable evidence of surfactant bioaccumulation has been collected and published, lower lethal toxicity associated with an increased BCF would argue in favour of the contention that it is not surfactant bioaccumulation *per se* which is of concern, but direct toxicity.

Biotransformation and biomagnification are processes that may occur once a chemical has entered an organism (bioaccumulated). Evidence for biotransformation of surfactants in aquatic organisms is scant, and limited to radiolabel studies. For the few surfactants investigated (e.g., $C_{14}EO_8$: Tolls ands Sjim 1999; C_{12}-LAS and C_{13}-LAS: Tolls et al. 1997), biotransformation was deduced to be the dominant factor in the elimination of these surfactants from the test organisms.

In order for biomagnification of a chemical to take place the compound must be stable in the environment for significant periods of time. Compounds which (bio)degrade relatively rapidly or which are readily metabolised (biotransformed) will not be biomagnified within the food chain. While the bioaccumulation of a chemical can still present a problem where exposure levels and uptake rates are sufficiently high in relation to depuration and metabolism rates, a high bioaccumulation potential does not automatically imply the potential for biomagnification. Indeed, for some chemicals, which are readily taken up by organisms near the bottom of the food chain, a capacity for metabolism is more likely in successively higher trophic levels. In some cases, calculated BCF values for surfactants in higher aquatic organisms (fish) were found to be 30-3000 times lower than values for algae (Ahel et al. 1993). The available information indicates that most commonly used surfactants do not have the properties required to exhibit biomagnification, i.e., they have a tendency to be rapidly degraded and metabolised and are not highly hydrophobic.

In conclusion, no evidence has been found to support concern with respect to the biomagnification of surfactants, although it is noted that most of the research effort has been devoted to a relatively small number of surfactant types. Bioconcentration factors in the aqueous phase are generally below the level of concern, and (for some nonionic surfactants at least) can be quantitatively related to the length of the hydrophobic and hydrophilic components. There is also evidence that overall molecular size may place constraints on biological uptake. The studies examined raise no concerns with respect to long-term retention of accumulated surfactant material in tissue, and indeed they present considerable evidence that many surfactants are metabolised. The fate of metabolites has not been thoroughly studied, however, and there is consequently a degree of uncertainty as to the fate and longer-term effects of some hydrophobic components (such as some alkylphenols) following partial metabolism.

5 ALTERNATIVE ANALYTICAL APPROACHES

In respect of the potential developments in analytical techniques the following questions should be addressed:

- Are the new methods likely to offer a better alternative to the existing ones?
- How practical and relevant are these new techniques to surfactants?
- Are surrogates to live animal testing reliable?
- Are the new methods suitable for standard tests?

Surfactant behaviour cannot be related to partitioning between two disparate liquid phases because of their inherent tendency to collect at phase interfaces or to form emulsions (micelles), placing the existing methods of estimating BCF in doubt. The lack of a widely-applicable, robust and simple method to assess bioaccumulation potential and sediment/water partitioning of surfactants has hindered the establishment of a rational and hence meaningful evaluation of the environmental hazards and risks that surfactants may pose. Surrogates to live animal testing are always preferable, and it is likely that the recently introduced MEEKC technique will provide a more valid result in the form of a pseudo-log P_{ow}. The technique has been used to investigate octanol-water partitioning of a wide range of organic compounds giving a good correlation with HPLC-generated values for simple organic molecules (Smith and Vinjamoori 1995). Salimi-Moosavi and Cassidy (1996) used the technique to separate long-chain surfactants and have further investigated the potential of the technique for surfactant applications. The newly developed techniques of MEEKC use the properties of surfactants to great effect in the analytical process. Currently in reverse-phase HPLC there is a tendency for irreversible adsorption of some compounds. This is not the case with MEEKC. It is a fact that products are often presented for testing as a mixture of substances, for which no useful (in analytical terms) information on the formulation is provided. There is therefore little possibility to apply a "correct" analytical technique. The MEEKC approach seems to offer a broader scope for a wider range of compounds even if a series of different conditions needs to be used on a formulation.

The indications from the literature are that the MEEKC technique would be very suitable as a standard method. It also seems feasible that the equipment could be used to determine log P_{ow}s of ordinary compounds, and there are references citing the use of diode array detection. While capillary electrophoresis is not as widely used as HPLC, there are at least two commercial models available at comparable cost to a HPLC system. Test costs are therefore likely to be similar to those for current log P_{ow} analysis.

The suitability of SPMDs as an alternative surrogate technique to live animal testing for estimation of BCFs for surfactants needs to be more closely investigated. Although a good relationship between BCFs for PAHs obtained using SPMDs and live animal tests on blue mussels, *Mytilus edulis*, (Røe et al. 1998), the intrinsic properties of surfactants may pose problems when interpreting data from the use of such devices. The justification for using SPMD is based on uptake and BCF for lipophilic chemicals, and the whole question centres on whether lipophilic descriptors are valid for surfactants – this seems illogical. The use of an SPMD requires analysis of the solvent inside the device – if surfactants sit on or in the semi-permeable membrane, there might possibly be very little material present in the solvent phase inside. BCF tests are considered to be prohibitively expensive, but the main cost element is the chemical analysis, not the 'biological' component. If it is necessary to analyse both the water and the content of the SPMD, then the cost of the work will not be very different from the cost of a BCF, and the primary advantage would be that a SPMD might equilibrate faster than an experimental animal. In BCF tests, actual uptake and depuration rates are measured, and the resulting estimate takes account both of passive depuration and metabolic transformation. SPMDs will model only passive processes.

A weakness of the OECD 117 method is that it does not always provide a reliable indication of the *quantity* of each component present – in fact, in some instances the peaks detected represent only trace components or solvents and active ingredients are not registered at all. Surfactants submitted for testing may often be complex mixtures, rather than pure compounds, and the analytical costs associated with alternative surrogate techniques may be multiplied accordingly. When adopting alternative approaches, it might be better to focus initially on a selected range of widely used 'generic' individual surfactant compounds, and use the resulting data as a form of range-finding exercise. In any case, the 'success' of the studies will depend critically on the precision of the chemical assays that are developed – even using the SPMD it will be necessary to analyse for individual compounds both in the internal solvent and in the exposure medium. The SPMD method seems to simply represent a technical improvement of the OECD 107 shake-flask method, but would still be subject to the same constraints when applied to surfactants, although the formation of emulsions would be avoided. For all its shortcomings, a practical advantage of the OECD 117 method is that it is possible to 'analyse' mixtures, without the need for compound-specific analytical methods (and without in most instances knowing which compounds are represented by the chromatography peaks).

Current developments in SPMD technology involve fairly large-scale test systems that would impose unacceptably high costs on current testing requirements, and in many cases practical restraints on a general widespread adoption of the method. There is obviously a need for 'laboratory scale' systems providing low-cost integrated methods suitable for use at realistic environmental concentrations. Small SPMDs suitable for laboratory use are under development, but their suitability for use with surfactants or other highly hydrophobic chemicals is currently unknown. However, in any program designed to develop an alternative surrogate technique for estimating surfactant BCFs, a sufficiently large number of chemicals will need to be examined in order to derive an independent QSAR. In view of the likely cost restraints, it is almost inevitable that there will be greater reliance on existing data. A thorough review of the literature with a view to defining exactly what (in terms or reliability and precision) could be achieved from existing data is therefore desirable. This review provides a sound basis on which to further develop this approach. Such an assessment can then be compared with estimates of what could be achieved from an acceptable (in terms of time and cost) experimental programme, and an assessment made as to whether such a programme would actually offer real, quantifiable benefit in terms of the quality of the QSAR. Pragmatically, there is no advantage in having a more thoroughly validated data set if it does not result in a tangible improvement in precision and reliability.

The main stumbling block to further development of the QSAR approach to BCF estimation is the substantial effort and cost that would be associated with establishing experimental BCF values with which to compare surrogate measures. A unified (harmonised) approach to BCF testing in live animals is currently lacking, reflected by the uncertainty of the reliability of existing BCF values. The time and cost of developing appropriate extraction and analytical methods for a suitably large number of surfactants would be high; before starting, it would be essential to set targets for recovery and precision, so that it would be possible to judge when sufficient work had been done to deliver a reliable and useable method. There would be no point in correlating an experimental measure with a surrogate measure if the confidence limits on the former were as high as ±100%. Setting such performance parameters should be an integral part of any such project.

6 GENERAL CONCLUSIONS

1. There is limited ecotoxicological data for surfactants in the marine environment.

2. BCFs for surfactants in the aqueous phase are generally below the level for concern. Many reported concentration factors are probably overestimates.

3. BCFs derived from current QSARs based on log Pow data for surfactants are not reliable.

4. Existing data does not indicate a specific generic problem with aquatic toxicity or persistence.

5. There is no evidence to support concerns with respect to biomagnification of surfactants.

6. There is no evidence to support concerns with respect to long-term retention of bioaccumulated surfactants.

7. Two surrogate partitioning techniques which may be usefully explored as alternative approaches to determining partition coefficients for surfactants are:

    • MEEKC (MicroEmulsion ElectroKinetic Chromatography)
    • SPMD (SemiPermeable Membrane Devices)

Acknowledgements

EOSCA wishes to thank the Associate Members who carried out this work.

References

1   Ahel M, McEvoy J, Gieger W (1993) Bioaccumulation of the lipophilic metabolites of nonionic surfactants in freshwater organisms. Environmental pollution 79:243-248

2   Bakke S, Vik EA, Stang P, Kelley A, Ravnestad M, Voldum K, Jensen RD (2000) Chemical mass balance model – a tool to predict present and future chemical discharges with produced water. Paper SPE 61199 Proceedings of the International Conference of Health, Safety and Environment in Oil and Gas E&P. Stavanger, 26-28 June (submitted)

3   Leo AJ, Hansch C (1979) Substituent constants for correlation analysis in chemistry and biology. Wiley, New York

4   Lewis MA (1990) Chronic toxicities of surfactants and detergent builders to algae: a review and risk assessment. Ecotoxicol Environ Safety 20:123-140

5   Lewis MA (1991) Chronic and sublethal toxicities of surfactants to aquatic animals: a review and risk assessment. Wat Res 25:101-113

6   Røe TI, Johnsen S, Nordtug T (1998) Bioavailability of polycyclic aromatic hydrocarbons in the North Sea. PhD thesis.

7   Salimi-Moosavi H, Cassidy RM (1996) Application of non-aqueous capillary electrophoresis to the separation of long-chain surfactants. Anal Chem 68:293-299

8   Scott Hall W, Patoczka JB, Mirenda RJ, Porter BA, Miller E (1989) Acute toxicity of industrial surfactants to *Mysidopsis bahia*. Arch Environ Contam Toxicol 18:765-772

9   Smith JT, Vinjamoori DV (1995) Rapid determination of logarithmic partition coefficients between n-octanol and water using micellar electrokinetic capillary chromatography. J.Chromatogr B 669:59-66

10 Rankin JC, Stagg RM, Bolis L (1982) Effects of pollutants on fish gills. In: Houlihan DF, Rankin JC, Shuttleworth TJ (eds) Gills. Society for Experimental Biology Seminar Series 16. Cambridge University Press, Cambridge, pp 207-219

11 Stagg RM, Rankin JC, Bolis L (1981) Effect of detergent on vascular responses to noradrenaline in isolated perfused gills of the eel, *Anguilla anguilla* L. Environ Pollut 24:31-37

12 Staples CA, Weeks J, Hall JF, Naylor CG (1998) Evaluation of aquatic toxicity and bioaccumulation of C8- and C9- alkylphenol ethoxylates. Environ Toxicol Chem 12:2470-2480

13 Sæten JO, Nordstad EN, Knudsen BL, Aas N (1999) Chemicals in the value chain. Paper presented on the 10th International Oil Field Chemicals Symposium, Fagernes, 28 February – 3 March

14 Tolls J, Sijm DHTM (1995) A preliminary evaluation of the relationship between bioconcentration and hydrophobicity for surfactants. Environ Toxicol Chem 14:1675-1685

15 Tolls J, Sijm DTHM (1999) Bioconcentration and biotransformation of the nonionic surfactant octaethylene glycol monotridecyl ether ^{14}C-$C_{13}EO_8$. Environ Toxicol Chem 18:2689-269

16 Tolls J, Sijm DTHM, Kloepper-Sams P (1994) Surfactant bioconcentration - a critical review. Chemosphere 29:693-717

17 Tolls J, Haller M, De Graaf I, Thijssen MATC, Sijm DTHM (1997) Bioconcentration of LAS: experimental determination and extrapolation to environmental mixtures. Environ Sci Technol 31:3426-3431

18 Tolls J, Haller M, Labee E, Verweij M, Sijm DTHM (2000) Experimental determination of bioconcentration of the nonionic surfactant alcohol ethoxylate. Environ Toxicol Chem 19:646-653

19 Ukeles R (1965) Inhibition of unicellular algae by synthetic surface active agents. J Phycol 1:102-110 [citation]

ALKYLPHENOL BASED DEMULSIFIER RESINS AND THEIR CONTINUED USE IN THE OFFSHORE OIL AND GAS INDUSTRY

P. Jacques[1], I. Martin[2], C. Newbigging[3] and T. Wardell[4]

[1]EOSCA ͨ/ₒ Baker Petrolite Ltd., Kirkby Bank Road, Knowsley Industrial Park (North), Liverpool L33 7SY
[2]EOSCA ͨ/ₒ T R Oil Services Ltd., Howe Moss Place, Kirkhill Industrial Esta , Dyce Aberdeen AB21 0GS
[3]EOSCA ͨ/ₒ Champion Technologies Ltd, Maitlands Quay, Sinclair Road, Aberdeen AB11 9PL
[4]EOSCA ͨ/ₒ Uniqema, Wilton Centre, Wilton, Redcar TS10 4RF

1 ABSTRACT

Due to their unique properties, alkylphenol/formaldehyde based resins are widely used in the offshore Oil & Gas industry for crude oil demulsification. For continued production at some oilfields there are no substitutes currently available, with resin chemistries forming an integral part in achieving crude oil specifications and economic production rates, at relatively low dosage levels - typically below 30 mg/l based on crude oil.

To date there is no conclusive evidence to suggest that the use of alkylphenol/formaldehyde based resins causes serious environmental problems in the aquatic marine environment. In certain North Sea markets, however, it has been suggested that these chemistries could possess the potential to cause environmental damage e.g. through endocrine disruption in marine organisms, as has been postulated for alkylphenol ethoxylates.

This paper will seek to illustrate the importance of alkylphenol/formaldehyde based resins in achieving economic oil production. It will also highlight, through the presentation of recent scientific studies sponsored by EOSCA, the health, safety and environmental profile of the chemistry, in order to illustrate that, to date, there is no strong evidence to suggest that alkylphenol based resins have a negative health, safety or environmental impact.

2 INTRODUCTION TO CRUDE OIL EMULSION PROBLEMS

During the lifetime of an oilfield, water will eventually be co-produced with the crude oil. As a field matures, formation water becomes co-produced with the oil and in many fields waterflooding adds a secondary source of water to the produced fluids. The presence of these oil and water mixtures can lead to the formation of emulsions, defined as the combination of two immiscible liquids, one dispersed in the other, and stabilised by an emulsifying agent[1]. Production equipment such as chokes or valves can restrict the natural flow of the produced fluids, thus causing them to shear and mix to form a dispersion of water in oil, which is stabilised by emulsifying agents such as paraffin waxes, asphaltenes or inorganic particles.

There are many factors that affect the stability of an emulsion[1], such as:

- the type of emulsifying agents present
- the agitation in the system
- the water content of the oil
- the viscosity of the oil
- the specific gravity of the oil
- the temperature of the system
- the age of the emulsion

3 THE NEED TO RESOLVE CRUDE OIL EMULSIONS

The resolution of emulsions is essential in order to maintain economic oil production, and hence a lot of emphasis is placed by oil producers upon the removal of most of the water from the oil. Pipeline operators and refineries also regard the treatment as an important step in reducing corrosion and erosion of their lines and process systems.

To summarise, the major reasons for resolving emulsions, and hence dehydrating and desalting crude oil, are[2]:

1   Crude oil purchasers specify maximum permissible contents of sediments and water, S&W, formerly called basic sediment and water, BS&W. Typically limits vary from 0.1 to 3 wt. %: e.g. 0.1% in some locations, 0.5% in the Gulf coast and Texas, and 3% for low-gravity California crudes.

2   Crude oil is bought and sold on a °API gravity basis and high-gravity oils command higher prices. Water lowers the °API gravity and reduces the selling price of oil.

3   Shipping emulsified oil wastes costly transportation capacities occupied by valueless water (i.e., S&W).

4   The viscosity of crude oil increases as the water content is increased. (Adding 1% more water typically produces a 2% viscosity increase in a 30°API crude and a 4% viscosity rise in a 15°API crude.)

5   Mineral salts present in oilfield waters corrode production equipment, tank cars, pipelines and storage tanks.

6   Refining of water-bearing crude can cause severe corrosion and plugging problems. Distillation of crude containing water-borne inorganic salts contributes to corrosion and fouling of refining equipment. Under some circumstances chlorides can hydrolyse to HCl, which is extremely corrosive.

7   There is a need to have a good oil-in-water interface in the separator. Failure to do so can result in process upsets which can lead to choking back wells in order to meet specifications. This obviously has cost implications.

8   Treated brine or produced water must be essentially oil-free to satisfy environmental discharge regulations (often 15 to 40 ppm) or to prevent re-injection problems such

as scale formation and/or reservoir plugging.   Note that the oil-in-water environmental regulation is far more severe than the water-in-oil transport/sales specification.

Any method of removing water, salt, sand, sediments, and other impurities from crude oil is called oil treating. Oil-treating methods have one common goal; namely, to provide a suitable environment for gravity to separate the brine from the crude. This paper will concentrate on chemical treatment as treating method, but others include:

1   Settling or providing low velocity (reduced turbulence and increased residence time to allow free water to separate).
2   Degassing or separating the gas from the liquid as it is released in the production equipment.
3   Washing or providing a continuous-phase water wash.
4   Heating to reduce oil viscosity and accelerate separation.
5   Electrical treating (i.e., applying electrostatic fields to enhance water coalescence).

4 CHEMICAL DEMULSIFICATION OF CRUDE OIL

Crude dehydration involves optimizing the synergistic use of up to four techniques; namely, demulsifying chemicals, retention time, heat and electricity, to produce clean oil and clean water[3].

All dehydration systems require settling time and demulsifiers; some require additional heating although most operate at production temperatures.   Electrostatic dehydration/desalting is used in certain areas. In almost all circumstances demulsifiers are a must-have chemical or the emulsion will almost certainly not be satisfactorily resolved.

Demulsifiers are surface-active chemicals, with four main actions[4]:

4.1 Strong attraction to the oil-water interface

They must be attached to and displace or neutralise the emulsifying agents already on the droplet interface.

4.2 Flocculation

They must neutralise any repulsive electrical charges between the dispersed drops and so allow the drops to touch and start to coalesce.

4.3 Coalescence

They must promote small droplets to combine into drops large enough to settle; this requires that the film surrounding and stabilizing the droplets be ruptured.

4.4 Solids wetting

They must prevent fines at the droplet interface from physically preventing coalescence. Clays, drilling muds, and iron sulfide fines can be water wet so that they leave the film interface and migrate into the water phase. Asphaltenes and waxes can be dissolved or oil wet to disperse them into the continuous crude phase.

A single chemistry type cannot provide all four of the required actions, so commercial demulsifiers are usually proprietary mixtures of these compounds (i.e., surfactants), blended with suitable solvents such as heavy aromatic naphtha (HAN) and isopropyl alcohol (IPA) to obtain a liquid that pours at the lowest expected temperature.

The formulated blended demulsifier is essentially oil soluble to promote rapid diffusion to the oil-water interface of the emulsified water droplet. At that site the various blended surfactants carry out the total demulsification process of flocculation, coalescence and solids wetting, leading to complete emulsion resolution and water separation.

Theories of how demulsifiers act are incomplete. They fail to explain the extreme specificity of the various types of chemicals on any particular crude-water mixture. One traditional theory about why demulsifiers work is that they "neutralise" the emulsifying agent. Another explanation is that the demulsifying chemical makes the film surrounding the water drop very rigid. When the oil drop expands on being heated, the film is ruptured. Alternatively, if the chemical makes the film contract, then heat is not required to burst the film.

5 ALKYLPHENOL/FORMALDEHYDE RESINS

Table 1 below presents a brief history of chemicals used to break water in oil emulsions since pioneering efforts in the early 1900s.

Table 1 *History of Chemical Demulsifiers (after reference 5)*

Period	*Dosage (ppm)*	*Type of Chemicals*
1920s	1,000	Soaps, salts of naphthenic acids, aromatic and alkylaromatic sulfonates, sulfated castor oil
1930s	1,000	Petroleum sulfonates, oxidized castor oil, and sulfosuccinic acid esters
Since 1935	100 - 500	Ethoxylates of fatty acids, fatty alcohols, and alkylphenols
Since 1950	50-200	Ethylene oxide/propylene oxide (EO/PO) copolymers, p-alkylphenol/formaldehyde polymers + EO/PO and modifications
Since 1965	30-100	Amine oxyalkylates
Since 1976	10 – 100	Oxyalkylated, cyclic p-alkylphenol formaldehyde polymers, and complex modifications, diepoxides
Since 1986	5 – 50	Polyesteramines, polyimines and blends

Amongst all the demulsifiers that have been used over the years, alkylphenol/formaldehyde resins stand out as one of the single most effective chemistries, in terms of economic and environmental considerations[5]. Firstly, they possess the ability to drop water from emulsions very quickly and, in addition, can water wet solids so that they leave the film interface and migrate into the water phase for subsequent removal.

Secondly, they have the ability to give good quality clean separated water for subsequent overboard discharge. This is important following the advent of ever tightening environmental legislation, and will be discussed later in the paper.

6 NORTH SEA RESIN USE

Data obtained from the SCOPEC Environmental Emissions Monitoring System (EEMS), a database operated on behalf of the DTI and UKOOA, which is set up for operators to report their chemical use and discharge data on a quarterly basis, indicates that during 2000 alkylphenol/formaldehyde resins were the single most widely used demulsifier chemistry in the UK sector of the North Sea, with over 50% of the total number of demulsifiers in use, containing resin chemistry[6].

The above fact indicates that a cessation in the use of resin chemistry, for example due to a ban based on environmental impact grounds, would undoubtedly cause serious operational difficulties and production system upsets in fields where crudes are extremely difficult to treat. This could have serious economic repercussions in terms of poor oil in water separation leading to crude oil being outside purchasing specification. It could also have negative environmental implications, in the form of increased overboard oil in water levels due to enforced use of less effective alternative demulsifier chemistries.

7 HSE PROFILE OF ALKYLPHENOL/FORMALDEHYDE RESINS

7.1 Health

Mammalian toxicity data indicates that alkylphenol/formaldehyde resins have an acute oral toxicity (LD_{50}) to rats of approximately 16,000 mg/kg[7]. Under current EU legislation[8], this experimental value indicates that resins would be classified as non-hazardous to man.

7.2 Environment

7.2.1 Harmonised Mandatory Control Scheme. The new Harmonised Mandatory Control Scheme[9] (HMCS) being introduced in the North Sea regions requires relevant member states to:

- Provide test data on all chemical substances and preparations used offshore to the Harmonised Offshore Chemical Notification Format[10].

- Subject chemicals to a Harmonised Pre-screening Scheme for Offshore Chemicals on a substance by substance basis, where possible[11].

- Rank chemicals according to their generic PEC:PNEC ratios using the Chemical Hazard Assessment and Risk Management (CHARM) 'hazard assessment' module as the primary tool[12].

- Make management decisions on chemical use based on environmental impact assessment, with one of the following outcomes[12]:

  - Permission
  - Substitution
  - Temporary Permission
  - Refusal of Permission

7.2.2 Pre-screening fate. Under the new HMCS Regulations, many alkylphenol/formaldehyde resins pass through the Pre-screening assessment stage without the requirement to be substituted.

This is illustrated in Table 2 below, and is based on the fact that, in general:

- many have a biodegradation rate of >20% in saltwater OECD 306 tests
- they are non-bioaccumulative due to their high molecular weight (>1000)
- their inherent aquatic toxicity is >10 mg/l against the most sensitive aquatic taxonomic group.

Table 2 *Ecotoxicological profile of typical Alkylphenol/formaldehyde Resin*

Aquatic Toxicity[13]	Criterion	Value[14]
Skeletonema Costatum	EC50, 72 hrs	15 mg/l
Acartia Tonsa	LC50, 48 hrs	670 mg/l
Corophium Volutator	LC50, 10 day	2149 mg/kg
Bioaccumulation Potential[15]	Log Pow Range	Surfactant *
Biodegradation[16]	% @ 28 days	39

* Log Pow is not valid for surfactants. Resin will not bioaccumulate due to very high molecular weight

7.2.3 CHARM Profile. When alkylphenol/formaldehyde resins are ranked according to their PEC:PNEC ratios using the CHARM "hazard assessment" module, they can achieve hazard quotient values substantially <1. The resin example in Table 2 is a good example of this, with a negative PEC:NEC[14] ratio of 6.01239E-08.

The above indicates that under HMCS criteria, resins may cause little environmental damage under standard application conditions. This is due overall to their environmental profile, high oil solubility - leading to little or no discharge with overboard water, and low application rates – typically 5 ppm.

7.3 Endocrine Disruption

Although resins are widely used in demulsification, concerns have been expressed in parts of the OSPAR region about the risks they pose to the marine environment[17]. These concerns have stemmed from the fact that because resins are produced from alkylphenols in a reaction with formaldehyde (and subsequent alkoxylation with ethylene and/or propylene oxide), it is assumed that they will display the same reported potential to cause hormonal disruption in marine organisms as alkylphenols and their ethoxylate chemistries.

In order to investigate these concerns EOSCA commissioned In Vitro Recombinant Yeast Assays at two independent third party Laboratories on a range of 12 "as used" resin bases, varying in their level of alkoxylation[18,19]. The assay methodology has been fully validated[20], and produces a measurable colourimetric response in the presence of compounds which interact with the human oestrogen receptor (hER).

The results from both laboratory tests showed conclusively that alkoxylated alkylphenol/formaldehyde resins do not display any endocrine disrupting properties when compared to the female hormone reference 17B-estradiol[18,19].

Both laboratories indicated that these results provide confidence that the resins are not oestrogenic in vivo. Hence, this demonstrates that resins used offshore in crude oil demulsification appear not to cause hormone disruption in the environment. It also illustrates that any residual alkylphenol that may be left in the resin base from the manufacturing process does not seem to have any negative effect.

As an additional piece of investigative work, EOSCA is currently evaluating testwork that will take into account the potential breakdown of resins in the environment, and subsequently examine the oestrogenicity of any metabolites that may be formed.

8 CONCLUSIONS

This paper has illustrated the importance of alkylphenol/formaldehyde based resins in achieving economic oil production in the North Sea. It has also highlighted, through the presentation of scientific data, that resins do not have an overtly negative health, safety or environmental impact.

The following key conclusions have been drawn:

- The loss of resin chemistry in the UK Sector of the North Sea would cause considerable operational and economic problems to the oil industry due to the lack of proven alternate chemistries available to treat difficult oilfield emulsions.

- Alkylphenol/formaldehyde resins have relatively low toxicity to man, an important factor to consider when reviewing the overall HSE impacts of oilfield chemicals.

- Under current environmental legislative requirements operating companies must justify the selection and use of demulsifier chemicals to the relevant competent authorities. A high priority part of this justification is the environmental impact assessment of the chemical relating to its use and discharge. Given this, it is crucial that this assessment considers the effects that demulsifier efficacy have on overall environmental impact.

- Alkylphenol/formaldehyde resins have the capacity to separate oil and water highly effectively at low chemical dosage levels, thus aiding in the achievement of low oil in water overboard levels. Hence, the substitution of resins for a less effective demulsifier chemistry, on environmental grounds, could in fact increase environmental harm.

- Alkylphenol/formaldehyde resins are almost entirely oil soluble. Consequently, almost all of this chemistry used offshore will remain with the crude oil, thus never contacting the marine environment, and resulting in minimal resin discharge in produced water.

- The recombinant yeast assays commissioned on alkylphenol/formaldehyde resins to date by EOSCA, prove a negative link between the chemistry and potential endocrine disruption in the marine environment.

References

1   B. Rowan, *The Use of Chemicals in Oilfield Demulsification,* Liverpool, Petrolite Ltd., (1992).

2   *Treating Oilfield Emulsions,* Third Edition, Oilfield Processing, Volume 2: Crude Oil.
3   S. Taylor, *Resolving Crude Oil Emulsions,* Chemistry & Industry, 19 October 1992.
4   H. Celius et al, *Separation Mechanisms and Fluid Flow in Oil/water,* Separation, 7th International Symposium, Oilfield Chemicals, Geilo, Norway, 1996.
5   F. Straiss et al, *SPE Production Engineering,* 1991, **334-8**.
6   EEMS Database, *Demulsifier use and discharge data from 2000,* obtained with the consent of DTI and UKOOA.
7   J. McMahon, Baker Petrolite, 1990, St Louis, Missouri, USA.
8   European Community: Directive 93/21/EEC, *Revised Annex VI to the Dangerous Substances Directive,* 67/548/EEC.
9   OSPAR Commission Meeting 26-30 June 2000, Summary Record Document OSPAR 00/20/1-E http://www.ospar.org/eng/html/welcome.html
10 OSPAR Recommendation 2000/5 on a Harmonised Offshore Chemical Notification Format (HOCNF).
11 OSPAR Recommendation 2000/4 on a Harmonised Pre-screening Scheme for Offshore Chemicals.
12 OSPAR Recommendation 2000/2 on a Harmonised Mandatory Control Scheme for the Use and Reduction of the Discharge of Offshore Chemicals.
13 OIC 01/6/2-E Draft OSPAR *Guidelines for Toxicity Testing of Substances and Preparations Used and Discharged Offshore* presented by Denmark at the Oil Industries Committee meeting, February 2001.
14 Baker Petrolite, proprietary ecotoxicological data, Liverpool, UK
15 *Partition Coefficients (n-octanol/water). High Performance Liquid Chromatography (HPLC) Method,.* OECD Guideline 117 for Testing of Chemicals, 30 March 1989.
16 OECD Guideline 306 for Testing of Chemicals, *Biodegradability in Seawater – Closed Bottle Test,* 17 July 1992.
17 Requirements for ecotoxicological testing and environmental assessment of offshore chemicals and drilling fluids, SFT, 1998.
18 Test Report, *Screen for the presence of oestrogenic activity in chemicals for use offshore,* CEFAS, 1999.
19 Test Report, *Screen for the presence of oestrogenic activity in chemicals for use offshore,* Brunel University, 2000.
20 E.J. Routledge and J.P. Sumpter, *Estrogenic activity of surfactants and some of their degradation products assessed using a recombinant yeast screen,* Environmental Toxicology and Chemistry 15, 1996, (3) **241-248**.

New Technology

USE OF ENZYMES FOR THE *IN-SITU* GENERATION OF WELL TREATMENT CHEMICALS

Ian D. McKay and Ralph E. Harris

Cleansorb Ltd, The Surrey Technology Centre, 40 Occam Road, The Surrey Research Park, Guildford, Surrey, GU2 7YG, UK.

1 INTRODUCTION

The effective chemical treatment of oil or gas wells requires the efficient delivery of the treatment chemicals to where they are needed.

Delivering certain types of chemicals from the surface is often problematic. Active chemical components may react with surfaces within the tubulars and downhole equipment or with the rock surface at the point of introduction into the wellbore or formation. As a consequence, some chemicals may not physically reach the target zone for which they are intended. In general terms, it is much easier to deliver a chemical treatment into a short section of a vertical wellbore than to effectively place a larger treatment evenly to the whole of a long horizontal or directional wellbore.

Recent trends in the oil industry such as the increasing use of extended reach wells, as well as existing problems with openhole wells and gravel packed completions have exacerbated the problems of achieving effective placement of a number of chemicals, particularly the placement of acid.

An alternative to delivering active chemicals from the surface is to produce the required chemical in the treatment fluid after the fluid is placed in the target zone downhole.

In-situ generation of chemicals is already used in some well treatment processes, for example the generation of hydrogen fluoride from ammonium bifluoride in mud acid. In the case of mud acid the hydrogen fluoride is generated in the mud at the surface. This substantially reduces the hazards associated with the operation compared to using hydrogen fluoride direct. In other cases the desired chemicals are generated downhole. Examples include (a) the setting of temporary plugs such as calcium silicate (produced from calcium chloride and sodium silicate) and guar-chromate (b) cross-linking of fracturing gels where the operation is designed to allow cross linking to occur as the fluid enters the formation and (c) cross linking of flow modifiers such as polyacrylamide for profile control.

A number of well treatment processes are based on the thermal hydrolysis of precursor chemicals in the treatment fluid that leads to a desired result. For example, thermal hydrolysis has been used in methods for increasing the pH of a treatment fluid in order to precipitate some classes of scale inhibitor. However, processes dependent on thermal hydrolysis are of necessity limited to hot wells, above say 100° C, as the rates of hydrolysis at low temperatures are generally too slow.

The use of enzymes as catalysts to facilitate the downhole production or deposition of a well treatment chemical is another approach to generating chemicals *in-situ* [1,2].

The ability of enzymes to catalyse specific reactions has made them very useful in a growing number of major industrial applications [2]. Enzymes are widely used in a range of major industries including the brewing, distilling, wine, animal feed, baking, dairy, cleaning, detergent, fats and oils, leather, personal care, pulp and paper, food and drink, pharmaceutical, fine chemical, diagnostic and textiles industries.

In these applications particular types of enzymes are used to either break down or produce specific chemicals. Because enzymes are catalysts they are only needed in small quantities. The use of enzymes can improve the efficiency of processes resulting in significant savings in water, energy or chemicals and can greatly reduce waste [48,50]. Examples of enzyme based processes which are much cleaner than their existing chemical counterparts include the use of enzymes in the production of medium density fibreboard (MDF) [49]. Use of enzymes to produce MDF avoids the use of formaldehyde. Another example is the use of enzymes in textile desizing. The enzyme-based process can reduce water consumption by 30 to 50% and environmental loadings by up to 40%. The textile desizing process recently received the prestigious Presidential Green Chemistry Challenge Award in June 2001 presented by the US Environmental Protection Agency [48,51].

Enzymes increasingly form the basis of environmentally clean technologies. They pose no significant environmental or health and safety problems. They are low hazard and eventually biodegrade to carbon dioxide, ammonia and water. Enzymes are more and more being used in processes which replace more hazardous or inefficient conventional chemical methods.

Using enzymes permits a chemical change to be produced in a treatment fluid in a predictable manner. It also allows a chemical change to occur under conditions where the non-catalysed rate would be insufficient to achieve the desired effect. The use of enzymes has potential to facilitate the delivery of a range of well treatment chemicals.

The approach is already being successfully used in the field for the *in-situ* production of acid. The enzyme-based process for producing acid and other processes for the delivery of other treatment chemicals, including the controlled deposition of scale inhibitors *in-situ*, are outlined below.

2 ENZYME-BASED *IN-SITU* PRODUCTION OF ACID

2.1 Background to use of acid for well treatment

Acid treatments of wells are one of the most frequent chemical treatments used in the oilfield. Acidizing is used to increase production rates in many situations. The most important applications include damage removal, completion and stimulation of horizontal wells, matrix acidizing, fracture acidizing and gel breaking.

Acidizing has been used to remove formation damage for over 100 years [3]. The acid most commonly used to remove formation damage in carbonate reservoirs is hydrochloric acid. In general, damage across the producing intervals of oil and gas wells is removed through injection of acid as a stimulation fluid [4].

Effective drilling fluid damage removal through efficient acidizing has the potential to deliver excellent financial benefits to operators. Reducing the skin value along the whole of a horizontal openhole section means that the well comes on production at maximum rates and stays on plateau for longer. In wells with production potentials of thousands of barrels per day, reducing the skin by even a modest amount, say from a skin value of 10 to 0 can result in an increase in the net present value of tens of millions of dollars [5,19]. In less prolific wells, significant financial benefits can still be obtained.

Acidizing treatments graphically illustrate the problems encountered when trying to successfully place a reactive chemical into wells in underground formations and the type of benefits which may be obtained from producing treatment chemicals such as acid *in-situ*.

All conventional acids, including hydrochloric acid and organic acids react very rapidly on contact with acid sensitive material such as calcium carbonate in the wellbore or formation. The rapid reaction means the acids do not penetrate very far into the formation before they are spent. Rapid fluid leak-off by wormholing may occur near the point of introduction. This can severely limit the effectiveness of reactive acids in a number of acidizing applications where deep acid penetration or uniform placement is needed.

In wells with relatively short production intervals the effectiveness of hydrochloric acid may be improved by using coiled tubing in combination with a diversion technique such as use of foam or gel. However, problems in placing acid are compounded when attempting to treat long horizontal or directional wells. Newly drilled horizontal wells normally require acidizing to remove drilling mud damage before being brought into production. The efficient placement of conventional acids is critical. It has proved to be very difficult to apply hydrochloric acid to uniformly remove drilling fluid damage in extremely long horizontal producing intervals [1]. This has been identified by several operators as a very serious problem with the result being disappointing well productivity [1,19]. In such wells uniform placement of acid along several thousand metres of wellbore may be required.

The rapid reaction rate of hydrochloric acid and organic acids with carbonate rock or scale prevents deep penetration into the formation and uniform placement. The industry has therefore actively sought and developed a range of methods by which the rate of reaction of an acid may be slowed or "retarded" in some way to permit penetration of the acid deeper into a carbonate formation or to achieve more uniform acidizing of filter cakes over long horizontal producing intervals.

Methods of retarding the reaction rate include:

- Emulsifying the aqueous acid solutions in oil (or solvents such as kerosene or diesel fuel) to produce an emulsion which is slower reacting [6-8].

- Dissolving the acids in a non-aqueous solvent [9-12].

- The use of non-aqueous solutions of organic chemicals which release acids only on contact with water [13-14].

- The use of a solution of methyl acetate which hydrolyses at high temperatures (preferred temperature of 160° C) to produce acetic acid [15].

- Gelling the acid [16,17].

The most common approaches to retarding acid include emulsifying the acid in a hydrocarbon, gelling of the acid or chemically retarding the acid in some way [1].

Emulsifying the acid is probably the most important of these methods. Emulsified acids may contain the acid as either the internal or the external phase of the emulsion. The former, which is more common, normally contains 10 to 30% v/v hydrocarbon as the external phase and 70 to 90% v/v of 15% hydrochloric acid as the internal phase. When acid is the external phase, the ratio of oil to acid is often about 2:1. Both the higher viscosity created by emulsification and the presence of the oil can retard the rate of acid transfer to the rock surface. This reduction in mass transfer rate, and its corresponding reduction in acid reaction rate, can increase the depth of acid penetration into the rock formation before the acid reacts with the rock or formation damage [4].

There are some problems associated with the use of oil external emulsified acids. For example there may be increased frictional resistance to flow of these fluids down well tubulars [4]. The presence of surfactants in the acidizing fluid, to produce the emulsion, can also affect the wetting characteristics of the rock formation i.e. may change a water wet rock surface into an oil wet surface. This can necessitate remedial post acidizing treatments to restore the rock surface to a water wet state if successful oil production is to be attained [4].

Gelled acids are used to retard acid reaction rate in treatments such as acid fracturing. Retardation results from the increased fluid viscosity reducing the rate of acid transfer to the fracture wall. Use of the gelling agents (normally water soluble polymers) is generally limited to lower temperature formations as most gelling agents degrade rapidly in acid solution at temperatures above 130° F (55° C) [4].

Gelling agents are seldom used in matrix acidizing because the increased acid viscosity reduces injectivity and may prolong the treatment with no net benefit i.e. the slower injection rate counters the benefit of a reduced reaction rate.

Chemically retarded acids are often prepared by adding an oil-wetting surfactant to the acid in an effort to create a physical barrier to acid transfer to the rock surface. In order to achieve this the additive must adsorb on the rock surface and form a coherent film. Use of these acids often requires continuous injection of oil during the treatment. At high flow rates and high formation temperatures, adsorption is diminished and most of these materials become ineffective [4].

2.2 *In-situ* acid production using enzymes

An alternative to retarding the rate of reaction of an acid introduced from the surface is to generate the acid *in-situ*. If the treatment fluid is initially non-reactive, when the fluid is placed into the wellbore the rapid reaction and resulting leak-off to the formation that occurs with conventional acids will not be encountered. If acid is then produced homogeneously throughout the treatment fluid, after the treatment fluid has been placed in the target zone, similar amounts of acid will be delivered to the whole of the intended target zone and the efficiency of the treatment will be high.

A process based on the use of enzymes to generate acid *in-situ* is now available and has been used in a number of different acidizing applications [1,2,18,19].

The patented acidizing process uses a solution of a suitable ester in combination with an enzyme capable of hydrolysing the ester. The enzyme hydrolyses the ester (the substrate for the enzyme) to produce an organic acid, normally acetic acid, *in-situ* (Figure 1).

Ester ⎯⎯⎯⎯⎯⎯⎯⎯⟶ Organic Acid + Alcohol
                ENZYME

Figure 1 *Basis of the enzyme-based acidizing process*

As the acid is produced, it becomes available for reaction with acid soluble materials such as calcium carbonate.

The amount of acid released, and the rate of release, can be controlled by using different formulations. The amount of acid delivered is determined by the initial concentration of the ester ("acid precursor") and the rate of production of the acid by the concentration of the enzyme catalyst.

The acid generating formulation is produced by blending the esters in water at the surface, then adding the enzyme catalyst. Mixing is continued for a few more minutes, then the treatment fluid is injected at the desired rate to fill the wellbore or the formation. Only simple mixing equipment is required and there is generally no need for high rate pumps. After placement of the fluid the well is shut in for the required period and acid is generated *in-situ*. The acid reacts with acid soluble material such as filter cake, calcium carbonate scale or a carbonate rock matrix, resulting in the production of water-soluble calcium acetate. The well is then put on production.

Almost all of the acid is produced after the fluid is placed. Because the fluid is essentially not reactive when placed, it does not suffer from the placement problems associated with pumping reactive acids such as HCl. In the case of a new well the presence of an intact filter cake will assist in the placement of the fluid by preventing leak off of fluid to the formation during pumping. Treatments are designed so that breakthrough of the filter cake takes significantly longer than the time required to place the treatment fluid. For example, enzyme based acidizing fluid has been displaced back from the toe to the heel of the well through the drill pipe in new open hole horizontal wells. This typically takes a few hours and once the horizontal section has been filled with the treatment fluid the well is shut in for 1 - 3 days to allow acidizing to occur downhole.

The relatively short time required to place the treatment fluid compared to the time required for in-situ acidizing to be completed allows acid to be effectively delivered to the whole of the target zone: across the whole of a long horizontal interval; deep into natural fracture networks in carbonate formations; to several metres radius of the wellbore in deep matrix acidizing.

Typically, the treatment fluid is formulated to produce 5-10% w/v acetic acid over one to three days. Greater than 95% of the acid produced is produced downhole after placement of the fluid. Enzyme-based acidizing has a number of advantages in a range of acidizing applications including those advantages that arise from the ability to produce a defined amount of acid at a controlled rate downhole (Table 1).

Enzyme based acidizing can be used for a large number of acidizing applications [1,2,18,19]. The most important is probably the removal of drilling fluid damage in new horizontal wells. As previously indicated, this offers very significant financial benefits to operators. The enzyme-based process can be used to remove damage caused by either water-based or oil-based muds in both carbonate and sandstone formations. When used for treatment of oil-based mud damage, formulations include a suitable surfactant. An additional advantage of enzyme-based acidizing is that when treating filter cakes containing biopolymers, the treatment fluid can be formulated to include polymer breaking enzymes or other types of polymer breakers to give a dual attack on the filter cake (carbonate dissolution and polymer breaking) using a single fluid. Indeed it has now been shown that the best results are obtained when both components of the filter cake are treated [20] but until now this has not been possible in a single treatment. The pH conditions created by the enzyme based acidizing system while acidizing is occurring downhole (i.e. the formation of acetic acid–acetate and carbonate buffers) are optimum for the activity of polymer breaking enzymes which may be incorporated into the acidizing fluid.

Table 1   *Advantages of enzyme-based acidizing*

Operational advantages

- Excellent zonal coverage may be obtained

- Uniform placement of acid along long wellbore intervals can be readily achieved

- Can achieve true deep matrix acidizing of carbonate formations

- Low hazard system compared to highly concentrated acids

- All components are very low hazard and toxicity

- Extent of acidizing (quantity of acid, rate of production) can be accurately controlled

- There is no need to pump at high rates or high pressure to obtain effective acid placement

- The components and the mixed fluid are easy to handle

- Ester components have a high flash point

- Fluid is low viscosity allowing rapid clean up post treatment

Environmental advantages

- Components (esters and enzymes) and products (calcium acetate, non-toxic alcohols) of the reaction all have a low environmental impact

- The use of toxic corrosion inhibitors is not required

Other applications for enzyme-based acidizing include the stimulation of horizontal wells which are already on production, deep matrix acidizing of carbonates (where the permeability of a formation is increased several fold uniformly to a radius of a few metres around the wellbore) treatment of carbonate scale, breaking of acid sensitive cross-linked gels such as guar-borate and stimulation of natural fracture networks.

The process may also have applications in fracture acidizing and acidizing of secondary mineralization in coal bed methane beds. Because the process delivers a predicable amount of acid per unit volume of fluid it could potentially also be used for acidizing poorly consolidated sandstone formations where carbonates are the cementing material without running the danger of collapsing the formation.

Laboratory evaluation has indicated that relatively low concentrations of acid need to be produced in-situ to achieve significant increases in the permeability of rock cores treated with the enzyme acidizing process. A series of core experiments were run in which the cores were exposed to formulations capable of producing relatively low concentrations of acid.

Experiments in which cores were treated several times with formulations capable of producing only 50 mM acetic acid had their permeability to 2% KCl increased several fold. A typical result using a San Andres dolomite core is shown in Figure 2. The permeability of the core increased from less than 0.7 to 3 mD during six successive treatments (Figure 2). In matrix acidizing field treatments 300 mM or more acid would be produced downhole using a single enzyme based acidizing treatment.

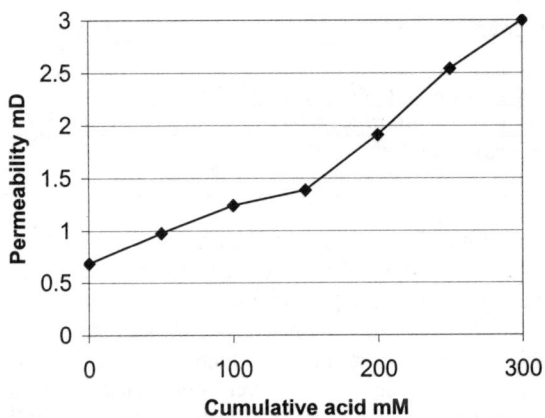

Figure 2   *Increase in permeability of a San Andres dolomite core during six sequential treatments with an enzyme-based acidizing formulation generating 50 mM acetic acid per individual treatment. Cores incubated for 24 h at 40° C and 500 p.s.i. at each stage*

The ability of the process to give good zonal coverage was also investigated in core experiments in which two Hassler sleeve cells were linked in series and acidized with either 15% hydrochloric acid or the enzyme-based acidizing process (Table 2). The treatment fluid reaching the mud damaged face of the second core had to flow across the mud damaged face of the first core i.e. simulating fluid flow across a filter cake on the wellbore face. The initial permeabilities (Ki) of the cores were measured with 2% KCl. The cores were then damaged with KCl/polymer water based mud. The resulting damage was then treated with either 15% HCl or an enzyme-based acidizing formulation which made 5% acetic acid while the core was shut in for 40 hours at 45° C and 500 p.s.i. The final regained permeabilities (Kr) were then measured with 2% KCl.

HCl breakthrough occurred almost immediately on contact with the first core. Further acid pumped to the first core leaked off through this core and consequently minimal acid reached the face of the second core. In addition to the Ki and Kr measurements, there was a lack of fluid flow from the second core outlet during the HCl treatment

The results of the treatment with the enzyme-based acidizing process clearly indicated that the treatment fluid had contacted both cores and increased the permeability of both.

It was apparent that although there was a significant increase in the permeability of the first core in the series that was treated with HCl (the large increase in permeability was due to formation of a wormhole) the HCl had not had a significant effect on the second core in

Table 2   *Effect on permeability of two cores mounted in series following treatment with 15% hydrochloric acid or enzyme-based acidizing fluid.   The enzyme based acidizing formulation produced 5% acetic acid over 40 hours*

Cores acidized with 15% HCl		
Test 1	Initial K (mD)	Final K (mD)
Core 1 (1st core in series)	4.66	837
Core 2 (2nd core in series)	3.37	4.78

Cores treated with enzyme-based *in-situ* acidizing process		
Test 2	Initial K (mD)	Final K (mD)
Core 3 (1st core in series)	6.89	25.07
Core 4 (2nd core in series)	3.97	68.5

the series. By contrast, both of the cores treated with the enzyme-based acidizing fluid had significantly increased their permeability. No wormholes were observed in either of the cores treated with the enzyme-based acidizing process.

Laboratory evaluation of the process has also indicated that highly effective damage removal can be achieved using formulations which are capable of producing relatively modest concentrations of acetic acid. Tests were conducted on carbonate core slabs to evaluate damage removal over a much larger area than is possible with conventional core plug apparatus using 2 types water based mud: KCl/polymer mud; CAT-1 mud.

Initial permeability to 2% KCl was measured before applying a mud squeeze at 500 p.s.i for 30 minutes. The core faces were flushed with 2% KCl to remove excess liquid mud and permeability to 2% KCl re-measured to determine the extent of mud damage. Arcasolve fluid was injected into the core slabs at 500 p.s.i. and 1 pore volume of fluid pumped through the core before the core slabs were heated to 45° C and shut in for 48 hours. Following shut in core slab permeability was re-measured using KCl to determine the extent of damage removal.

The results are shown in Table 3. The treatment gave regained permeabilities of 84% and 93% of the initial permeability respectively. With other core material, muds and formulations higher regain permeabilities have been achieved. In the case of some carbonate materials, regained permeabilities of several hundred percent of initial permeability have been obtained as mud damage has been completely removed and the matrix permeability of the carbonate has been increased by the enzyme-based acidizing treatment. It should be noted that the treatment formulation used in this experiment did not include any polymer breaking enzymes. Combining acidizing with polymer breaking appears desirable [20]. The slightly acidic conditions present in the enzyme-based acidizing fluid are generally near optimum for polymer breaking enzymes. Formulations containing polymer breaking enzymes as well as acid producing enzymes are now being used in the field and yielding excellent results.

Enzyme-based acidizing has already been proved effective in the field for cleaning up new horizontal wells, stimulating mature horizontal wells and deep matrix acidizing. The first field application for the process was in deep matrix acidizing. Increases in production or injection rate of up to 80% have been observed in formations believed to be undamaged [1,19]. The increases exceeded the benefit expected from increasing the permeability of the formations alone and suggested that the treatments had also removed some near wellbore damage.

Table 3   *Water-based mud damage removal using the enzyme-based acidizing process*

Permeability mD	Water-based mud	
	Cat 1	KCl-polymer
Initial	24.0	51.0
Damaged	0.48	1.02
Regained	20.2	47.4

A new horizontal well in a sandstone formation drilled with a carbonate containing polymer mud and treated with the enzyme-based acidizing process to remove drilling mud filter cake damage has recently been reported by the operator to have a productivity ten times that of vertical wells in the same field. Sustained increases (of several fold) in the production rate of a mature horizontal well in a carbonate formation after treatment with the enzyme based acidizing process have also been reported [18].

Use of an acidizing system with a reaction rate so highly retarded that essentially no reaction takes place during the time that the acid is being pumped into the reservoir has been considered in mathematical terms. This hypothetical system has been proposed as the ultimate for a matrix treatment [4]. Enzyme-based acidizing approaches such a system.

3 ENZYME-BASED PROCESSES FOR IN-SITU DEPOSITION OF OTHER MATERIALS

Other enzyme-based processes for *in-situ* generation of chemicals other than acid have been described [2]. Enzyme-based processes may be used to generate minerals, gels or resins *in-situ*.

Hydrolase and oxidoreductase enzymes are the enzymes most likely to be useful in oilfield applications [2]. Hydrolases do not require co-enzymes and are the class of enzyme already being used for the earlier oilfield applications of enzymes (polymer conditioning and polymer breaking).

Hydrolytic enzymes break a covalent bond in their substrate adding a water molecule in the process. The enzymes have a very high degree of specificity, meaning that a given enzyme will only attack specific types of bond in specific types of compound. Hydrolytic enzyme mediated processes for the production of gels, resins and minerals, based on changes in the pH or on the *in-situ* production of specific chemicals have been described [2]. Oxidoreductase enzymes may be used to generate a range of resins [2].

Enzyme based processes for generating resins and minerals have potential for use in sand consolidation. Consolidation of initially completely unconsolidated sand by enzyme-based calcium carbonate deposition has already been demonstrated in the laboratory [2]. Unconfined compressive strengths of up to 100 p.s.i. have been obtained. The process for calcium carbonate deposition is based on urease which hydrolyses urea to carbon dioxide and ammonia. In the presence of a calcium salt, calcium carbonate is deposited (Figure 3).

The enzyme-based process is able to deposit up to 100 g of calcium carbonate per litre of treatment fluid. The yield is proportional to the initial calcium concentration, where there is sufficient urea also present. Where the fluid is made up in seawater, the calcium and magnesium present in the seawater increase yield (Figure 4). Even at the highest levels of yield, the volume of the produced calcium carbonate is only a few percent that of the treatment fluid, so when treating a poorly consolidated sandstone no significant

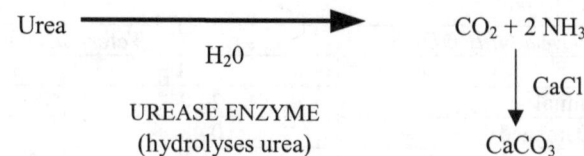

Figure 3 *Enzyme based process for calcium carbonate deposition*

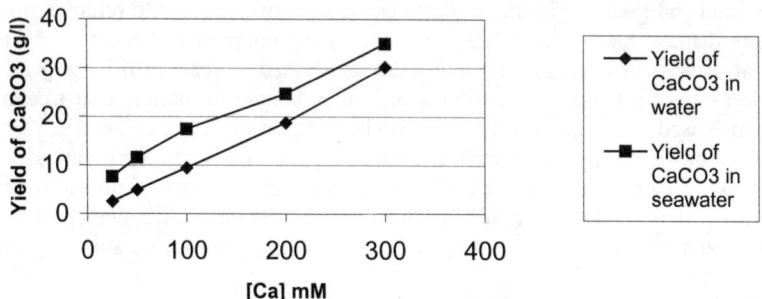

Figure 4   *Effect of initial calcium concentration on yield of calcium carbonate in urease-based process for deposition of calcium carbonate. End point yield at 25° C (24 hours using a known amount of urease)*

reduction in the porosity of the sandstone is expected. Work is currently progressing on using the process to strengthen low unconfined compressive strength sandstones.

Enzyme-based processes for deposition of minerals have potential for water shut-off if sufficient of the mineral can be produced. Enzyme-based processes for generating gels also have potential for water shut-off.

4 ENZYME BASED PROCESS FOR DEPOSITION OF SCALE INHIBITORS

A further possibility which has been proposed is the use of enzymes to deposit inhibitors.

Scaling of wells is a widespread oil industry problem. In many parts of the world, including the North Sea, Alaska, Gulf of Mexico and parts of South America and the Middle East, there is a tendency for wells to scale up during oil production. Scaling occurs due to the particular chemistry of the formation water. Carbonate scaling can occur as a consequence of pressure and temperature changes as the fluid approaches the wellbore.

Susceptible wells need to be dosed with suitable scale inhibitors on a regular basis or scaling will occur and the wells may suffer a large and often irreversible decline in productivity. A very small amount of scale inhibitor in solution can prevent the deposition of a large amount of scale from the same volume of solution.

In the absence of scale inhibitor at a concentration equal to or above the minimum inhibitor concentration, scaling can occur very rapidly; in some cases production from a well can cease in a few days or even a few hours.

There are currently two main approaches to placing scale inhibitors in underground formations following the pumping of a treatment fluid downhole:

(1) Adsorption of the inhibitors from the bulk treatment fluid onto the rock surfaces.
(2) Precipitation of the inhibitors from the bulk treatment fluid caused by a change in the conditions such as pH or temperature downhole [21-47].

Once placed, the inhibitor then desorbs or dissolves into the water co-produced with the oil over a period of time (weeks to a few months) to inhibit scale deposition. When the concentration of scale inhibitor in the produced water falls to a predetermined level, with a safety margin above the minimum inhibitor concentration, a repeat treatment is needed.

Adsorption based methods of placing scale inhibitors are wasteful as most of the inhibitor is not adsorbed. Over 90% of the inhibitor may immediately return to the surface once the well is returned to production and so does not contribute to long term scale inhibition. Precipitation methods are recognised as superior to adsorption methods for placing scale inhibitors in the formation. Deposition of the inhibitor in a less soluble form results in longer periods between treatments.

The basis for some current commercially successful high temperature precipitation processes is the thermal hydrolysis of at least one chemical component present in the scale inhibitor formulation after placement of the treatment fluid downhole. This causes an increase in the pH of the solution. The pH increase results in the precipitation of the scale inhibitor as a metal salt. The pH change occurs homogeneously throughout the treatment fluid. This is far preferable to precipitation methods based on the sequential injection of incompatible chemicals, which may or may not mix adequately downhole since mixing usually has to occur within the rock matrix. Problems with poor mixing of two solutions containing incompatible chemicals, due to plug flow for example, are not encountered. Other processes based on pH decrease to precipitate scale inhibitors in less soluble forms have been proposed but do not appear to be in commercial use.

In the current commercial processes based on thermal hydrolysis, the hydrolysis of the pH increasing substances does not occur at a sufficient rate at low temperatures to facilitate scale inhibitor precipitation. A need therefore exists for more effective, predictable methods of precipitating scale inhibitors at temperatures less than about 100° C.

Recently, an enzyme-based system for the deposition of scale inhibitors has been evaluated. The basis for this process is shown in Figure 5.

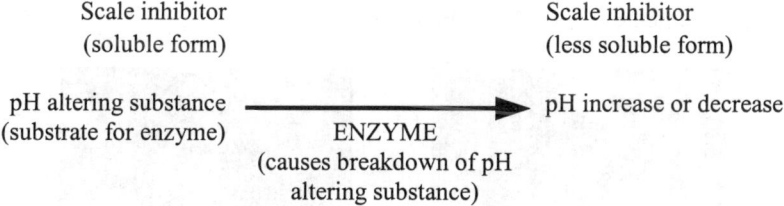

Figure 5 *Basis of scale inhibitor deposition processes based on enzyme catalysed hydrolysis of pH altering substances*

This is similar to commercial processes based on thermal hydrolysis but hydrolysis of the pH altering components is achieved using enzymes instead of thermal hydrolysis. Again, the pH change is homogeneous throughout the treatment fluid.

In principle, enzyme-based scale inhibitor deposition can be achieved using either pH increase or pH decrease, depending on the inhibitor.

We have investigated the possibility of using a combination of urea and a urease (a urea hydrolysing enzyme) to produce alkaline conditions due to the production of ammonia

(Figure 6). Under alkaline conditions, in the presence of suitable metal salts, a range of scale inhibitors may be precipitated.

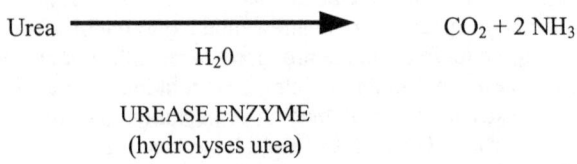

$$Urea \xrightarrow{\quad\quad\quad\quad} CO_2 + 2\,NH_3$$
$$H_2O$$

UREASE ENZYME
(hydrolyses urea)

Figure 6 *Enzyme based process for increasing the pH*

Results obtained to date using the enzyme-based process for increasing the pH are outlined below. A pilot study has proven the feasibility of the approach. It has shown the following:
- Urease enzyme can be used to break down urea in scale inhibitor formulations and increase the pH, resulting in deposition of scale inhibitors.
- The enzyme can be used in the temperature range of about 20° C to about 80° C which is below temperatures where thermal hydrolysis of urea is effective.
- Urease may be used to deposit a range of scale inhibitors.
- The time taken for deposition to occur can be varied by adjusting the amount of urease enzyme incorporated into the formulation.
- Deposition takes place over a period of several hours after addition of the enzyme to the formulation.
- Urease enzyme is required in order to achieve scale inhibitor deposition at any temperature below that at which thermal hydrolysis is effective.

Similar yields of scale inhibitor can be deposited using the enzyme-based process at low temperature as are obtained at higher temperatures with thermal processes. The appearance of the precipitated material is the same. Figures 7 and 8 show the appearance of a typical formulation of scale inhibitor A before and after addition of a known amount of urease enzyme (incubation for five hours at 50° C).

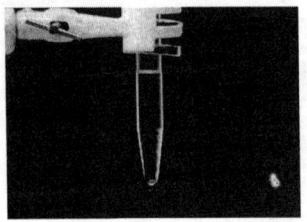 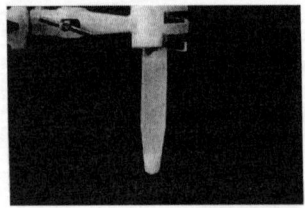

Figures 7 and 8 *Appearance of scale inhibitor A formulation before (left) and after (right) addition of urease*

It can be seen that the solution before adding the urease and increasing the pH is clear. All components are completely soluble. After 5 hours of incubation with an optimum amount of urease enzyme, the pH of the solution has risen from about pH 4.0 to pH 7.5 and the scale inhibitor has been deposited as the less soluble calcium salt. There is no pH change or scale inhibitor deposition in control tubes without enzyme. A typical time course of pH change and scale inhibitor deposition with this particular formulation at 50° C is shown in Figure 9 below.

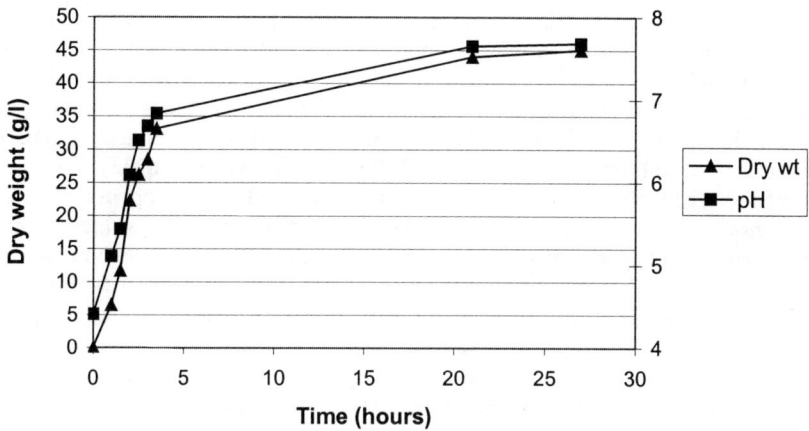

Figure 9   *Yield of precipitated scale inhibitor and pH vs time following addition of a known amount of urease enzyme to a solution of scale inhibitor A*

The next phase of this project is currently under way and is a detailed evaluation of the process under reservoir conditions of temperature and pressure. This will ensure that the process operates as expected.   The results of the study will be reported in due course.

5 DISCUSSION

Enzymes are now being successfully used for the controlled in-situ generation of well treatment chemicals. An enzyme-based process for the *in-situ* generation of acetic acid is being used in the field for a variety of acidizing applications. This process has a number of advantages over conventional acidizing methods. In particular excellent zonal coverage can be achieved.   Good results in applications including filter cake removal may be obtained using formulations which produce only 5% w/v acetic acid, in contrast to the trend in conventional acid treatments towards using higher acid concentrations, faster pump rates and larger volumes, which are needed to counter the rapid reaction rate of such acids with carbonate formations. More efficient acidizing treatments result in maximum production rates and significant economic benefits. As the fluid is essentially non-reactive when placed it is very low hazard and can be pumped through tubulars, pumps etc. without encountering the corrosion and compatibility problems associated with conventional acids such as hydrochloric acid. When being used in new wells the enzyme-based acidizing process can be placed using the drill string and mud pumps and the intact filter cake regulates leak off and assists in placement of the treatment fluid.

Other enzyme-based processes are being developed for use in the oil industry, including processes for the *in-situ* production of minerals, gels and resins. These processes have potential for sand consolidation or ·water shut-off applications. Sand consolidation does not appear to require the deposition of a large volume of cementing material such as calcium carbonate. Current work should provide information on the extent of porosity and permeability changes taking place when poorly consolidated sandstone is strengthened and the changes in the unconfined compressive strength. Another process under development uses enzymes for the controlled deposition of scale inhibitors. Initial indications are that the enzyme-based process behaves in the same way as processes based on thermal hydrolysis. Again, this is currently being investigated in tests under reservoir conditions.

A common feature of all of the enzyme-based processes is that they normally consist of a single treatment fluid in which the enzyme-induced change is homogeneous throughout the fluid. This can increase the efficiency of the treatments compared to conventional chemical treatments which may have more than one treatment fluid.

Enzyme-based well treatments offer environmental in addition to operational benefits. The behaviour of enzyme containing systems is predictable with a high degree of confidence and they can be configured to give the desired result over different time frames.

The enzymes used in oilfield processes are selected for their compatibility with the reservoir conditions. They are robust enzymes active over a wide range of temperatures and salinity.

A major driver in the oil industry at the moment is the need to reduce the environmental impact of chemical treatments. On this basis it would appear that the use of enzymes in chemical treatment processes has growing potential for the industry.

6 CONCLUSIONS

1. Enzymes can be used to produce a range of useful oilfield chemicals *in-situ*.
2. An enzyme-based process is already being used in the field for highly effective acidizing.
3. Novel enzyme-based processes have been shown to be capable of depositing minerals, gels, resins and scale inhibitors.
4. There is potential for enzyme-based water shut off, sand consolidation and scale inhibitor deposition applications.
5. Enzyme-based systems are generally very simple and highly predictable.
6. Significant environmental benefits may result from switching to and using enzyme-based processes.

References

1. Almond, S.W., Harris, R.E. & Penny, G.S. (1995). SPE 30123. Utilization of biologically generated acid for drilling fluid damage removal and uniform acid placement across long formation intervals. pp.465-478 *in* Proceedings of the European Formation Damage Control Conference held in The Hague, The Netherlands 15-16 May 1995.
2. Harris, R.E., & McKay, I.D. (1998). SPE 50621. New Applications for Enzymes in Oil and Gas Production. pp. in Proceedings of the SPE European Petroleum Conference held in The Hague, The Netherlands, 20–22 October 1998.
3. H. Frasch. Increasing the Flow of Oil Wells. US Patent 556,669 March 17, 1896.
4. Williams B.B. *et al.,* (1979). Acidizing Fundamentals SPE Monograph No. 6. SPE New York and Dallas.
5. Daneshy, A. (1995). Economics of Damage Removal and Production Enhancement. SPE 30234. *in* Proceedings of the European Formation Damage Control Conference held in The Hague, The Netherlands 15-16 May 1995.

6. US Patent No. 1,922,154 De Groote.
7. US Patent No. 2,050,932 De Groote.
8. US Patent No. 2,681,889 Manaul et al.
9. US Patent No. 2,059,459 Hund et al.
10. US Patent No. 2,206,187 Herbsman.
11. US Patent No. 2,863,832 Perrine.
12. US Patent No. 2,910,436 Fatt et al.
13. US Patent No. 2,059,459 Hund *et al.*
14. US Patent No. 2,863,832 Perrine.
15. Abrams, A. et al. (1983). Higher pH acid stimulation systems. Journal of Petroleum Technology, 2175-2184.
16. Mukherjee, H. & Cudney, G. (1993). Extension of acid fracture penetration by drastic fluid loss control. Journal of Petroleum Technology Feb 1993. pp 102-105.
17. Jones, A.T., Døvle, M & Davies, D.R. (1995). SPE 30122. Using acids viscosified with succinoglycan could improve the efficiency of matrix acidizing treatments. pp.453-463 in Proceedings of the European Formation Damage Control Conference held in The Hague, The Netherlands 15-16 May 1995.
18. Harris, R.E. et al. (2001). SPE 68911. Stimulation of a Producing Horizontal Well Using Enzymes that Generate Acid In-Situ –Case History. Proceedings of the SPE European Formation Damage Conference held in The Hague, The Netherlands, 21–22 May 2001.
19. www.cleansorb.com\page7.html
20. SPE 65405 New Treatment for Removal of Mud-Polymer Damage in Multilateral Wells Drilled Using Starch-Based Muds.
21. SPE 60222 Combined Scale Removal and Scale Inhibition Treatments.
22. SPE 60217 Potential Application Of Amine Methylene Phosphonate Based Inhibitor Species In HP/HT Environments For Improved Carbonate Scale Inhibitor Performance.
23. SPE 60198 Increasing Squeeze Life on Miller with New Inhibitor Chemistry.
24. SPE 58725 Development and Deployment of a Scale Squeeze Enhancer and Oil-Soluble Scale Inhibitor To Avoid Deferred Oil Production Losses During Squeezing Low-Water Cut Wells, North Slope, Alaska.
25. SPE 60190 Ester Cross-Linking of Polycarboxylic Acid Scale Inhibitors as a Possible Means to Increase Inhibitor Squeeze Lifetime.
26. SPE 37790 Encapsulated Scale Inhibitor Treatment.
27. SPE 37275 The Design of Polymer and Phosphonate Scale Inhibitor Precipitation Treatments and the Importance of Precipitate Solubility in Extending Squeeze Lifetime.
28. SPE 31125 The Correct Selection and Application Methods for Adsorption and Precipitation Scale Inhibitors for Squeeze Treatments in North Sea Oilfields.
29. SPE 30106 Scale Inhibitor Adsorption/Desorption vs. Precipitation: The Potential for Extending Squeeze Life While Minimising Formation Damage.
30. SPE 29001 Mechanistic Study and Modelling of Precipitation Scale Inhibitor Squeeze Processes.
31. SPE 28998 Solubility and Phase Behaviour of Polyacrylate Scale Inhibitors and their Implications for Precipitation Squeeze Treatments.
32. SPE 26605 The Effect of pH, Calcium, and Temperature on the Adsorption of Phosphonate Inhibitor Onto Consolidated and Crushed Sandstone.
33. SPE 25164 Precipitation and Dissolution of Calcium-Phosphonates for the Enhancement of Squeeze Lifetimes.

34. SPE 25163 The Modelling of Adsorption and Precipitation Scale Inhibitor Squeeze Treatments in North Sea Fields.
35. SPE 17008 Scale Inhibitor Precipitation Squeeze for Non-Carbonate Reservoirs.
36. US Patent No. 6,123,869 Precipitation of scale inhibitors (26-Sep-00).
37. US Patent No. 5,840,658 Process for the controlled fixing of scale inhibitor in a subterranean formation (24-Nov-98).
38. US Patent No. 5,604,185 Inhibition of scale from oil well brines utilizing a slow release composition and a preflush and/or afterflush (18-Feb-97).
39. US Patent No. 5,346,010 Precipitation of scale inhibitors in subterranean formations (13-Sep-94).
40. US Patent No. 5,346,009 Precipitation of scale inhibitors (13-Sep-94).
41. US Patent No. 5,211,237 Precipitation of scale inhibitors (18-May-93).
42. US Patent No. 4,602,683 Method of inhibiting scale in wells (29-Jul-86).
43. US Patent No. 4,393,938 Treating wells with ion-exchange-precipitated scale inhibitor (9-Jul-83).
44. US Patent No. 4,357,248 Treating wells with self-precipitating scale inhibitor (11-Feb-82).
45. Carlberg, B.J. (1983) Precipitation squeezes can control scale in high volume wells. Oil & Gas Journal pp. 152-154.
46. Bourne, H.M., Williams, G.D.M., Ray, J., & Morgan, A. (1997). Extending squeeze lifetime through in-situ pH modification - laboratory and field experience. Proceedings of the 8th International Oil Field Chemical Symposium, Geilo, Norway, 2-5 March 1997.
47. Bourne, H.M. Collins, I.R., Cowie, L.G. Nicol, M. & Strachan, C. (1997). The role of Additives on inhibitor precipitate solubility and its importance in extending squeeze lifetimes. IBC Conference Proceedings, Solving Oilfield Scaling, 22-23 January 1997, Aberdeen.
48. www.novozymes.com
49. Cunnah, P. (2000). Laccase – "An Industrious Enzyme" In Brief. 20. Published by Biocatalysts Ltd. (www.biocatalysts.com).
50. Enzymes Industry and the Environment 3rd Edition (1995). Published by Novo Nordisk A/S. Available as PDF download from www.novozymes.com
51. www.epa.gov/greenchemistry/index.htm

ELECTRICALLY CONDUCTIVE OIL-BASED MUD

M.A. Tehrani [1], C.A. Sawdon and S.J.M. Levey

[1] M-I Drilling Fluids UK (LTD), Research and Technology Centre, Ashleigh House, 1, Abbotswell Road, Aberdeen AB12 3AD.

1 INTRODUCTION

Drilling operations for the production of oil and gas are performed with the aid of a drilling fluid or mud. The drilling fluid cools and lubricates the drill string and bit, removes the drill cuttings from the wellbore, and provides sufficient hydrostatic pressure to prevent the ingress of formation fluids into the wellbore. There are a variety of fluids used for this purpose. These drilling fluids or "muds" fall mainly into two categories; water-based drilling fluids (WBM) and oil-based (mineral or synthetic) drilling fluids (OBM). To impart the many required properties to the drilling fluids, a variety of additives are included in the formulations. Water-based drilling fluids consist of a brine phase with various additives for density, rheology, fluid loss, and lubricity control. They perform adequately where downhole temperatures are not too severe, where reactive shales (water-sensitive clays) are not encountered, or where corrosion is not a major problem. The alternative in those circumstances is an OBM, where the liquid phase may be entirely oil or consist of a brine-in-oil invert emulsion. These fluids too contain various additives for the control of density, rheology and fluid loss. Table 1 gives an example of the components of a typical OBM.

Table 1   *Typical formulation for an oil-based mud (s.g. 1.70, oil/water: 70/30, water activity: 0.80)*

Component	Concentration Kg/m^3
Base oil	375.4
Invert emulsifiers	42.8
Lime	17.1
Filtration control additive	17.1
Viscosifier	14.3
$CaCl_2$ salt	75.2
Freshwater	210.2
Weighting material (barite)	940.3

To perform satisfactorily, a drilling fluid must have the following characteristics:

- minimise pressure drop in surface pipes and drill string
- provide a yield stress suitable for supporting/transporting mud solids and drill cuttings
- be chemically, thermally and mechanically stable
- provide hole stability
- provide good lubricity
- produce a filter cake to prevent excessive fluid loss to the formation
- aid signal transmission (e.g. electrical logging).

Although oil-based drilling fluids are more expensive than water-based fluids, it is on the basis of their added operational advantage and superior technical performance that they are often used for drilling operations. Historically, however, a limitation to their use has been the environmental concerns over the discharge of contaminated cuttings in some parts of the world, e.g. the North Sea. But, in recent years, advances in re-injection technologies as well as novel collection, transportation and recycling techniques, have effectively proven that discharge is not the only option for disposal of cuttings and waste waters from drilling operations using oil-based drilling fluids.[1, 2]

An area where oil-based muds have been at a technical disadvantage is in high quality wellbore imaging. The use of finely detailed formation imaging logs is an increasingly critical component in evaluating the full potential of a field prior to initiation of the development phase. Wellbore image logging is aimed at acquiring detailed geological and petrophysical data of formations crossed by the wellbore. Acquisition and analysis of these images allow recognition and quantification of geological features, such as orientation and type of fractures and faults, and may be used to optimally locate the zone where the reservoir is perforated to allow the inflow of hydrocarbons to the wellbore. The level of detail available for subsequent processing is driven primarily by the resolution and azimuthal coverage provided by borehole imaging devices.

Borehole imaging techniques include high-resolution resistivity devices such as the formation micro imaging (FMI) and ultrasonic borehole imaging tools. Resistivity imaging or logging devices are particularly effective at unravelling geological features because of their inherent high-resolution capability and the wide dynamic range of the formation properties being measured. These devices send alternating current radially from one or more electrodes into the formation (Figure 1). The current returns via the mud column to the metallic housing of the tool. This technique necessitates a current path through the drilling fluid and the filter cake present between the electrodes and the borehole wall. Consequently, the use of resistivity tools for borehole imaging has been historically restricted to water-based drilling fluids.

In water-based drilling fluids, the resistivities of the fluid, filter cake and filtrate are low because of the naturally high conductivity of water and, to a greater extent, saline solutions. Because high conductivity provides relatively unrestricted current flow from a logging tool, signal response from the formation is generally of the highest quality when using WBMs. Conversely, conventional oil-based fluids are designed to provide oil-wet surfaces and water-free filtrates. These fluids consist of an internal brine phase that is strongly emulsified into a non-polar oil phase. Thus, the fluid, filter cake, and filtrate are non-conductive and block the flow of electrical current. Consequently, responses from resistivity logging tools in conventional OBM are poor or non-existent.[3]

This paper describes the development and application of a conductive oil-based mud (COBM) that allows high quality resistivity imaging while maintaining the performance advantages of invert emulsion fluids. The focus of the paper is on the chemistry that

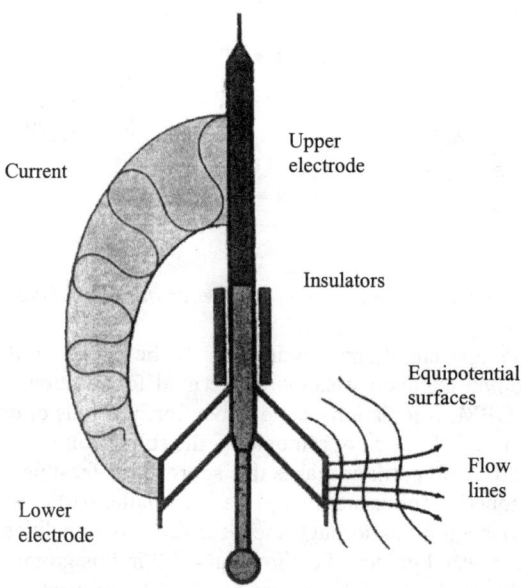

Figure 1 *Schematic of the Formation Micro Imager (FMI) tool and the current path*

produces an electrically conductive continuous phase and provides a high performance fluid with a conductive mud, filter cake and filtrate. The paper also presents the results of yard tests and a field trial of the COBM where the fluid was used to drill a highly deviated well through complex geological structures.

2 METHOD

As shown in Figure 2, water-based drilling fluids have a conductivity that is more than one million times higher than that of conventional OBM. However, knowledge of the FMI tool design and some preliminary modelling work at the outset indicated that a 10^3-10^4 fold increase in OBM conductivity may be adequate for successful resistivity logging. As seen in Figure 2, this is considerably higher than any gain achievable by decreasing the oil-water ratio of the OBM (i.e., increasing the water content) or by increasing the frequency of the applied voltage. There are different routes to making an oil-based mud electrically conductive, some of which may not be practical solutions for field application. Several options have been considered and evaluated which have resulted in a number of application patents.[4-8] The various options are described below:

2.1   Conductive Solids

Conductive solids can create a conductive path through a non-polar fluid and fall in two groups: non-interacting, interacting.[5,6] Non-interacting particles that are uniformly shaped must be present in high concentrations (near to close packing) in order to have any effect on conductivity (Figure 3a). Examples include metal-coated particles and graphite. The high concentration of the solids will have a dramatic effect on rheology. Non-interacting, non-uniformly shaped particles with large aspect ratios can form a conductive network at

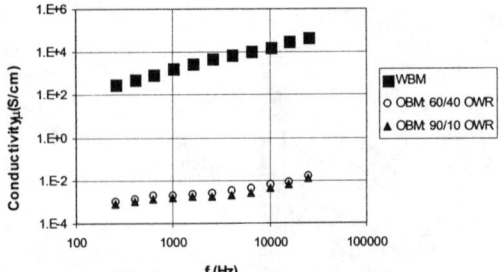

Figure 2   *Comparison of oil-based mud and water-based mud conductivities*

considerably lower concentrations (Figure 3b). To be effective, these should be able to bridge around the non-conductive solids in the mud formulation, e.g. weighting material such as barite. In OBM formulations, conductive fibres or rods of one millimetre length or longer may be required. The risk of removal by the solids control equipment utilised at the surface to remove the drill cuttings makes this approach unfeasible.

Interacting conductive solids are micron size or smaller particles that form a percolating network at low concentrations through their surface functionalities or even by Brownian motion, Figure 3c. Special grades of carbon black fall in this group. However, the presence of impurities in the mud, especially strong emulsifiers, reduces the effectiveness of such particles in boosting the conductivity in invert emulsion muds. A further drawback of the conductive solids approach is that the filtrate will be completely non-conductive.[5,6]

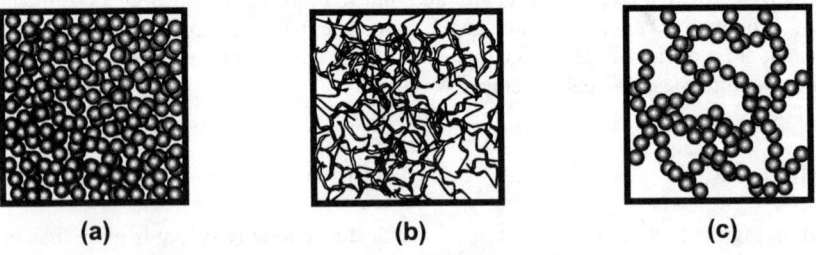

Figure 3   *Conductive solid particles; (a) non-interacting, regular-shaped, (b) non-interacting, irregular-shaped, (c) interacting.*

2.2   Brine Phase Contribution

The brine phase of an invert emulsion mud has very high conductivity but the dispersed droplets are normally so effectively emulsified that they behave as non-conductive solid particles. One option to increase OBM conductivity is to decrease the stability of the emulsion by either using less or weaker emulsifiers. This will cause local agglomeration and perhaps coalescence of brine droplets, thereby producing conductive paths through the mud.

The partial destabilisation of the invert emulsion has to be carefully controlled. If gone too far, complete emulsion breakdown may occur and the mud system can fail. Generally, given the thermal and chemical changes occurring in the mud during the drilling operation, such control is often impossible to achieve. Further, the presence of large brine droplets in the mud can lead to water in the filtrate with undesirable effects on reactive shale and formation wettability.

Another route to using the high conductivity of water is to form a micro-emulsion. Solvation of water in the oil phase by means of an appropriate solvent and/or surfactant can convert the oil phase into a micro-emulsion with high conductivity. However, such systems are often sensitive to variations in chemistry and temperature, and are difficult to control.

2.3   Oil Phase Contribution

Making the continuous oil phase conductive has the advantage over the previously discussed options of producing a conductive filtrate. This approach requires an ionic material that dissociates into charge carrying species in the oil phase. Many salts will dissolve in non-polar solvent environments but without the ability to ionise readily to allow electrical conduction, thus the polarity of the organic phase is a crucial element. Since the base fluids (i.e. the oil phase) used in OBMs are generally non-polar, it is necessary to include an appropriate polar solvent that can facilitate the dissolution and ionisation in the continuous phase.[4]

Solvent - The performance requirements of the polar solvent include full miscibility with the oil phase, and boiling and flash points comparable to the base oil. Water solubility of 5--10% can be tolerated. Further, as the drilling fluid is in prolonged contact with various elastomers, downhole and in surface equipment, compatibility with a wide range of elastomers is also required. Finally, the solvent must be acceptable in terms of health, safety and environmental performance. The dielectric constant of the solvent should be sufficiently high to promote ionisation. Table 2 compares the conductivity of drilling fluids formulated with an organic nitrogen compound and several glycol ethers.

Table 2   *Conductivity of muds formulated with an organic nitrogen compound and different glycol ethers*

Solvent	Number of propylene glycol repeat units (n)	Mud conductivity (μSm^{-1})
Propylene glycol n-butyl ether (PnB)	1	13000
Dipropylene glycol n-butyl ether (DPnB)	2	6300
Tripropylene glycol n-butyl ether (TPnB)	3	3900
UCON LB65*	6	260
UCON LB135*	23	100

* These are polymeric materials and the "n" is an average number

Ionic additive (Salt) – There is a wide range of compounds which may be considered as ionic additives for the oil phase. Some inorganic metal salts exhibit good solubility in organic solvents, among them are the halides of Group I alkali metals. For others, it may be possible to improve the solubility by forming organometallic complexes. For example, more covalent metals such as calcium, iron, molybdenum and copper may have better ligand co-ordination properties than the alkali metals.[9-11] Organic salts are another group of compounds which may have good solubility in organic liquids. Some of these possibilities have been examined with differing degrees of success.

2.3.1 Inorganic salts. Some inorganic salts show a strong tendency to dissolve in organic solvents. The bromide salts of the alkali metals show promise in that the larger anion offers better solubility in lower polarity solvents. This also suggests that the higher up the alkali metal series the less ionic in nature (i.e. less electropositive) the metals become and, therefore, more covalent and more soluble in solvents. An example of this is shown in Table 3 where the conductivity of a 5% by weight solution of the salt in dipropyleneglycol *n*-butylether is given at 25°C:

Table 3   *Conductivity values at 25 °C for alkali metal bromides in a glycol ether*

Salt	Solubility in glycol ether	Conductivity (μSm^{-1})
LiBr	Soluble with heat	18000
NaBr	Slightly soluble with heat	150
KBr	Very slightly soluble with heat	25

Thus, electrical conductivity is related to the solubility and the number of ions in solution. For the above salts the values vary as LiBr > NaBr > KBr, which is in line with classical theory that suggests that the degree of ionic character increases with decreasing electronegativity of the metal cation.

The real test of the effectiveness of the ionic additive (or salt) and the solvent is the addition of base oil and water or brine. With most salt-solvent combinations (e.g. the above salts and solvent), the addition of the base oil and water reduces the polarity of the continuous phase and forces the salt into the aqueous phase, causing a subsequent loss of conductivity in the oil phase. To overcome this problem the solvent has to either increase the polarity of the oil phase to the extent that the salt remains (at least partially) in the continuous phase, or complex the metal ions into the oil phase. The use of highly polar solvents may have the drawback of causing severe elastomer incompatibility. Appropriate solvents (e.g. alkyl ethers of oligomers of ethylene glycol) may be useful for their co-ordinating characteristics. Such molecules are expected to wrap their co-ordinating ether linkages around the metal cation, leaving the hydrophobic alkyl groups to enable solvency in the low-polarity continuous phase.

With most inorganic salts, increasing the polarity of the solvent does not eliminate the problem of conductivity loss upon addition of base oil and water. An exception to this is lithium bromide which, with a high polarity solvent such as diethylene glycol *n*-hexyl ether (Hexyl Carbitol), may form a single phase of high conductivity. The evolution of a second phase of low conductivity depends on the concentration of the components.

Studies have shown that it is possible to solvate inorganic salts in an organic solvent to raise the conductivity of the solution. Furthermore, oligomeric ethylene oxide groups have a role in holding water in the organic phase. Since metals which are more covalent in nature have increased solubility in the organic phase, Group II metal halides such as calcium bromide and magnesium chloride should exhibit better performance. These salts were tested in dipropyleneglycol n-butyl ether (DPnB) and Hexyl Carbitol, two solvents with different polarities. Similar to its performance with the Group I metal salts, DPnB did not retain any conductivity in the oil phase upon the addition of base oil and water. With Hexyl Carbitol, calcium bromide behaved in a similar way to lithium bromide, producing a high conductivity single phase. As with lithium bromide, a correct balance of components is needed to achieve an optimum single-phase system.

Iron and many of the transition metals show a high degree of covalency in their bonding, allowing the dissolution of their salts in organic solvents. However, addition of the base oil and water to the mixture results in phase separation into a high conductivity aqueous phase and a non-conductive oil-rich phase. Increasing the concentration of the polar solvent reduces this problem.

In summary, inorganic salts require high concentrations of highly polar or co-ordinating organic solvents in order to remain partially in the oil phase. In general, such solvents tend to be incompatible with the elastomers used in oilfield equipment.

2.3.2 Organic salts. Organic salts to be considered as charge carriers are based around an organic cation-anion pairing which might ionise in organic solvents. Organic salts with high water solubility pose the same problem as inorganic salts, i.e. transfer to the aqueous phase. Thus salts with lower water solubility are preferred.

Certain organic salts are less susceptible to phase transfer to the water phase, but will form ion pairs which cannot be readily separated by non-polar solvents such as mineral oils and, therefore, need more polar solvents. This may be improved by the use of bulky anions and cations to hinder the close approach of ions, but polar solvents are still needed to allow free movement of ions for electrical conductivity. In a complex system such as an oil-based mud, there is also the possibility that free ions can encounter any other counterions which may be present in the system. This can be a major limitation if the counterions encountered form stable ion pairs.

Due to their organic phase solubility and low pKa values, sulfonic acids and their salts may have potential for ionisation in a low-polarity medium. A number of sulfonates with inorganic and bulky organic cations were examined. In all cases the structure of the sulfonate molecule, particularly its asymmetry, allowed an interfacial interaction between the oil and water phases, thus limiting ion mobility. Consequently, no major increase in oil phase conductivity was observed.

Another group of additives with potential as charge-carrying species are the quaternary phosphonium and ammonium salts. These are used as phase transfer catalysts in industrial processes for their ability to dissolve in both polar organic and aqueous media whilst retaining their salt characteristics. Various phosphonium salts with different combinations of aromatic and aliphatic chains were investigated with similar results to those of the sulfonates. Tetrabutyl phosphonium bromide showed the highest level of conductivity, around 6000 μSm^{-1} at ambient temperature. However, according to the information supplied by the manufacturer, this salt is more toxic than its ammonium counterparts. This lead to the evaluation of a number of quaternary ammonium compounds with predominantly $C_4 - C_8$ chains. Five per cent by weight solutions of the salts in a semi-polar solvent (SPS) gave the following conductivities at ambient conditions:

Table 4   *Conductivity of several quaternary ammonium salts in a semi-polar solvent (SPS) and in 40/60 (v/v) mixture of oil and SPS*

Quaternary ammonium salt (Predominant chain length)	Conductivity in SPS (μSm^{-1})	Conductivity in oil/SPS: 60/40 (μSm^{-1})
C_4	4000	660
C_5	2600	560
C_6	2300	480
C_8	1700	410

It is clear that the conductivity is inversely related to the molecular weight and chain length. When the above tests were repeated with equal molar quantities of the salts, equal conductivities were obtained. This suggests that the ion concentration is crucial to the level of conductivity rather than the size of the cation and its ionic mobility. However, the larger the alkyl group the lower is its tendency to partition into the aqueous phase.

One of the advantages of taking the ionic dissolution approach was that the conductivity of the COBM is proportional to temperature. Hence, downhole temperatures assist in improving conductivity sufficiently to generate good FMI response. Figure 4 shows the relationship between conductivity and temperature for the newly developed COBM.

The ultimate test for both additives was the ability to produce a stable drilling fluid with relatively high conductivity, good rheology, fluid loss control, shale inhibition, and high contamination tolerance. To allow adequate flexibility in controlling rheology and fluid loss, the conductive formulations were prepared with products normally utilized in conventional oil-based drilling fluids.

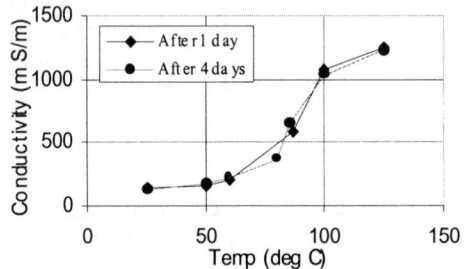

Figure 4   *Effect of temperature on conductivity of COBM*

3   MEASUREMENTS

3.1   Laboratory Tests

It is standard procedure to hot roll (or age) the mud after mixing, at a temperature similar to that expected down-hole, to allow the clays and polymeric materials to fully yield and swell. The mixing time and temperature may vary, but 16 hours and 121°C, respectively, are normally used. The properties of the mud, e.g. rheology and conductivity, are measured before and after ageing.

During laboratory testing, mud conductivity was measured at ambient conditions with a dip probe conductivity meter (Model CM-35 manufactured by WPA, Cambridge, UK) operating at a fixed frequency of 500Hz. The instrument could measure conductivities in the range $1\text{-}10^6$ μSm^{-1}. In addition, higher temperature measurements were made in a specially designed high-temperature, high-pressure (HTHP) cell operating at up to 500 psi and 180°C. These measurements were made over a range of frequencies up to the 15kHz, the operating frequency of the FMI imaging tool.

3.2   Yard Tests

After optimising the properties of the conductive mud in the laboratory, the logging performance of the fluid was evaluated with an actual FMI tool in a specially constructed short test well (Figure 5). The test well was a 9.5m long hole of 200mm diameter and 66° deviation from the vertical, and was made through several layers of cement and concrete. The cement layers, each about 50cm deep, were sandwiched between two non-conductive

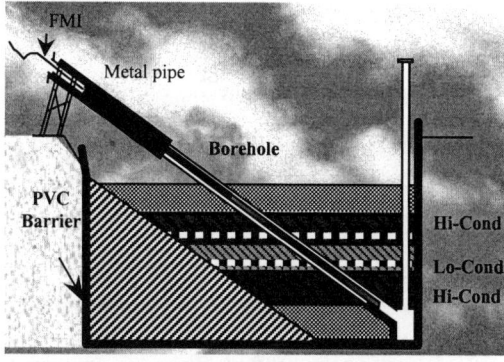

Figure 5   *Schematic of test well*

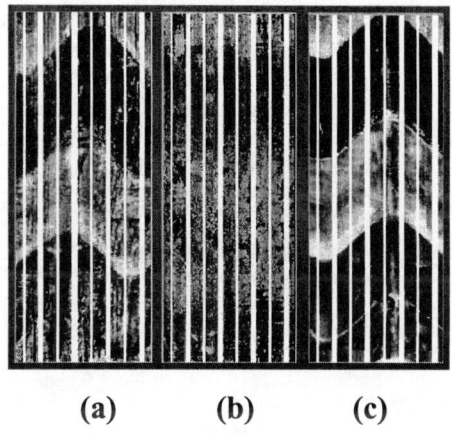

(a)   (b)   (c)

Figure 6   *FMI logs produced: (a) in brine, (b) in conventional OBM, (c) in COBM*

concrete layers. Resistivity of the cement layers varied from 4 to 40 ohm-m. Within this range, the resistivities of the top, middle and bottom layers of cement were ranked as low, high and medium, respectively. Fullbore FMI logs were obtained in 2% NaCl brine, conventional OBM, and conductive oil-based muds designed with conductivities in the range 77-10,000 μSm⁻¹. An example of the logs produced is shown in Figure 6.

Figure 6a shows an FMI image produced in brine. The quality of the image is similar to that achieved by typical FMI in WBM with real rocks. The layers with different resistivities are clearly delineated in the figure. Figure 6b shows an FMI log in an OBM. The only visible feature is the highly diffused variation in contrast from top to bottom of the image. No usable information can be extracted from this image. Figure 6c shows an FMI log produced in a COBM with a conductivity of 5500 μSm⁻¹. The imaging of the layer boundaries and other details is as clear as that achieved in brine. A more detailed analysis of the images obtained in the yard tests can be found elsewhere.[12]

Results of the validation tests indicated that good FMI logs could be obtained over a range of conductivities. This knowledge, together with the positive effect of temperature on conductivity as illustrated in Figure 4, was later used to optimize the COBM for field application.

4   FLUID OPTIMISATION FOR FIELD APPLICATION

To optimise the fluid for the drilling and logging operations, detailed testing of the formulations were conducted over a range of fluid densities, 1.1 - 2.0 s.g., and with different base oils, including diesel, low toxicity mineral oil and synthetic materials. Resistance to contaminants was also examined with a variety of fluid contaminations such as drilled solids, seawater and cement. The results showed that the fluid could be engineered without major reformulation to perform as an effective invert emulsion fluid while maintaining the required level of conductivity.

The HTHP rheology, potential for barite sag, elastomer compatibility and shale inhibition were examined and the fluid was found to perform similarly to a conventional or synthetic OBM.[12]

Limitations on the fluid system were set with respect to wellbore temperature. The fluid was expected to perform well between 75° and 150° C. Below 75°C, higher quantities of the conductive additives would be required to produce adequate conductivity for optimal logging. This could adversely affect elastomer compatibility and increase fluid costs. At temperatures exceeding 150°C, the salt component of the conductive additives can undergo thermal degradation according to the Hofmann elimination (Figure 7), where the quaternary salt decomposes to trialkyl amines and alkenes under heat and in the presence of water.[13] There is no indication as to the reactivity in relation to chain length, but it can be assumed that the less water-soluble salts would be less prone to this degradation.

$$RCH_2CH_2NR_3 + OH^- \longrightarrow RCH=CH_2 + R_3N + H_2O$$

Figure 7   *Hofmann elimination reaction*

5   FIELD TRIAL

The field trial of the COBM took place in the Norwegian sector of the North Sea. The primary objective of this well was to acquire very detailed geological information, including structural dip, sedimentary dip, faults, fractures and deformation patterns. Recognition of such geological features was considered important for a better understanding of the productive capacity of the reservoir.

Because of the resolution and azimuthal coverage required, FMI was selected as the only tool capable of delivering high quality results. Acquisition of FMI data was so critical that, in the absence of a conductive OBM, water-based fluid would have been used in order to log with the FMI tool. However, because of the long open-hole section that would be required (2,400 m with approximately 1,250 m in the reactive overburden shales), drilling with an aqueous fluid was considered high risk, particularly given the high probability of bad weather conditions which would intermittently suspend drilling operations during well construction. Thus, there was a clear need for a drilling fluid that exhibited the drilling performance of an OBM and the logging performance of a WBM.

The COBM was used to drill a long 8½-in. (21.6cm) hole through an upper shale interval into the reservoir section. Drilling with the COBM started at a hole angle of 23° (at a depth of 2178m) which was then built up to 69° from the vertical and held through the reservoir to a total depth of 4,620m. Approximately 1,250m of reactive shale was drilled prior to reaching the reservoir.

The drilling program called for FMI logging over the entire reservoir section. In addition, the FMI was to be run in the lower part of the shale section (just prior to entering the reservoir) to assess the level of FMI response, and to determine whether or not there would be a need to displace to a water-based drilling fluid before entering the reservoir where FMI logging was of utmost importance.

Upon completion of fluid optimisation for field application, the final components of the COBM were determined to deliver the drilling performance of a standard OBM. The COBM formulation and fluid properties are given in Tables 5 and 6.

Table 5   *Field trial COBM formulation*

Product	Conc. ($Kg\ m^{-3}$)
Low toxicty mineral oil	444.1
Invert emulsifiers	50.7
Co-solvents	89.5
Ionic additive	20.3
Lime	6.8
Filtration control additives	23.9
Viscosifier	24.0
CaCl$_2$ salt	26.5
Freshwater	86.9
Barite	618.6

Table 6   *Properties of COBM used in the field test*

Mud s.g.	1.5 - 1.75
Plastic Viscosity (Bingham model)	< 65 cP (mPas)
100 rpm reading (173 s^{-1})	< 18 Pa
HTHP fluid loss at 150°C	< 4 mL at 150°C
Conductivity at ambient conditions	> 50 µSm^{-1}

Based on the results of the yard tests, a conductivity of about 800 µSm^{-1} was thought to be adequate for logging at a bottom hole temperature of about 120°C. Given the positive effect of temperature on conductivity, the fluid was designed to have a conductivity greater than 50 µSm^{-1} at ambient conditions. Figure 8 illustrates the HTHP conductivity of the field mud samples at different times during the drilling operation.

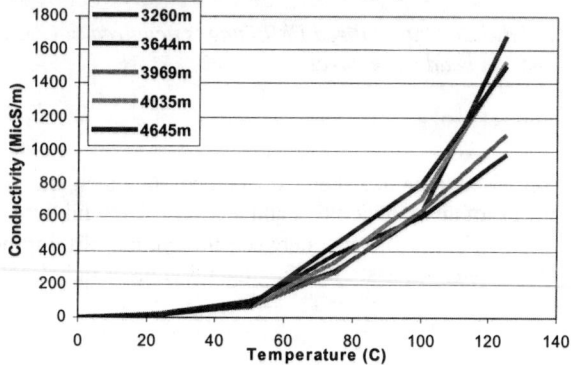

Figure 8   *Effect of temperature on conductivity from field samples of COBM taken from various depths*

5.1   Field Trial Results

The conductive oil-based mud as described above proved to be a stable and easily maintained fluid, and behaved like a standard OBM system. The conductivity was easily maintained and adjusted without compromising the other properties of the fluid or wellbore stability.

Generally, the quality of resistivity logs depends on the contrast between the conductivity of the mud and the formation. The larger this contrast (i.e. the lower the mud conductivity or the higher the formation conductivity), the poorer the image quality. From a logging standpoint, the conductivity of the COBM in this field trial was adequate for obtaining good quality image logs, even in formations with a conductivity as high as 500,000 μSm^{-1}.

Figure 9 is an examples of the FMI processing and interpretation from the field trial which shows faults and geological features - a fault dipping West with 50° inclination is recognised at 3,642 m. This event caused interruption of formation bedding. Overall, FMI image resolution and feature identification was good and within the range of quality that would be expected from an FMI logged in WBM.

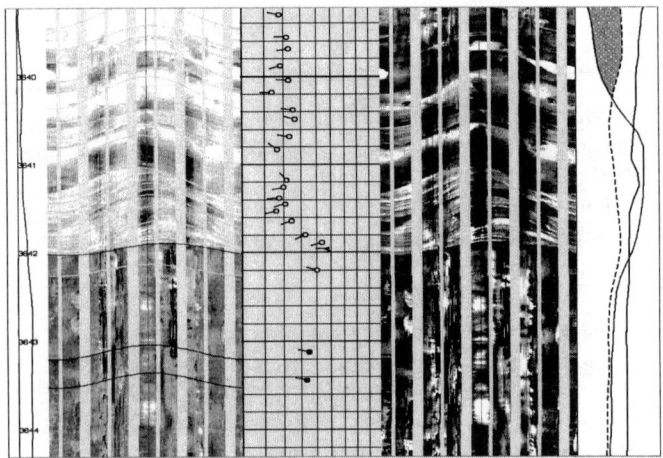

Figure 9   *Static and dynamic normalised FMI images acquired in COBM showing faults and formation bedding features*

6 CONCLUSIONS

- A conductive oil-based mud has been developed in which the oil phase is made conductive by means of an ionic additive and a polar solvent mixture.
- The fluid strikes an effective balance between the optimal drilling performance of an OBM and the high-quality resistivity logging previously only acquired with WBM systems.

Acknowledgements

The authors thank the management of M-I *L.L.C.* for permission to present this paper.

References

1   Ekeli Ø, *et al.*, "Phillips Drills Entire OBM Exploratory Well With Vacuum-Cuttings System," *Oil & Gas J.*, Dec. 20, 1999, 129.
2   Pruett, J.O.II and Walker, D., "Solids Control Performance Measurement – The Key to Minimising Drilled Cuttings Waste Generation," presented at the IBC Conference on Minimising Drilling Waste Generation, Aberdeen, Scotland, March 24-26, 1999.
3   Mercer, B., *et al.* "Detect Fractures When Drilling With Oil-Based Mud," *Petroleum Engineer International*, Sept. 1994, 31.
4   Maitland, G., Sawdon, C and Tehrani, A., "Electrically Conductive Non-Aqueous Wellbore Fluids," PCT patent application WO 99/14286.
5   Sawdon, C.A., *et al.*, "Electrically Conductive Non-Aqueous Drilling Fluids," PCT patent application WO 00/41480.
6   Schlemmer, R.P. and Schultz, J.F.A., "Electrical Well Logging Fluid and Method of Using Same," US Patent 6,006,831 (Dec. 28, 1999).
7   Patel, A.D.: "Conductivity Medium for Open-hole Logging and Logging While Drilling," US Patent No. 6,029,755 (Feb. 29, 2000).
8   Patel, A.D. "Water Soluble Invert Emulsions," US Patent No. 5,990,050 (Nov. 23, 1999).
9   F.A. Cotton and G. Wilkinson; *Advanced Inorganic Chemistry*; Fourth Edition, Wiley Interscience, 1980, 264.
10 Climax Molybdenum Company; "Organic Complexes of Molybdenum," *Molybdenum Chemicals Bulletin Cdb-9*, June 1956.
11 Jones, M.M., "A New Method of Preparing Some Acetylacetonate Complexes," *J. Am. Chem. Soc.*, July 5, 1959, **81**, 3188.
12 Laastad, H., *et al.* "Water-Based Formation Imaging and Resistivity Logging in Oil-Based Fluids – Today's Reality," *SPE 62977, presented at the SPE 2000 Annual Technical Conference, Dallas, Texas,* Oct. 1-4, 2000.
13 Fessenden, R.J. and Fessenden, J.S., *Organic Chemistry*, Second Edition, PWS Publishers, 1982, 729.

THE USE OF SURFACTANTS TO GENERATE VISCOELASTIC FLUIDS

Ralph Franklin[1], Michael Hoey[2] and Raman Premachandran[1]

[1] Akzo Nobel Chemicals Inc, 1 Livingstone Avenue, Dobbs Ferry NY 10522
[2] Currently with Infineum USA L.P., 1900 East Linden Avenue, Linden, NJ 07036

1 INTRODUCTION

Amphiphilic molecules in aqueous media generally form spherical micelles above a critical concentration[1]. With increasing surfactant concentration, these aggregates can undergo structural transition to form spherical to rod like micelles under appropriate conditions of salinity, temperature or the addition of some organic compounds[2-12]. Rod like / worm like/cylindrical micelles, obtained from surfactant systems and certain salts have attracted considerable interest because of their unique viscoelastic properties[13-21]. These viscoelastic fluids possess both viscous, fluid-like properties and elastic, solid-like properties. In simple terms, viscoelastic fluids have the ability to both flow and support solid materials. Such systems can be optimised for performance in variety of applications including oilfield[22-28], cleaning[29, 30], and transportation (drag reduction)[31-33]. Viscoelastic surfactants also find application in lubricants[34], anti-misting[35], detergents[36], personal care[37], foam fluids[38], mobility control[39], and as heat exchange fluids[40].

The viscoelasticity of a surfactant system is attributed to the formation of worm-like, thread-like, or rod-like micelles. These elongated structures look and behave like (living) polymers in the way they associate. The intrinsic flexibility of the micelles depends on the persistence length of the aggregated micelles, which can vary from 100 to 1000 angstroms, while the elasticity of the system is related to the degree of micellar entanglement. However, unlike polymers they are composed of relatively low molecular weight components that can break away from one another and then re-associate. Thus, they exhibit reversible shear thinning over numerous cycles, unlike polymers, which degrade with repeated shear. The ability to be able to pump a fluid without any significant loss of its solid suspending character is a key requirement for fracturing purposes involving proppants. Thus, the attractiveness of surfactant systems is obvious.

For oilfield applications, the structural and flow behaviour of brine gels is important. The flow and structural properties of these aqueous surfactant gels depend on the concentration and nature of the brine, additives, surfactants, temperature, and shear. Usually, high concentrations (>10%) of surfactants are required to produce viscoelastic gels. However, for economic viability, it is essential to identify surfactants that form gels at very low concentrations. Rheological characterisation of a formulation with respect to concentration, temperature, shear and time is important in optimising performance for specific applications. In this paper, we discuss the rheological characteristics of erucylmethylbis(hydroxyethyl)ammonium chloride, Ethoquad®E/12-75, and dimethylaminopropyltallowamide oxide, Aromox®APA T, systems at various temperatures and shear rates relevant to their performance in oilfield stimulation.

2 DISCUSSION

2.1 Experimental details

Ethoquad® E/12-75 (75% active in isopropanol) and (Aromox® APA T (50% active in propylene glycol) were supplied by Akzo Nobel Chemicals Inc. KCl, MgCl$_2$ and other salts were purchased from Sigma Chemicals and used without further purification. Viscosity of different gels was measured using a Brookfield DV II Cone and plate rotational Viscometer at 0.5 rpm using CP51 spindle, or a Fann 35A viscometer at 100 rpm. Rheological characterisation of the gels was done using Rheometric SR 5000 stress controlled rheometer using a 40-mm, 4° cone and plate. Six percent by weight of Aromox® APA T was added to a 3% w/w brine solution and the sample stored in oven at 50°C until it formed a clear gel. Unless otherwise mentioned all measurements were made at room temperature. Sample preparation was done by dispersing the surfactant in the brine and then storing the sample at 75°C, for the Ethoquad® E/12-75, or 50°C, for the Aromox® APA T, until the sample formed a clear gel. The spiking experiments with Aromox® APA T were done by adding either tallow fatty acid (Uniqema) or Armeen® APA T (Akzo Nobel).

2.2 Characterisation of Ethoquad® E/12-75

Our initial work with Ethoquad® E/12-75 utilised a Fann 35A viscometer to help characterise the product. This type of rotational viscometer is commonly used in the oilfield chemicals industry. The evaluations were performed using 2% v/v of surfactant in 3% ammonium chloride brines. A good fluid was considered to be one that exhibited limited fall off in viscosity between 110°F and 150°F at 100 rpm and a minimum viscosity of 90 cps. A typical viscosity profile is shown in Figure 1.

However, in working with different batches of surfactant, it was observed that some would form gels that would unexpectedly and dramatically fail the Fann test. A rigorous analysis of the Ethoquad® samples and precursors identified several minor compositional differences. What had been underestimated was the significant impact that trace impurities had on the system. It became clear that this system is greatly influenced by the presence of impurities, and for fracturing purposes the effect is highly undesirable. While it has not been possible to attribute the loss in performance to any single component some general guidelines were developed. Two key factors were identified, the chainlength distribution of the components, and the levels of manufacturing intermediates or impurities.

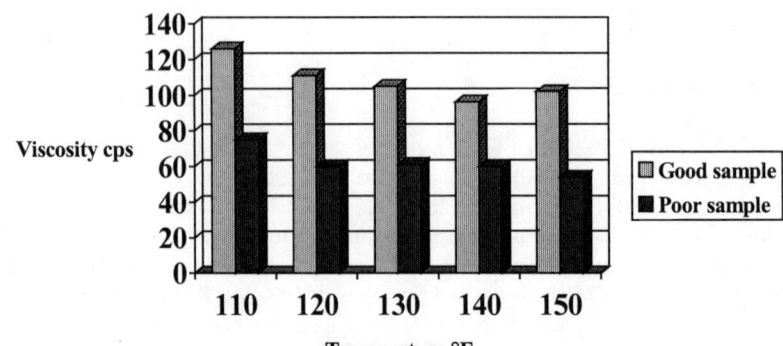

Figure 1   *Viscosity profiles of good and poor performing samples of Ethoquad® E/12-75 as measured on a Fann 35A viscometer using 2%v/v of Ethoquad® E/12-75 in a 3%w/w ammonium chloride brine*

Manufacturing intermediates would include materials such as amides, nitriles, fatty acid, dialkylamine adducts, and the like. The dominating factor impacting the product performance is the chainlength distribution. It was found that by limiting the level of C18 and less components to 1-2% the impact of the other trace contaminants from the manufacturing processes was moderated[23]. However, it is preferred that these should also be kept to <1%, Table 1.

Table 1   *Effect of impurities on the viscosity of Ethoquad®E/12-75 fluids*

Fann 35A test result	% intermediate products	% C18 and less chains
Pass	0.30	2
Fail (marginal)	0.18	4
Fail	0.87	9
Pass	1.4	1
Fail	0.31	10

More recently, we have characterised the product using a stress-controlled rheometer. This allowed us to better characterise differences between an acceptable viscoelastic gel and a poor one. Figure 2, shows a typical profile of a good viscoelastic gel produced with Ethoquad® E/12-75 and KCl brine. The gel was characterised using the dynamic temperature sweep mode of the instrument. It can be seen that the sample exhibits low flowability (G") and high elasticity (G') up to the crossover point. At the crossover point, the gel melts and the elasticity falls off. The elasticity of the fluid after melting is important for proper settling of the proppants. Thus, it is important that the crossover temperature is not too low and the fall off in elasticity not too sudden.

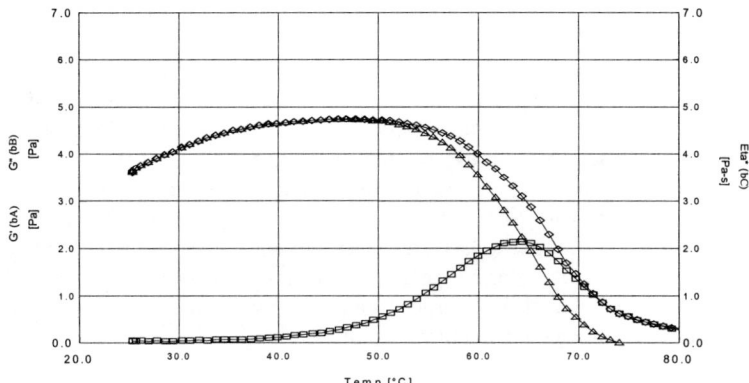

Figure 2   *Dynamic temperature sweep measurement of 3% (actives) Ethoquad® E/12-75 in 3% KCl brine.   Δ represents G' (elastic component), ☐ represents G"(viscous component), ◊ represents Eta (viscosity)*

Poorly performing gels tended to exhibit lower G' values that would fall off sharply as the temperature increased. Correspondingly, the viscous component, G", increased significantly with temperature leading to a crossover point significantly lower than for the good sample. Our results to date indicate that samples with a low flowability index (G"/G') at 25°C and a crossover temperature of > 58°C should offer acceptable performance for fracturing purposes. The rheometric dynamic analysis illustrates more effectively than a rotational viscometer the degree to which the rheological character of the system is affected by surfactant impurities.

2.3   Characterisation of Aromox® APA T

Aromox® APA T is another product that exhibits excellent viscoelastic properties in solutions of electrolytes. The product is an amidoamine oxide derived from N-[3-(dimethylamino)propyl]tallowamide and is biodegradable. The initial samples of this product were made using isopropanol as a solvent and in performance testing the results were generally disappointing. By changing solvent to propylene glycol we were able to significantly improve the performance, but at the same time we created some manufacturing issues. Using propylene glycol as a solvent, we observed an undesirable level of hydrolysis of the amide function during the oxidation stage. No explanation is offered for this but it was a problem that had to be addressed to obtain a consistent product. An improved procedure was subsequently developed[41].

The initial characterisation work with Aromox®APA T used a Brookfield rotational viscometer to look at the flow behaviour of potassium chloride brines gelled with 6% w/w of the Aromox®APA T.   To see just how sensitive the system is to the presence of isopropanol, a spiking experiment was done using a reference sample of the Aromox®APA T.   The isopropanol was spiked into the surfactant prior to making the brine gels.

Isopropanol levels as low as 1-2% in the surfactant composition significantly impacted the viscosity profile of the resulting gels.

Figure 3 shows the viscosity profiles produced by fluids prepared from different batches of the Aromox®APA T. The results clearly illustrate that while some samples exhibited good viscosity over a wide temperature range others failed to show acceptable levels of performance. Good samples were considered to be those that generated fluids that exhibited a viscosity maximum of about 15,000 to 20,000 centipoise at 55-60°C.

Analysis of all of the tested samples revealed differences in the levels of residual amine (unoxidized dimethylaminopropyl tallowamide) and fatty acid resulting from the preparative process. Contrary to our expectations the performance of the purest material tested, one that was low in both residual amine and fatty acid, exhibited inferior performance to reference materials and showed no viscosity maximum. Reference materials were samples that had undergone customer evaluation and were considered to offer the desired performance for fracturing applications.

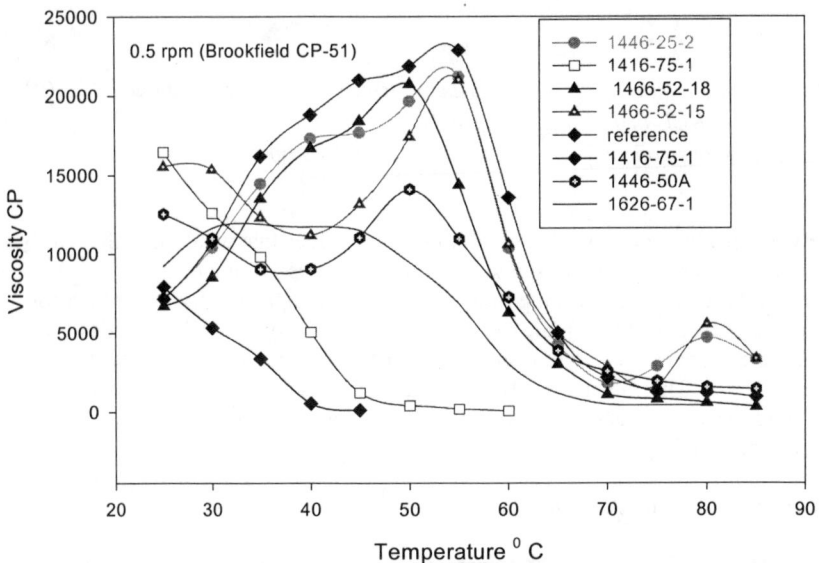

Figure 3   *Viscosity profiles of Aromox® APA T fluids showing batch to batch variations*

To determine the effect of amine and fatty acid on the performance of the product, spiking experiments were conducted using the "pure" material as a base. The effect of added amine and acid can be seen in Figure 4. The addition of amine to about 3.5% - 5% w/w increased the high temperature viscosity. However, increasing the amine loading much beyond 5% w/w resulted in phase separation. The addition of fatty acid to the pure material generally lowered viscosity over all temperatures and at levels about 2.5% w/w phase separation was observed. The negative effect of fatty acid could be compensated for to some extent by the use of additional amine.

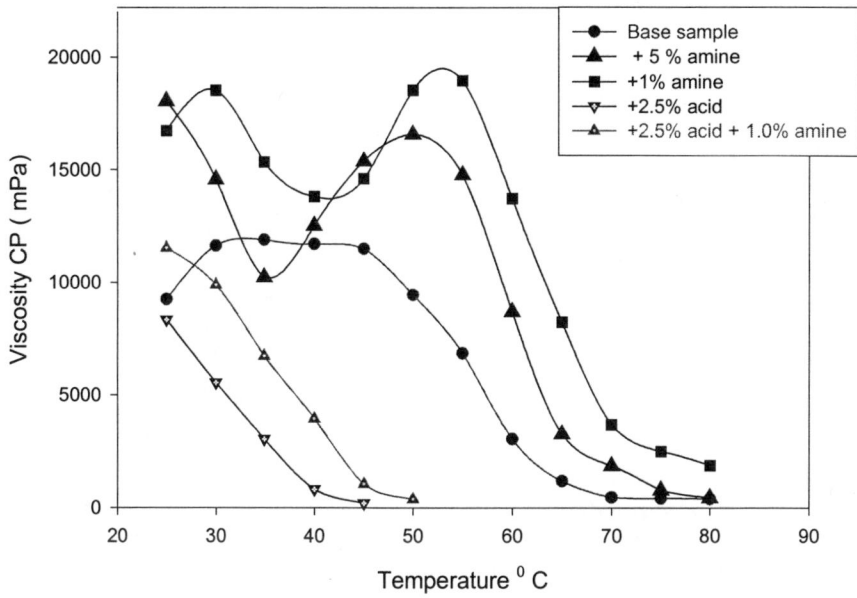

Figure 4 *Impact of added amine and fatty acid on the viscosity profile of fluids generated with Aromox® APA T*

The effect of surfactant concentration with 3% KCl brines was evaluated, and at loadings of less than 4% w/w thickening of the brine was observed. At loadings greater than 4% viscoelastic gels were observed. The effect of brine concentration and type was also investigated using a 6% w/w loading of Aromox®APA T. For KCl, the viscosity maximum shifted to higher temperatures on increasing the concentration to 3% and then decreased beyond there; sodium chloride showed similar behaviour. With monovalent ions a trend in performance was noted such that $Li^+>NH4^+>Na^+=K^+$. Divalent systems such as calcium chloride brines produced good viscosity profiles at loadings of <2%, but at higher loadings the systems would phase separate at elevated temperatures.

2.3.1. Dynamic temperature sweep analysis. In this method the samples were subjected to a constant stress of 0.5 Pa and the parameters, G', G'', phase angle, and complex viscosity recorded. In examining the performance of Aromox®APA T using the rheometer in this mode it was decided to take three samples that had been defined on the rotational viscometer as having modest, good, and poor performance. The product exhibiting a modest performance was low in amine and fatty acid, the good performer was low in fatty acid but with a higher level of amine, and the poor performer was low in amine but high in fatty acid. The results are illustrated in Figures 5, 6 and 7.

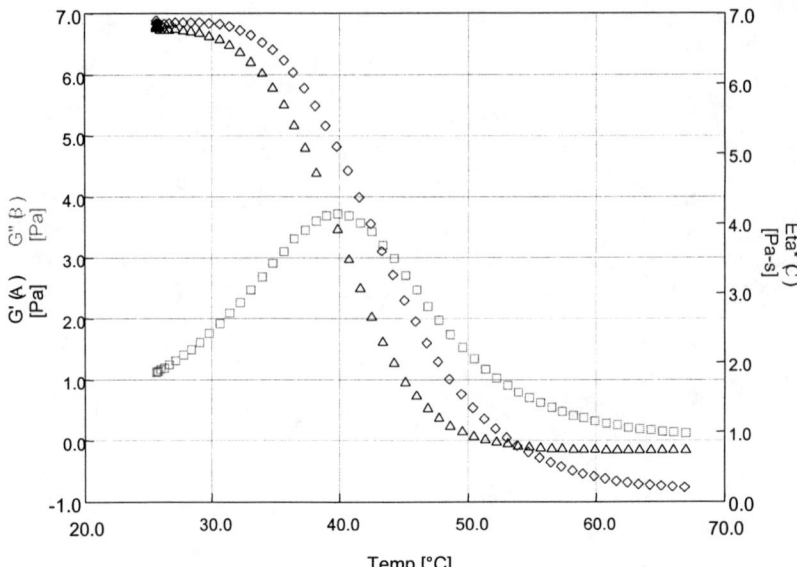

Figure 5   *Dynamic temperature sweep diagram for Aromox® APA T exhibiting modest performance . Δ represents G' (elastic component), □ represents G"(viscous component), ◊ represents Eta (viscosity)*

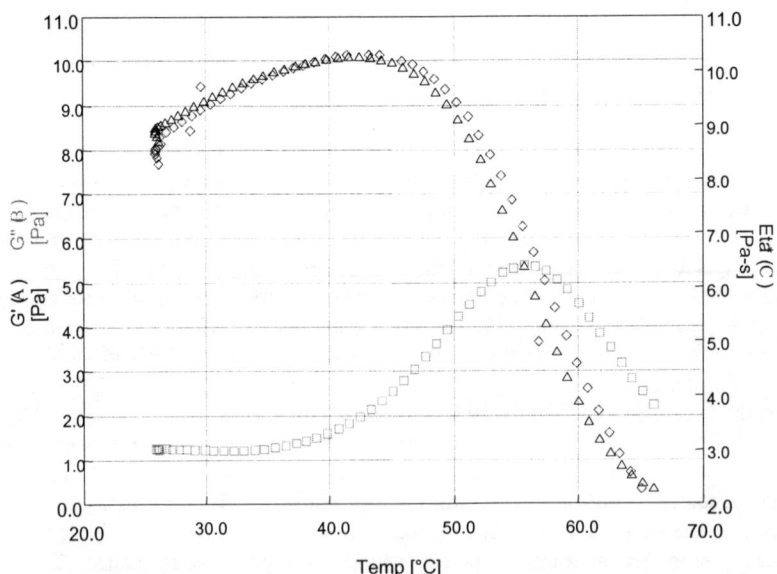

Figure 6   *Dynamic temperature sweep diagram for Aromox® APA T exhibiting good performance. Δ represents G' (elastic component), □ represents G"(viscous component), ◊ represents Eta (viscosity)*

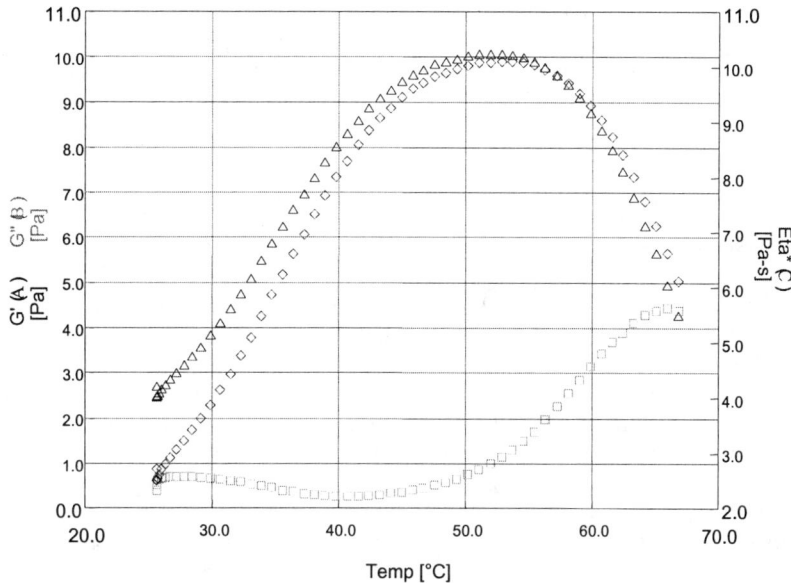

Figure 7 *Dynamic temperature sweep diagram for Aromox® APA T exhibiting poor performance. Δ represents G' (elastic component), ☐ represents G"(viscous component), ◊ represents Eta (viscosity)*

The product low in both residual amine and acid shows a relatively low crossover point, ~40°C, and low elasticity at room temperature; G' is approximately 6.8 Pa. As the sample melts, beyond the crossover point, the outer phase viscoelasticity is rapidly lost and its ability to suspend proppants diminished. In contrast, the good sample with a higher level of free amine has a G' of about 8 Pa and a crossover temperature in the region of 55°C. The loss of viscoelasticity beyond the crossover temperature is less dramatic in this case and thus the deposition of suspended solids would be more controlled. The most surprising result came from what was considered to be the poorest sample of the three. Our experiments using a rotational viscometer had indicated that high levels of fatty acid had a very negative effect on performance. However, the results from the rheometer show that while G' has a low value at 25°C it increases significantly as the temperature is increased. The crossover point is not reached until nearly 70°C and the high yield strength exhibited at the crossover point (~50Pa) indicates the ability to suspend materials such as proppants at quite high temperatures. The discrepancy between the observations from the rotational viscometer and the rheometer has been attributed to a wall slip effect with the rotational viscometer. The slip effect is most pronounced with highly elastic gels.

Results obtained from these three samples demonstrate that both the amine content and the fatty acid content significantly impact the performance of this product. The amine impurities increase both the low and high temperature viscosity and elasticity whereas fatty acid reduces the elasticity and viscosity at low temperatures but increases them at higher temperatures. Thus, one might anticipate that the performance characteristics of the product could be optimised for specific applications by making appropriate adjustments in amine and acid levels. To evaluate this we examined the performance of a batch in which the fatty acid was at a low level and the amine was within what we had assigned as an acceptable range. The values of G' and G" for this sample were somewhat lower than our reference material. Encouraged by earlier results that had indicated that additional fatty acid could boost the high temperature performance the sample was spiked with additional fatty acid. The result of the spiking was to produce a product that closely matched the performance of the reference. Thus, we were able to demonstrate that the rheological profile can be modified with some degree of predictability by the simple addition of amine or fatty acid.

3   SUMMARY AND CONCLUSIONS

The two surfactants described illustrate the impact on performance of manufacturing impurities. In the case of Aromox® APA T the effect can be used to advantage in varying the performance of the product in a predictable way to meet the needs of the specific application. The work presented has focussed on brine solutions and the requirements for oilfield fracturing purposes but clearly these systems can be adapted to other applications such as personal care and cleaning formulations. The Aromox® APA T will gel both water and brine systems thus providing additional versatility. For applications involving suspended solids the crossover temperature, outer phase density, and elasticity are important in controlling the release of the solids.   From our observations with Aromox®APA T, the crossover temperature and elasticity of the system can be adjusted by use of low levels of additives. The additives may be impurities resulting from the manufacturing process and thus incorporated into the product during manufacture. Understanding how the impurities impact the rheological properties allows for the subsequent adjustment of process conditions and specifications to achieve optimum performance. Since some surfactants are made in solution, consideration has to be given to the solvent used. Clearly, for Aromox® APA T monohydric alcohols such as IPA have an adverse effect on the rheological properties of the gels produced, whereas Ethoquad® E12 –75 is relatively unaffected. Surfactant manufacturers should realise that their process conditions and product consistency can significantly impact the application performance of viscoelastic surfactants. By understanding how the composition of the product impacts its performance, controls can be put in place to optimise performance and eliminate the need for batch to batch formulation adjustments by the formulator.

References

1   C. Tanford, *The hydrophobic effect,* 2nd ed.: Wiley New York, 1980.
2   P.A. Hassan, J.V. Yakhmi, *Langmuir* 2000, 16, 7187.
3   H. Rehage, H. Hoffman, *Mol. Phys.*1991, 74, 933.
4   M.E. Cates, S.J. Candau, *J.Phys: Condens. Matter* 1990, 2, 6869.
5   H. Hoffman, U. Munkert, C. Thunig, M. Valienta, *J. Colloid Interface Sci.* 1994, . 163, 217.
6   T. Shikita, H. Hirata, T. Kotaka, *Langmuir* 1989, 5, 398.
7   F. Kern, R. Zana, S.J. Candau, M.E. Cates, *Langmuir* 1992, 8, 437.
8   M. Tornblom, U. Henrikson, M. Ginley, *J. Phys. Chem.* 1994, 98,7041.
9   S.Q. Wang, *J. Phys. chem.* 1990, 94, 8381.

10 H. Hoffman, A. Rauscher, M. Gradzielski, S.F. Schulz, *Langmuir* 1992, 8, 2140.
11 S.J. Candau, E. Hirsch, R. Zana, M.Adam, *J. Colloid Interface. Sci.* 1988, 122, 430.
12 F. Kern, R. Zana, S.J. Candau, *Langmuir* 1991, 7, 1344.
13 T.M. Clausen, P.K. Vinson, J.R. Minter, H.T. Davis, T. Talmon and W.J. Miller, *J. Phys. Chem.* 1992, 96, 474.
14 T. Shikita, Y. Morishima, *Langmuir* 1996, 12, 5307.
15 T. Shikita, Y. Morishima, *Langmuir* 1997, 13, 5229.
16 J.F.A. Soltero, J.E. Puig, O. Manero, P.C. Shculz, *Langmuir* 1995, 11, 3337.
17 Soltero, J.F.A.; Bautista, F.; Puig, J.E.; Manero, O.; *Langmuir*, 1999, 15, 1604.
18 Hassan, P.A.; Valaulikar, B.S.; Manohar, C.; Kern, F.; Bourdieu, L.; Candau, S.J.; *Langmuir* 1996, 12, 4350.
19 Hassan, P.A.; Manohar C.; *J. Phys. Chem. B* 1998, 102, 7120.
20 Hassan, P.A; Candau, S.J.; Kern, F.; Manohar, C.; *Langmuir*, 1998,14, 6025.
21 Narayanan, O.R.; Hassan, P.A.; Manohar, C.; Salkar, R.A.; Kern, F.; Candau, S.J.; Langmuir 1998, 14, 4364.
22 Whalen, R.T; *Viscoelastic surfactant fracturing fluids and method for fracturing fluids and a method for fracturing subterranean formations,* US6035936, 2000.
23 Gadberry, J.F.; Hoey, M.D; Franklin, R; Vale, G. D.; and Mozayeni, F.; *Surfactants for hydraulic fracturing compositions,* US5979555, Akzo Nobel NV, 1999.
24 Farmer, R.F.; Doyle, A.K.; Vale, G. D.; Gadberry, J.F.; Hoey, M.D.; Dobson, R.E.; *Method for controlling the rheology of an aqueous fluid and gelling agent therefor,* US6239183, Akzo Nobel NV, 2001.
25 Nehmer, W.L; *Viscoelastic surfactant gravel carrier fluids,* CA1298697A1, 1992.
26 Rose, G. D.; *Water based hydraulic fluids,* US4806256, The Dow Chemical Company, 1989.
27 Tibbles, R.J; Parlar, M.; Chang, F.F.; Fu, D.; Davison, J.M.; Morris, E.A.; Wierenga, A.M.; Vinod, P.S.; *Fluids and techniques for hydrocarbon well completion,* US6140277, Schlumberger Technology Corporation, 2000.
28 Brown, J.E.; Card, R.J.; Nelson, E.B.; *Methods and compositions for testing subterranean formation,* US5964295, Schlumberger Technology Corporation, 1999.
29 J.E. Rader, W.L. Smith, *Methods for opening drains using phase stable viscoelastic cleaning compositions,* US5833764, 1998.
30 G.D. Rose, A.S. Teot; K.L. Foster, *Aqueous bleach compositions thickened with a viscoelastic surfactant,* US4800036, The Dow Chemical Company, 1989.
31 A.S. Teot, G.D. Rose, G.A. Stevens, *Friction reduction using a viscoelastic surfactant,* US4615825, The Dow Chemical Company, 1986.
32 M. Hellsten, I. Harwigsson, *Use of alkoxylated alkanolamide together with ethoxylated alcohol as a friction reducing agent,* US5979474, Akzo Nobel NV, 1999.
33 M. Hellsten, I. Harwigsson, *Use of a betaine surfactant together with an anionic surfactant as a drag-reducing agent,* US5902784, Akzo Nobel NV, 1999.
34 R.L. McLean, *Vibration dampers utilizing reinforce viscoelastic fluids,* US3640149, Houdaille Industries, Inc., 1972.
35 G.D. Rose, K.G. Seymour, A.S. Teot, *Shear stable antimisting visocelastic formulation,* US4770814, The Dow Chemical Company, 1988.
36 R.G. Welch, D.N. Githuku, L.J. Hollihan, Jackson, Charles, A.; *Process for producing detergent agglomerates from high active surfactant pastes having non linear viscoelastic properties,* US5574005, The Procter and Gamble Company, 1996.
37 D. Balzer, *Aqueous viscoelastic surfactant solution for hair and skin cleaning,* US5965502, Huels Aktiengellschaft, 1999.

38 J.E. Bonekamp, G.D. Rose, D.E. Schmidt, A.S. Teot, E.K. Watkins, Viscoelastic *surfactant based foam fluids*, US5258137, The Dow Chemical Company, 1993.
39 W.B. Gogarty, *Use of viscoelastic fluids for mobility control*, US3822746, Marathon Oil Company, 1974.
40 G.D. Rose, *Method for heat exchange fluids comprising viscoelastic surfactant compositions*, US4534875, The Dow Chemical Company, 1985.
41 M.D. Hoey, et al; US patent application, Akzo Nobel NV, 2000.

FUNCTION AND APPLICATION OF OILFIELD CHEMISTRY IN OPEN HOLE SAND CONTROL COMPLETIONS

Liz George[1], L. Morris[2], S. Daniel[3], B. Lungwitz[3], *M. E. Brady[3] and P. Fletcher[3]

[1] Shell U.K. Exploration and Production, 1 Altens Farm Road,Nigg, Aberdeen. AB12 3FY. U.K.
[2] Client Support Laboratory for Europe, CIS and Africa, Schlumberger Oilfield Services, Enterprise Drive, Westhill Industrial Estate, Westhill, Aberdeenshire, AB32 6TQ, UK
[3] Schlumberger Oilfield Chemical Products, Sugar-Land Product Centre, 110 Schlumberger Dr, Sugar-Land, Texas, TX 77478, USA.

* Corresponding author

1 INTRODUCTION

The number of oil and gas wells completed open hole has increased significantly in recent years. Many of these applications require sand control to aid wellbore stability and for sand exclusion. This paper describes the requirements of fluids associated with these applications and the chemistries involved.

We describe surfactants which are capable of forming viscoelastic fluids stable over a wide range of temperatures and salinities, and that also impart a low level of formation impairment. These viscoelastic surfactant (VES) systems may also be employed in a variety of oilfield applications such as matrix and fracturing fluid treatments but in this paper we focus on sand control applications. The tolerance of these viscoelastic surfactants to high brine concentration is attributable to their unique chemical properties, which we describe in this paper. The physico-chemical properties of this surfactant broaden the scope of field applications previously attainable only by mechanical means (isolation tools). The ability to form viscous higher density solids free fluids makes them ideal for circulating gravel pack placements where well control is an important issue and sand transport is critical. In addition we describe the compatibility of these materials with near well-bore clean-up treatments.

We will also describe the relationship of this chemistry to environmental compliance issues, which are gaining ever-increasing importance in the oilfield sector. A comparison with other gelling agents with respect to biodegradability and toxicity is outlined where continuous improvement is the main goal.

2 BACKGROUND

There are two principal techniques for gravel packing; the so called "high rate water packing" (HRWP) technique employs a high velocity transport of sand slurries at relatively low concentrations of sized sand (0.5-1 lbs/ gal). A critical velocity is required to ensure that the sand packs the lower side of the wellbore during the alpha wave portion of the treatment. When the sand slurry reaches the toe of the well the void in the upper part of the wellbore is filled from toe to heel during the beta wave[1]. This method can employ gelled carrier fluids but more common practice today is the use of non-viscous carrier fluids. The second technique uses alternate path technology where viscous carrier fluids are formulated to transport sand at high concentrations (2-8 lbs/gal). When losses occur into the formation, due to filter cake erosion[2,3] or swabbing effects after setting the packer, the sand slurry becomes dehydrated very quickly and forms a node or bridge preventing the transport of sand toward the toe of the well. In HRWP applications the occurrence of this screen out or bridge normally marks the end of the job and only part of the screen-wellbore annulus becomes gravel packed[4,5]. In the alternate path technique pressure build up behind the sand bridge diverts sand slurry into transport and packing tubes situated around the screen. The slurry bypasses the bridge and exits through nozzles into the screen-wellbore annulus. This process continues from the heel to the toe of the well.

Common gelling agents for gravel packing have been hydroxyethyl cellulose (HEC) and biopolymers such as xanthan gum. It is well known that these polymeric systems have several intrinsic disadvantages such as a significant degree of irreversible formation damage and high friction pressures. More recently VES base fluids have been utilised as gelling agents in a variety of applications including fracturing, matrix stimulation and sand control[6-11].

3   CHEMISTRY OF VISCO-ELASTIC SURFACTANTS

When the concentration of individual surfactant molecules is high enough they aggregate to form structures called micelles[12]. The minimum concentration required to form such aggregates is known as the critical micelle concentration (cmc). In aqueous media free of salts and inorganic species, the cmc decreases by a factor of 2 for each carbon atom increase in the hydrophobic alkyl chain length. The presence of salts in the solution phase can decrease the cmc values of ionic surfactants dramatically. The geometry of the micelles formed may also change in some cases to form worm or rod-like micelles[13-15] (Figure 1) (instead of the classical spherical micelle structures), which in semi dilute concentrations generate relatively high viscosities due to entanglement of micelles: these structures can grow to extremely large dimensions (approximately 1 micron when viewed using cryoscopic transmission electron microscopy) (Figure 2), which produces a high propensity for the micelle structures to interact with each other forming a gel network. The head group on viscoelastic surfactants can associate strongly with water via hydration spheres and hydrogen bonding which also contributes to the hindrance of fluid movement and increase viscosity. Such fluids exhibit very interesting rheological behaviour. The fluids produced are termed viscoelastic because they have the capability of returning to their original viscous state or form when an applied stress is released.

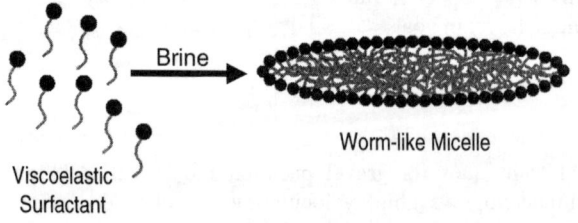

Figure 1 *The addition of salt to a viscoelastic surfactant produces a worm-like micelle*

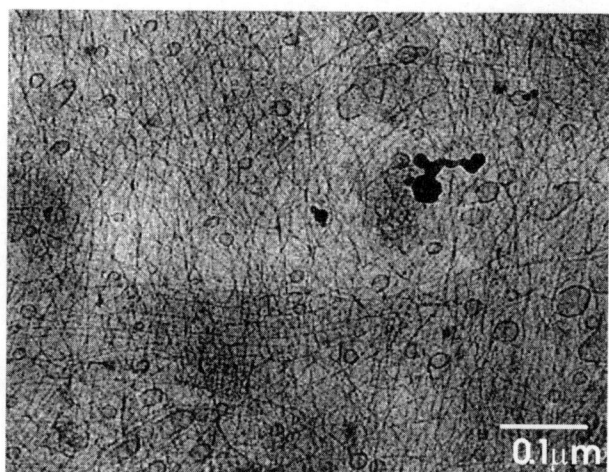

Figure 2 *Cryoscopic Transmission Electron Microscopy shows numerous micelle structures, formed from a quaternary amine with a long hydrophobic tail group, which can grow to 1-2 microns*

As a result of this unique rheological behaviour, viscoelastic surfactant (VES) based fluids are excellent particle suspension media. Some of the principal advantages of VES fluids are ease of preparation, low formation damage and high retained permeability in the sand/ proppant pack[6]. A simple dilution of the surfactant in brine, with agitation, is all that is required to prepare the fluid.

Currently several VES surfactants are known: 1) cationic surfactants, mainly quaternary amines; 2) anionic surfactants, based on fatty acids and 3) zwitterionic surfactants. The principle classes of surfactants are shown (Figure 3). The quaternary amine and fatty acid surfactants do have some limitations in oilfield applications, predominantly that electrolyte or salt concentrations are limited to approximately 6 wt% (0.84-1.07 mole l^{-1}) where carrier fluids are formulated with sodium or potassium chloride. This salt intolerance can be addressed partially by modifying the tail or chain length of the surfactant. One feature of longer chain quaternary amine surfactants is that a competent gel structure, involving the formation of worm like micelles, is maintained with divalent/ alkaline earth metal salts such as $CaCl_2$ up to densities of 9.5 lbs/ gal (1.65 mole l^{-1}). A limitation to date of fluids formulated with high electrolyte concentrations is that they have limited temperature stability (<150 °F): electrostatic interactions and Van der Waals forces have a large role in the reformation of rod-like micelle entanglements after the fluid system has been perturbed[13]. An alteration in micelle shape may occur by changes in molecular environment such as pH, ionic strength, temperature and the presence of other additives in solution.

The chemical structure of a zwitterionic surfactant is shown (Figure 3c). We believe that the head group, having both a positive and negative charge, produces micelle properties that are less sensitive to electrostatic interactions with other species in solution such as salt anions. The individual micelle structures may also have stronger interactions with each other via dipole-dipole interactions. The mechanism of these micelle interactions at higher temperature and electrolyte concentrations is the subject of on-going work.

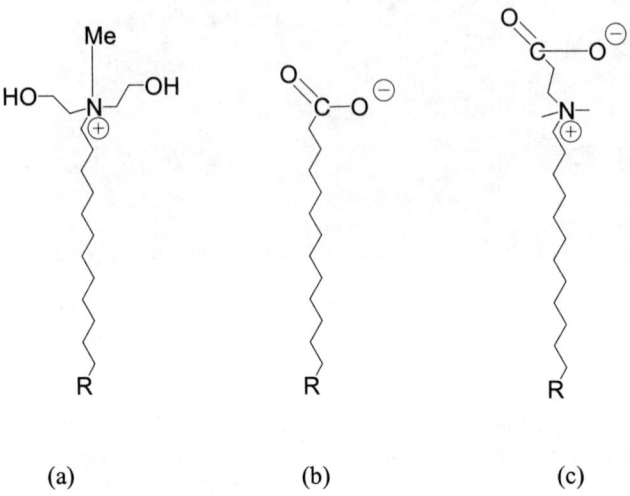

Figure 3 *Several types of surfactants capable to form wormlike micelles; (a) quaternary amine; (b) fatty acid and (c) zwitterionic surfactant*

These surfactants are compatible with high density brines up to 14 lbs/ gal and fluids prepared with them have unique thermal stability up to >270 °F. Viscosity-temperature profiles (Figure 4) for several formulations incorporating different concentrations of calcium bromide are shown. These solutions have been gelled with a zwitterionic surfactant and a co-surfactant (sodium alkyl-aryl sulphonate) that interacts with the surfactant chains to aid in the formation of worm like micelles. The fluctuations in viscosity with temperature are difficult to ascribe to specific surfactant properties. We know from the literature[12] that the cmc is decreased by the dehydration of the head group that occurs with increased temperatures. Hydrogen bonding disruption also occurs with increased temperature, which increases the cmc. These changes work in opposite directions and may not be equivalent in magnitude at a specific temperature. With this zwitterionic surfactant it is apparent that higher salt concentrations are indeed beneficial with respect to viscosity, particularly at temperatures in excess of 270 °F. This observation is the reverse of normal observations with an ionic VES because too much counter-ion shields the charge on the surfactant head group, changing several properties such as head group area and radius of curvature. This can lead, for example, to the formation of spherical micelles that do not impart viscosity at the concentrations normally used (<5% v/v). The limit to the salt tolerance for the zwitterionic surfactant evaluated here is 13.5-14 lbs/ gal (1.68 moles l^{-1}) which may be ascribed to the amphoteric head group being no longer capable of screening charges: disruption of the micelle structure occurs resulting in lower viscosity. The rheological temperature stability of these fluids approaches that of many commonly used viscosifiers such as xanthan gum biopolymer and hydroxyethyl cellulose stabilised with pH buffers and radical scavengers.

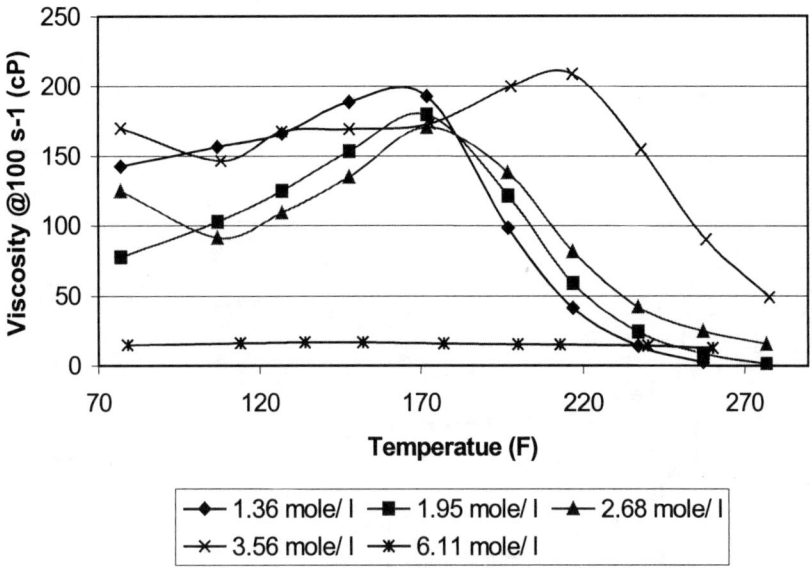

Figure 4 *Viscosity as a function of temperature for fluids gelled with 3% v/v zwitterionic surfactant (w/ co-surfactant) in various concentrations of CaBr₂ brine giving densities ranging from 9-14 lbs/gal (1.08-1.68 sg)*

4   APPLICATIONS

In this section we introduce gelled fluids formulated with a zwitterionic surfactant that may be applied in circulating gravel-packing applications. We focus on the placement of sand using an alternate path technique (described above).

4.1 Zwitterionic Surfactant and HydroxyEthyl Cellulose (HEC) Based Viscous Gravel Pack Slurries

We have demonstrated that gels formed with certain zwitterionic surfactants possess both greater salt and temperature stability, approaching that of polymers such as hydroxyethyl cellulose (a common viscosifier for completion brines). Another important property of VES micelles is that they are "soluble" in produced hydrocarbons. This facilitates inflow when the well is brought on production. In fact what occurs is that the micelles change shape from rod like to spherical and the fluid loses its viscoelastic properties. Alternative carrier fluids gelled with HEC often require breaker technology (oxidizers causing scission of polymer chains) to ensure restored permeability of the gravel pack during hydrocarbon flow back. The disadvantage of this approach is that the viscosity of the carrier fluid, and hence its sand carrying capacity, may be compromised particularly if the well treatment is delayed after the inclusion of the breaker or if pumping is interrupted. The viscosity retained in HEC based carrier fluids with and without an oxidiser (1.66 sg CaCl₂/ CaBr₂ brine) at 250 °F (121 °C) is shown (Figure 5). Both fluids lose viscosity relatively quickly (50% reduction within 2-4 hours). Whereas the base fluid still maintains enough viscosity to securely transport sand (based on laboratory and field practice with alternate path technology a viscosity > 50 cp at 100 s^{-1} suffices), the same fluid with oxidiser has lost sand carrying capacity, at flow rates typical in alternate path technology (2-6 bbls/ min), within 1-2 hours. In many cases this characteristic is still suitable to complete the gravel pack operation without failure. However, there is little room for error if the operation is delayed.

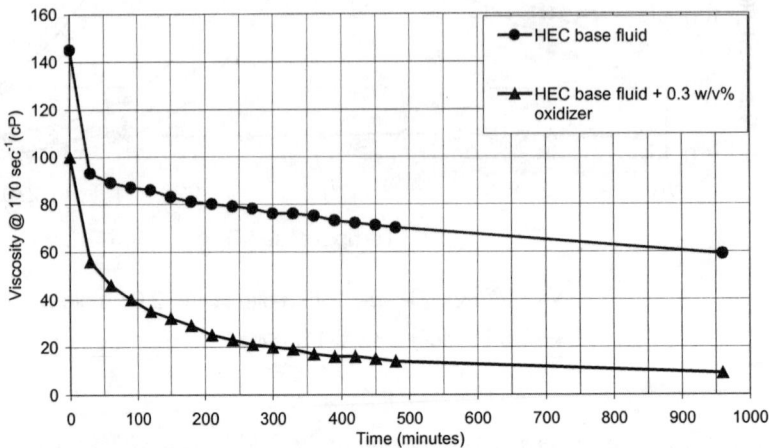

Figure 5 *Reciprocating capillary viscometer measurements for a gelled carrier fluid comprising 8% w/v HEC mixed in 13.8 lbs/ gal (1.66 sg) CaBr$_2$/CaCl$_2$ brine with and without 0.3% w/v of an oxidiser (polymer breaker) at 250 °F (121 °C)*

The viscosity-temperature profile for a 1.66 sg (13.8 lbs/ gal CaBr$_2$) brine base fluid with 20 % v/v of a chelating agent (Hydroxyalkylethylenediaminetriacetate) gelled with 3% v/v of a zwitterionic surfactant is shown (Figure 6). The chelant has been incorporated in the carrier fluid to dissolve the CaCO$_3$ within the reservoir drilling fluid (RDF) filter cake[16-18]. CaCO$_3$ is used as a bridging material to prevent excess mud, solids and filtrate invasion. The chelating agent chemistry (Figure 7) is fully compatible with the zwitterionic surfactant and has been described for other well treatments[18]. Lower density carrier fluids used both to transport sand and to place clean up chemistry in a single stage process have been described in previous papers[16-18,20]. The use of high density gels attainable with zwitterionic surfactants, that are also highly temperature stable and compatible with low corrosivity clean up chemistry, is something that has not been published in detail to date. In gravel packing operations the screen and tubulars are exposed to corrosive environments for many years. It is crucial that the installed screens maintain their integrity for the economic life of the well as intervention costs are generally high. Although strong acids may be used to remove RDF filter cake damage they are also much more corrosive than chelating agent solutions[10,11]. Even when inhibited the corrosion rate for strong acids will increase after several days which has long term implications for the integrity of the screen hardware during the life-time of the well.

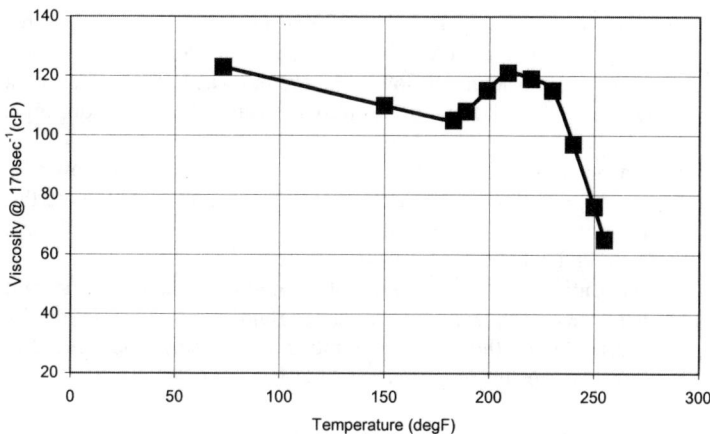

Figure 6 *The viscosity as a function of temperature of a carrier fluid incorporating 3% v/v active zwitterionic surfactant and 20% v/v of a chelating agent solution prepared from Hydroxyalkylethylenediaminetriacetic acid. The fluid also incorporates 0.1% v/v of a corrosion inhibitor.*

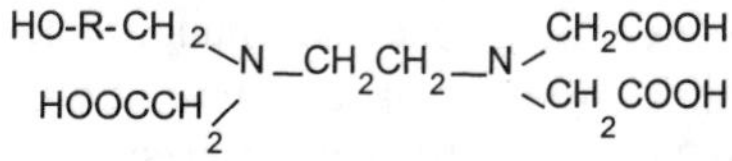

Figure 7 *Structure of Hydroxyalkylethylenediaminetriacetic acid, a precursor for reagents prepared for the dissolution of CaCO₃ present in reservoir drilling fluids and their filter cakes*

4.1.1 Carrier Fluid Impact on Core, and Core Gravel Pack Composites

Preserved North Sea field core samples were vacuum saturated with simulated formation brine before being loaded into a permeameter fitted with the capacity to accommodate a gravel pack and screen. The core was confined in a modified HTHP cell, with one end converted to hold a 1-inch diameter sample. The sample was cleaned by alternatively flooding Isopar oil and simulated formation brine at 360 ml/hr. The sample was judged clean when the effluent from the core was clear. When the sample was clean, the back-pressure on the system was raised to 100 psig and Isopar oil was flooded through the sample at 360 ml/hr. The flood was continued until the sample had attained irreducible water saturation and a stable pressure drop had been reached. The permeability of the sample was calculated by measuring the pressure drop at 5 different flow rates (60 to 360ml/hr). A linear regression of the data was performed and the calculated gradient inserted into Darcy's equation. The data was only accepted if the r- squared coefficient was better than 0.99 and the intercept was between 50 and −50 mBar. On completion of the base undamaged core permeability measurement the sample's injection direction was reversed and the temperature of the cell increased to 180 °F (82.2 °C). The top of the cell was opened and the gravel pack carrier fluid, pre-heated to a temperature of 180°F, was poured onto the "wellbore face" of the rock sample. The cell was sealed and the temperature and pore pressure were raised to 250 °F (121 °C) and 100 psig respectively. The gravel pack fluid was then injected slowly through the sample, until 5 pore volumes had been collected from the system out flow. The system was left under test conditions for 12 hours to allow any internal fluid breaker to activate, before performing the return permeability.

On completion of the required period for breaker activation, the flow direction was reversed and Isopar oil was flooded through the sample at 360 ml/hr. Note that the VES fluid does not require an internal breaker but its gel structure is deteriorated via production of reservoir fluids, mutual solvents or by brine dilution.

The flood was continued until the sample had attained a stable permeability and the system effluent was free of any gravel pack carrier fluid. The return permeability of the sample was measured using the same conditions and method as used for the base permeability reading. The results are discussed later in section 4.1.3.

4.1.2   Retained Permeability Through 20/40 Gravel Pack

The gravel pack slurries were prepared with 20/40 mesh sand, held together in 1.5 inch diameter Teflon jackets with fine and coarse stainless steel gauzes over the end faces. Two gravel pack slurries were evaluated. One with the HEC carrier fluid and the other with the zwitterionic surfactant based fluid. The gravel pack sand samples were vacuum saturated with Isopar oil before being loaded into the testing apparatus. The Teflon jacket containing the sand was confined hydrostatically. Base line permeability was measured using the same technique as above. The gravel pack fluid, pre-heated to a temperature of 180 °F was then injected into the 20/40 mesh sand. The cell was sealed and temperature and pore pressure were raised to 250 °F and 100 psig respectively. The gravel pack carrier fluid was then injected slowly through the sample, until 5 pore volumes had been collected from the system out flow. The gravel pack slurries were shut in for the same period as before and the return permeability of the sample was measured using the same conditions and method as used in the base undamaged gravel pack permeability measurements.

The results are shown in table 1. It is evident that the level of formation and gravel pack damage induced by the HEC carrier fluid is significantly greater than that imparted by the VES slurry.

Table 1   *A comparison between HEC and VES fluids and their impact on rock and gravel pack permeability impairment*

Fluid type	Formation Base Permeability (mD)	Formation Return Permeability (mD)	Formation Permeability Retained (%)	Gravel Pack/ Core Base Permeability (mD)	Gravel Pack/ Core Return Permeability (mD)	Gravel Pack Permeability Retained (%)
HEC	482	260	54	3010	1880	63
VES	474	471	99	2800	2810	100

4.2 Inclusion of Clean up Chemistry During Gravel Packing Operations

In this section we describe the application of high-density gravel pack carrier fluids containing a chelating agent for simultaneous RDF filter cake removal using the alternate path technique. We describe the rheological profile of the fluid and its ability to facilitate cake flow back through a completion comprising a 20/ 40 gravel pack sand with a 12 gauge (300 micron opening) wire wrapped screen.

4.2.1 Rheological Compatibility

The clean up fluid composition used was a 25 % v/v of a chelating agent solution (CAS) in 12.5 lbs/ gal (1.50 sg) NaBr brine, reduced in pH to 4.0 with 36 wt% hydrochloric acid. This solution was prepared in a Waring Blender and 3% v/v of a zwitterionic surfactant added at high shear rate. 0.1% v/v corrosion inhibitor was added to the gel. The resulting fluid density was 11.7 lbs/ gal (1.41 sg). The viscosity-temperature profile (Figure 8) is typical for this fluid with the highest viscosity being generated between 120 and 140 °F with another peak at 220 °F. Such fluctuation in viscosity has not presented problems during the pumping of sand slurries via alternate path methodology. It is evident that the rheological stability at temperatures greater than 200 °F is significantly greater than gels formed with quaternary amine surfactants or anionic surfactants. We believe that the rod like micelle formed with a zwitterionic surfactant is less dependent on electrostatic or Van der Waals interactions that occur with quaternary amine and anionic surfactants[13]. Instead, we believe that stronger dipole-dipole interactions may participate in reforming the rod like micelle structure when shear or perturbation of the fluid occurs (shear at higher temperature).

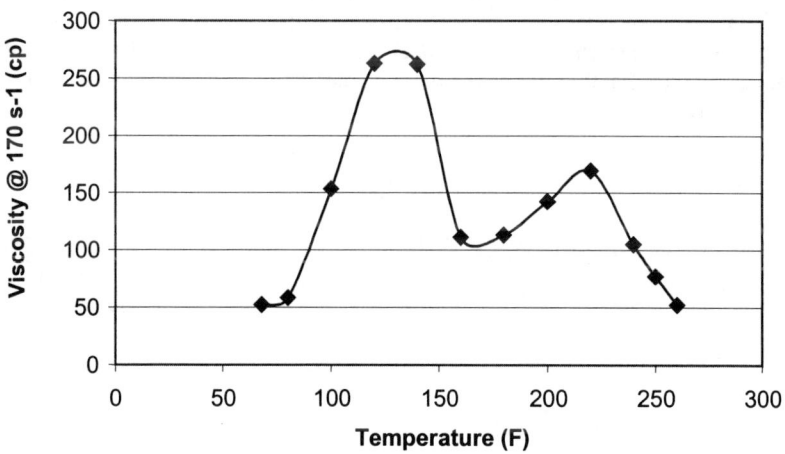

Figure 8 *Variation of viscosity at 170 s^{-1} with temperature of a fluid comprising 25% v/v chelating agent solution in 12.5 lbs/ gal NaBr brine with 3% v/v zwitterionic surfactant*

4.2.2 Simultaneous Filter Cake Removal during Gravel Packing Operations

A similar fluid system to that above (3% v/v Zwitterionic VES with a 40% v/v solution of a chelating agent in 12.8 lbs/ gal NaBr, reduced to pH 4) was utilised in further tests to evaluate its effect on the flow back properties of a polymer-carbonate based RDF. The RDF comprised a sodium bromide brine, with approximately 48 lbs/ bbl sized calcium carbonate (the particle size was selected to pass through 20/ 40 mesh sand), 1.5 lbs/ bbl xanthan gum and 6 lbs/ bbl derivitised starch. Twelve pounds per barrel (lbs/ bbl) simulated drill solids were added to the RDF (60% sand and 40% Kaolinite- hymod prima clay). The density of the RDF was 1.66 lbs/ gal (1.4 sg). A baseline was evaluated as follows: the water based RDF was injected into a 400 md sandstone core plug for forty-eight hours dynamically followed by a forty-eight hour static period at the reservoir conditions typical for the North Sea field of interest (124 °C (255 °F) and 2760 psi overbalance).

The pressure required to initiate flow (Figure 9) through the cake and the 20/ 40 gravel pack screened completion was extremely high (450 psi at 220 ml/ min). The retained permeability was less than 0.25% (this translates to a skin factor greater than 350). Another test was performed where the RDF filter cake was laid down in the same manner as before and the clean up carrier fluid, with a composition outlined above, was displaced at 100ml/min and 200 psi overbalance. The displacement was performed for 5 minutes with shut in for a further 30 minutes. This was repeated for 2 litres of clean up fluid, with a final shut in soak lasting 4 hours. The fluid with the VES and chelating agent solution (CAS) in NaBr was applied to the filter cake at 124 °C for a total of 6 hours. The pressure measured to initiate flow this time was much lower (11 psi) and the retained permeability measured at this point was 29% (Figure 10). Further crude oil was passed through the core, mimicking draw down and a final retained permeability was measured at 48.2%. This translates to a skin of 1.0. It is important to consider what the implications of these results are on well productivity. We utilised Hawkins equation to calculate the skin factors above assuming a 8.5" diameter wellbore and a 4.89" internal diameter base pipe inside the sand screen. Using the skin factors we utilised nodal analysis to look at impact on well productivity. We used an example well case where the reservoir interval was 500 ft with a reservoir thickness of 100 ft and a vertical to horizontal permeability of 1/ 10. Reservoir pressure was 2400 psi and the gravel pack permeability was 100 D. The nodal plots (Figure 11) showed that the base case, where no clean up was performed, produced 800

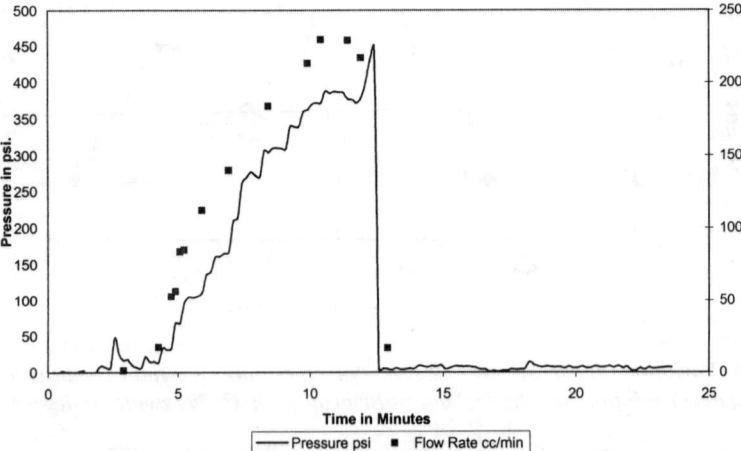

Figure 9 *Pressure-flow rate responses during flow back in the production direction of a North Sea crude oil through a sandstone core plug and water based drilling fluid filter cake that had been laid down at 2760 psi for 96 hours at 124 °C*

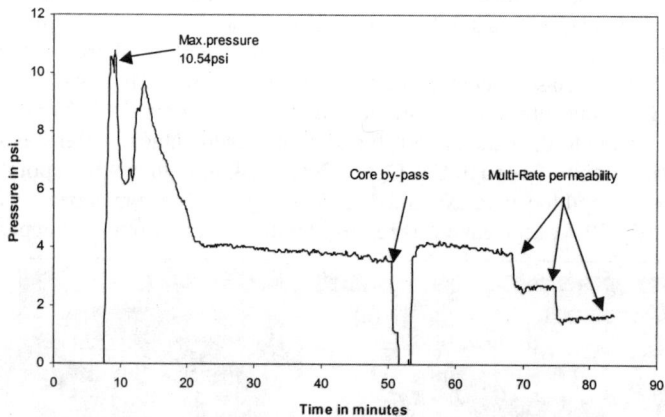

Figure 10 *Pressure responses during flow back of a North Sea crude oil through a sandstone core plug and water based drilling fluid filter cake treated for 6 hours with a Viscoelastic High Density Gravel Pack carrier fluid including CAS*

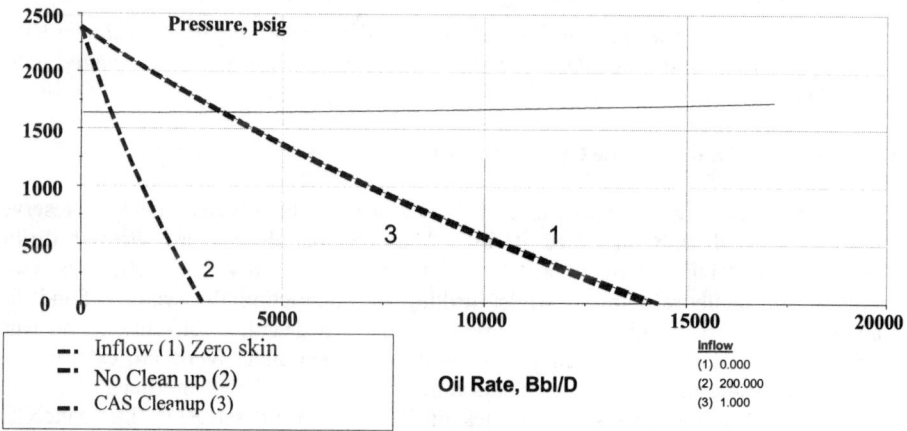

Figure 11 *Nodal Analysis for experiments described in section 4.2.2*

barrels of oil per day (bopd) whereas the clean up treatment produced 3449 bopd. A gravel-packed well with zero damage would produce 3547 bopd. The clean up treatment would make a significant impact on production with an extremely short pay back period with respect to covering additional costs for the well treatment.

4.2.3 Solids Removal During Clean Up of RDF Filter Cakes

A water based RDF was prepared with a 13.6 lbs/ gal $CaCl_2$/ $CaBr_2$ brine and 6 lbs/ bbl modified starch derivatives formulated for fluid loss and to provide adequate cuttings carrying capacity. Thirty lbs/ bbl $CaCO_3$ pore-throat bridging agents and 12.5 lbs/ bbl of kaolinitic clay (simulated drill solids) were added. The RDF fluid was injected into an Aloxite disc with a pore diameter of 35 microns. The injection period was 16 hrs at 500 psi overbalance and 250° F (121° C). A filter cake of 3-4 mm thickness with a compacted layer of 1 mm was produced (Figure 12a). A filter cake formed in the same manner and subsequently treated with a solution comprising 26% v/v CAS (pH 4) in 12.5 lbs/ gal NaBr brine gelled with 3% v/v of a zwitterionic surfactant is shown (Figure 12b). The corrosivity of this class of chelant, partially neutralised with HCl, is extremely low[10,11] and so only a low concentration of corrosion inhibitor is required (0.1% v/v). The clean up fluid was applied to the cake at 250° F and 50 psi overbalance. After 3 hours the clean up solution had dissolved enough CaCO₃ to effect breakthrough at which point the experiment was terminated and the cake examined. A striking difference in appearance was observed: approximately 50% of the cake was removed with numerous pinholes apparent.

Figure 12   *A water based RDF filter cake appearance before and after treatment with a gravel pack carrier fluid including a modified chelating agent solution and gelled with a zwitterionic surfactant*

4.3 Clean up of a Reversible Oil Based Filter Cake

There is an increasing number of extended reach and offshore wells where the reservoir section is entered with oil based RDFs. The principal drivers are superior drilling performance and reduced rig time. Also shaly streaks, sensitive to water, often cause problems for wellbore stability whilst drilling or completing the well. Sand face completion installation, after the pay zone has been drilled with a conventional oil based RDF, has proven difficult: the emulsions used in conventional inverts are extremely oil wetting thus rendering clean up treatments aimed at removing CaCO₃ inefficient[21].

We applied a gravel pack carrier/ cleanup fluid to an oil based RDF filter cake. The oil based RDF was formulated with emulsifier and oil wetting surfactants whose physico-chemical properties are altered after exposure to low pH (<7) fluids. The head group of these oil wetting surfactants becomes larger when protonated and renders the interfaces they reside at water wettable or hydrophilic. This allows acids and chelating agents to attack the cake during or after gravel packed completions. An example of the impact in using such oil based RDFs with a high density clean up system is shown (Figure 13). The 20/ 40 mesh sand (at 8 lbs/ gal) slurry comprised a 50% v/v of a chelating agent solution, (20 wt% CAS reduced to pH 4 in NaBr brine to a density of 10.6 lbs/ gal) gelled with 3% v/v of a zwitterionic surfactant. This slurry was placed above the aforementioned oil based RDF filter cake. Two experiments are shown where the filter cake has been created for periods of 1 and 16 hours at 300 psi overbalance and 170 °F. In the case where a 16 hour injection was carried out the overbalance during the carrier fluid soak period was reduced to 25 psi after 1 hour. This was performed to simulate completion of the gravel pack using alternate path technology and closing a flapper valve that isolates the open hole horizontal from a large hydrostatic pressure in the column of fluid above the packer. In both experiments break through occurred despite the presence of 25 lb/ bbl simulated drill solids (calcium montmorillonite clay) and 85 lbs/ bbl CaCO₃ weighting/ bridging agent. This phenomenon is not observed when such clean up treatments are applied to oil based RDF filter cakes incorporating conventional emulsifying or oil wetting agents (imidazoline, tall oil fatty acid/ amide chemistry). The increase in permeability and injectivity of the cake requires longer soak times when a greater period of drilling fluid injection has occurred.

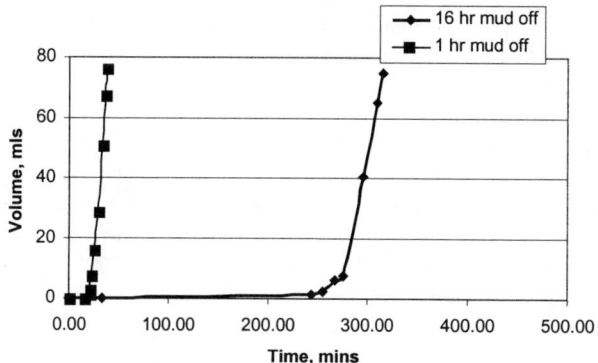

Figure 13 *Cake breakthrough experiments utilising an invertible oil based mud: carrier fluid was 50% v/v CAS w/ 3% v/v Zwitterionic gelling surfactant in NaBr brine (10.6 ppg)*

5   ENVIRONMENTAL ISSUES

We have demonstrated several examples where high-density carrier fluids can be effectively applied during gravel packing open hole horizontal wells. We have also demonstrated that clean up chemistry is compatible with the zwitterionic surfactants used to viscosify fluids with high electrolyte concentration.

As important as identifying technical solutions, consideration of the environmental impact plays a large role in applying new technology. Table 2 shows the results of common toxicity and biodegradation tests performed to assess the environmental impact of chemicals used in oilfield applications as per OSPARCOM protocol. As with most surfactants commonly used in oilfield applications the ecotoxicity to algal and sedimentary reworker species is high compared to non surface-active chemistries. Quaternary amines have been identified for some time as reagents that are problematic in this respect. As a result discharge of such chemistry has very strict limitations. It is the goal of the industry to reduce the overall environmental footprint in all oilfield activities. Some of the most stringent protocols are applied in the North Sea sectors. Zwitterionic surfactants also, by their very nature, will adsorb onto many different surfaces including those of common living marine species. In considering the toxicity of a chemical to several aquatic organisms it is evident that zwitterionic surfactants exhibit a marginally lower impact (Arcartia tonsa) whilst for other species a similar effect is observed (Skeletonema): the higher the EC50/ LC50 value the lower the toxicity. The majority of surfactants do possess this disadvantage (cf. polymeric moieties), however the overall impact must be assessed. In this regard the biodegradability of zwitterionic surfactants is considerably higher than that of quaternary amines. This may be ascribed to the carboxylate functionality present in the head group. Fatty acids with the same functionality are also highly biodegradable. A significant step towards the goal of continuous improvement is the reduced longevity (by virtue of greater biodegradability) of zwitterionic surfactants in marine and aqueous environments. This is one instance where both the technical performance enhancement and improved environmental properties have increased the scope of application of VES fluids to wells requiring higher density and where the conditions are more extreme such as higher temperature. The goal remains to improve this still further particularly from an environmental perspective.

Table 2 *Toxicity, bioaccumulation and biodegradability data for a Quaternary amine (R-N$^+$(OH)$_2$Me) and a zwitterionic surfactant (R-NH$_2^+$(CH$_2$)$_2$COO$^-$)*

Chemistry	R-N$^+$(OH)$_2$Me	R-NH$_2^+$(CH$_2$)$_2$COO$^-$
Skeletonema EC50 72 hr	< 0.5 mg/ L	0.43 mg/ L
Acartia tonsa LC50 48 hr	1 mg/ L	15 mg/ L
Corophium volutator (10 days)	1533 mg/ Kg	-
Scophtalmus maximus (OECD203) LC50, 96 hr	1.4 mg/ L	2.2 mg/ L
Biodegradation (28 days-OECD 306)	<15% (OECD 302)	66%
Log Pow	<0.7	4.93-5.93
HOCNF* category	A	B

* HOCNF- Harmonised Offshore Notification Format

6 CONCLUSIONS

We have presented new viscoelastic surfactants with zwitterionic functionality that form stable rod-like micelle structures in the presence of high salt concentrations. The gels formed still have retained viscosity during exposure to temperatures as high as 270 °F. We have demonstrated that zwitterionic surfactants expand the scope of gravel packing operations utilising alternate path technology that employs shunts located on the sand screens. This technology facilitates gravel packing at densities as high as 14 lbs/ gal. This has important implications in circulating gravel packs in deep water and where well control is critical.

The zwitterionic surfactants also impart stable gels in the presence of low corrosivity clean up chemistry (chelating agents such as hydroxyalkylethylenediaminetriacetic acid salts) which remove the CaCO$_3$ bridging agents commonly added to both water based and oil based reservoir drilling fluids. Indeed the zwitterionic surfactant evaluated here does not require a co-surfactant when the chelating agent is present. These clean up solutions have a large impact on both the flow initiation pressure and retained permeability when filter cakes have been generated under extreme conditions of overbalance and for extended periods. In many cases it is necessary to remove major filter cake components in open hole gravel packed completions. This has been evaluated by representative laboratory experiments and quantification of the impact on well productivity.

We have demonstrated that clean up fluids incorporating a chelant and zwitterionic surfactant in high-density sodium bromide brine are efficient at removing CaCO$_3$ within pH reversible oil based drilling fluid filter cakes.

Finally we have stressed the need of identifying solutions that also show improvement toward reducing the environmental footprint.

Acknowledgements

The authors would like to acknowledge Michael Patey and Carol Shiach for much of the experimental work and contributions to this publication. We would also like to thank the management of both Shell and Schlumberger for permission to publish this paper.

References

1   W.L. Panberthy, Jr., K.L. Bickham, H.T. Nguyen and T.A. Paulley: "Gravel Placement in Horizontal Wells," *SPE 31147* presented at the SPE Formation Damage Symposium, Lafayette, LA, Feb. 14-15, 1996.

2   L.G. Jones, R.J. Tibbles, L. Myers, D. Bryant, J. Hardin and G. Hurst: "Gravel Packing Horizontal Wellbores with Leak-Off Using Shunts," *SPE 38640* presented at the SPE Annual Meeting, San Antonio, TX, Oct. 5-8, 1997.

3   L.G. Jones, R.J. Tibbles, L. Myers, S. Crowder and M.J. Kaberlein: "Fracturing and Gravel Packing with Alternate Paths," *SPE 46255* presented at the SPE Western Regional Meeting, Bakersfield, CA, May 11-15, 1998.

4   T.E. Becker and N.H. Gardiner: "Drill-in Fluid Filter-Cake Behaviour During Gravel-Packing of Horizontal Intervals- A Laboratory Simulation," *SPE 50715* presented at the 1999 SPE International Symposium on Oilfield Chemistry, Houston, TX, Feb. 16-19, 1999.

5   M.H. Johnson, J.P. Ashton and H. Nguyen: "The Effects of Erosion Velocity on Filter-Cake Stability During Gravel Placement of Openhole Horizontal Gravel-Pack Completions," *SPE 23773* presented at the SPE International Symposium on Formation Damage and Control, Lafayette, LA, Feb. 26-27, 1992.

6   J.E. Brown, L.R. King, E.B. Nelson, and S.A. Ali: "Use of a Viscoelastic Carrier Fluid in Frac-Pack Applications," *SPE 31114* presented at the SPE Formation Damage Symposium, Lafayette, LA, Feb. 14-15, 1996

7   B. Rimmer, C. MacFarlane, C. Mitchell, Henry Wolfs, Mathew Samuel: "Fracture Geometry Optimization: Designs Utilizing New Polymer-Free Fracturing Fluid and Log-Derived Stress Profile / Rock Properties", SPE 58761 presented at the SPE International Symposium on Formation Damage Control held in Lafayette, LA, 23–24 February 2000.

8   M. Samuel, D. Polson, D. Graham, W. Kordziel, T. Waite, G. Waters, P.S. Vinod; D. Fu and R. Downey; "Viscoelastic Surfactant Fracturing Fluids: Applications in Low Permeability Reservoirs", SPE 60322 presented at the 2000 SPE Rocky Mountain Regional/Low Permeability Reservoirs Symposium and Exhibition held in Denver, Colorado, 12–15 March 2000.

9   F. Chang, Q. Qu, W. Frenier: "A Novel Self-Diverting-Acid Developed for Matrix Stimulation of Carbonate Reservoirs", SPE 65033 presented at the SPE International Symposium on Oilfield Chemistry held in Houston, Texas, 13–16 February 2001.

10 M. Parlar, R.J. Tibbles, F.F. Chang, D. Fu, L. Morris, M. Davison, P.S. Vinod and A. Wierenga, "Laboratory Development of a Novel, Simultaneous Cake-Cleanup and Gravel-Packing System for Long, Highly-Deviated or Horizontal Open-Hole Completions", SPE 50651 presented at the SPE International Symposium on Oilfield Chemistry held in Houston, Texas, 16-19 February 1999.

11 M.E. Brady, S.A. Ali, C. Price-Smith, G. Sehgal, D. Hill and M. Parlar, "Near Wellbore Cleanup in Openhole Horizontal Sand Control Completions: Laboratory Experiments", SPE 58785 presented at the SPE International Symposium on Formation Damage held in Lafayette, Louisiana 23–24 February 2000.

12 M. Rosen, M. Dahanayake, "Industrial Utilisation of Surfactants-Principles and Practice", AOCS Press, Champaign, Illinois, 2000

13 H. Hoffmann, G. Ebert, "Surfactants, Micelles and Fascinating Phenomena", Agnew. Chem. Int. Ed. Engl., 1988, **27,** 902-912

14 F. Lequeux, "Structure and Rheology of Wormlike Micelles", Current Opinion in Colloid and Interface Science, 1996, **1**, 341-344.

15 S. Kumar, D. Bansal, Kabir-ud-Din, "Micellar Growth in the Presence of Salts and Aromatic Hydrocarbons: Influence of the Nature of Salt", Langmuir, 1999, **15**, 4960-4965.

16 M.E. Brady, A.J. Bradbury, G. Sehgal, F. Brand, S.A. Ali, C.L. Bennett, J.M. Gilchrist, J. Troncoso, C. Price-Smith, W.E. Foxenberg, and M. Parlar, "Filtercake Cleanup in Open-Hole Gravel-Packed Completions: A Necessity or A Myth?", SPE 63232 presented at the SPE Annual Technical Conference and Exhibition held in Dallas, Texas, 1–4 October 2000.

17 M. Parlar, C. Bennett, J. Gilchrist, F. Elliott, J. Troncoso, C. Price-Smith, M. Brady, R.J. Tibbles, S. Kelkar, B. Hoxha and W.E. Foxenberg, "Emerging Techniques in Gravel Packing Open-Hole Horizontal Completions in High-Performance Wells", SPE 64412 presented at the SPE Asia Pacific Oil and Gas Conference and Exhibition held in Brisbane, Australia, 16–18 October 2000.

18 S.D. Mason, O.H. Houwen, M.A. Freeman, M.E. Brady, W.E. Foxenberg, C. J. Price-Smith and M. Parlar, "e-Methodology for Selection of Wellbore Cleanup Techniques in Open-Hole Horizontal Completions", SPE 68957 presented at the SPE European Formation Damage Conference held in The Hague, The Netherlands, 21–22 May 2001.

19 W.W. Frenier, David Wilson, Druce Crump, and Ladell Jones," Use of Highly Acid-Soluble Chelating Agents in Well Stimulation Services", SPE 63242 presented at the SPE Annual Technical Conference and Exhibition held in Dallas, Texas, 1–4 October 2000.

20 S. Kelkar, M. Parlar, C. Price-Smith, G. Hurst, M. Brady, and L. Morris, "Development of an Oil-Based Gravel-Pack Carrier Fluid", SPE 64978 presented at the SPE International Symposium on Oilfield Chemistry held in Houston, Texas, 13–16 February 2001.

21 L.N. Morgenthaler, R.I. McNeil, R.J. Faircloth, A.L. Collins and C.L. Davis: "Optimization of Stimulation Chemistry for Openhole Horizontal Wells," *SPE 49098* presented at the SPE Annual Meeting, New Orleans, LA, Sep. 27-30, 1998.

EFFECTIVE TOPSIDE CHEMICAL DETECTION VIA A NOVEL ANTIBODY ENGINEERING TECHNIQUE

K.A.Charlton,[1] G.Strachan,[1] A.J.Porter,[1] S.M.Heath,[2] and H.M.Bourne,[2]

[1] Remedios Ltd, MacRobert Building, King Street, Aberdeen, AB24 5UA
[2] TR Oil Services Ltd, Howe Moss Avenue, Dyce, Aberdeen, AB21 0GP

1 INTRODUCTION

The accurate analysis of any one of the cocktail of oilfield chemicals that may be present in produced water can be problematic. Most analyses involve a certain amount of sample stabilisation and clean up prior to analysis of a particular component of the chemical. The methods used are mainly chromatographic (gas or liquid), utilising a range of detectors coupled to other spectroscopic detection techniques such as Inductive Coupled Plasma Spectroscopy. These methods are often time consuming involving multiple stages, subject to interference, require skilled operators and are not appropriate for on-site applications. An analytical result may be reported a week (or longer) after the sample was taken due to the logistics in getting the sample from the platform to the laboratory and preparing it for analysis. This delay does not allow the operator to effectively manage production chemistry issues on the facility and a more rapid but accurate technique is required for on-site measurement.

The use of antibodies for immunodetection of chemicals offers a superior alternative. Immunoassays have been used for the last thirty years in medical diagnostics and have more recently been applied to the detection of chemicals in environmental matrices. The development of genetically engineered antibodies, consisting of the part of the antibody that recognises its targets is slowly replacing the traditional use of monoclonal antibodies. These antibody fragments can be generated against even the most difficult chemical targets (eg. haptens or specific parts of larger chemical molecules) faster and more reliably than via traditional antibody technology. The resulting molecules show higher sensitivity in all assay formats and can be produced in quantity and more cost-effectively than via existing antibody production systems. In addition, these antibody fragments can be improved upon further to make them more stable than traditional antibodies, in the non-physiological conditions associated with offshore samples.

This paper will describe the generation of engineered antibodies to small chemical haptens including a specific scale inhibitor and demonstrate its potential as a robust, reliable and cost-effective chemical diagnostic tool for the offshore oil and gas industry.

2 METHODS AND RESULTS

2.1 Phage Display Technology

The discovery that functional antibody fragments could be displayed on the surface of bacteriophage, allowing the selection of antibodies to any antigen of choice, has revolutionised the field of molecular immunology.[1,2] This ability to display the entire functionally active antibody repertoire of a suitable host on the surface of phage allows the isolation and production of monoclonal antibodies without recourse to immortalisation of B-cell lines required for hybridoma technology.[2-4] Phage display libraries commonly consist of the heavy and light chain variable domains of an antibody separated by a flexible peptide linker, fused to the phage minor coat protein gpIII and displayed as single-chain Fv (scFv) fragments. All such libraries can be classified as being either naïve or immune. Naïve libraries can be constructed from the naturally expressed repertoire of a host not previously exposed to the target(s) of interest, or be assembled with one or more of the complementarity determining regions (CDRs), which form the antigen binding site, being generated synthetically (synthetic libraries). Immune libraries in contrast are assembled from the expressed repertoire of a donor which has been immunised against the target antigen. The isolation of high affinity antibodies to hapten antigens requires naïve libraries containing $>10^{10}$ individual clones. In contrast such antibodies can readily be generated from immune libraries of 10^7-10^8 clones. Immune libraries are however biased towards the immunising antigen, and are unsuitable for use with other targets. A single large naïve library can be used with any antigen with an equal chance of success for each.

We have generated diagnostic antibodies to a variety of small molecular weight hapten target molecules using both naïve synthetic and immune libraries. The method employed for selection is illustrated in Figure 1. The libraries of phage antibodies were maintained in *E .coli* cells and rescued using a suitable helper phage strain. Phage were applied to an immunotube coated with the target chemical antigen conjugated to the carrier

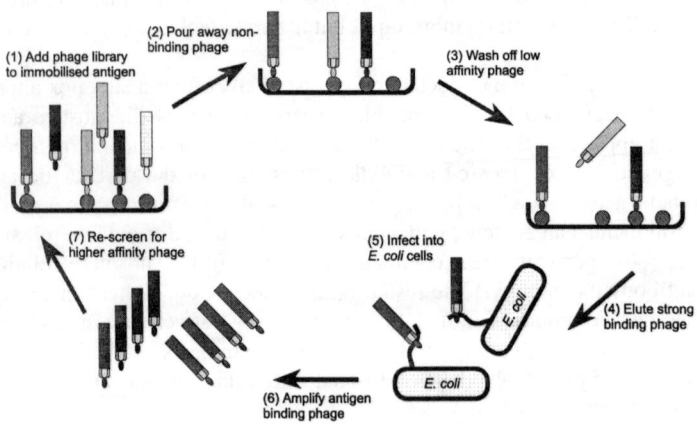

Figure 1   *Flow diagram of phage panning showing the cycle of binding (1), selection (2-3), recovery (4-5) and amplification of antigen specific phage antibodies*

protein Bovine Serum Albumin (BSA).[5] Unbound phage were poured off and those bound weakly or non-specifically removed by extensive washing with PBS. Specific phage were then eluted by incubation with 10 μM free antigen solution, and infected into *E. coli*. The enriched library was then rescued as before and the process repeated for 3-4 rounds.

2.2 Target Antigens

In order to demonstrate the technology, and its applicability to offshore oil and gas production chemical sensing, we selected a number of small molecular weight hapten antigens as target molecules (Figure 2). Phthalic acid was chosen because, with a molecular weight of just 166 Da, it is close to the lower size limit for which antibodies can reasonably be generated. Atrazine is a triazine herbicide applied extensively throughout Europe and the USA in agriculture prior to its use being banned, and now represents the major pesticide pollutant of river and drinking water in those regions.[6] CM40-026 is a

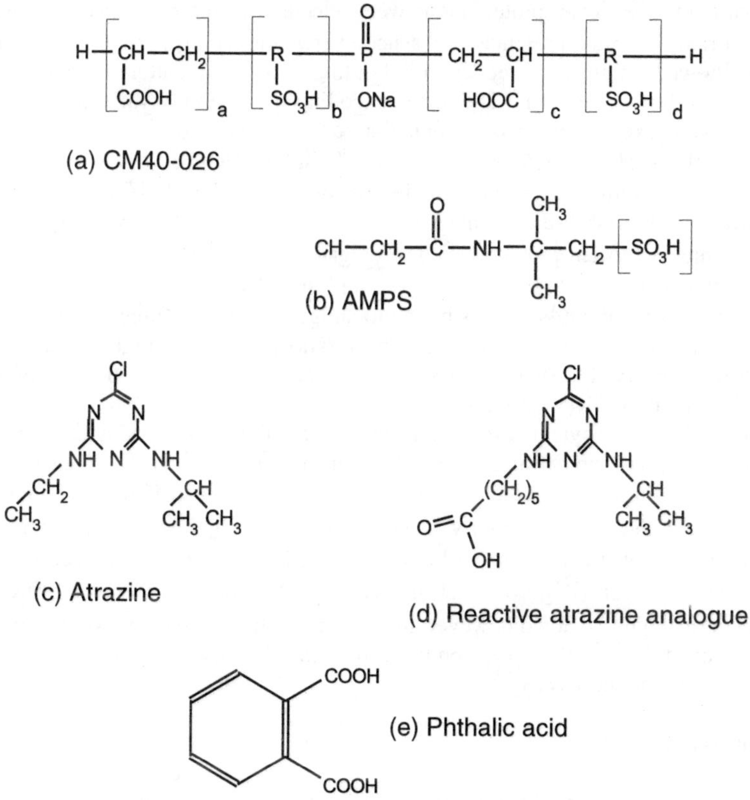

(a) CM40-026

(b) AMPS

(c) Atrazine

(d) Reactive atrazine analogue

(e) Phthalic acid

Figure 2 *Structures of hapten antigens used in this study; (a) CM40-026 is a statistical polymer where the number of groups a.b.c and d, and their positioning either side of P are random; (b) AMPS, represented by R in structure (a); (c) the triazine herbicide atrazine and (d) the reactive analogue used for conjugation; (e) phthalic acid*

high molecular weight polymer used in North Sea oil extraction to prevent accumulation of scale and subsequent blockage of production wells. A low molecular weight derivative (ave. mw 810 Da) comprising the central phosphate and a small number of the flanking acrylic acid and AMPS groups was produced for antibody selection. Haptens were conjugated to BSA using the *N*-hydroxysuccinimide method.[7] This approach requires the presence of a free carboxylic acid group, which was added to the hapten via a hexanoic acid where necessary (Figure 2d).

2.3   Assessment of Antibody Binding Characteristics

Following three or four rounds of bio-panning (Figure 1) individual phage-antibody clones were rescued in 96 well tissue culture plates using standard methods.[8] Culture supernatants, each containing monoclonal phage antibodies, were analysed initially by Enzyme Linked Immuno Absorbent Assay (ELISA). Those able to bind to hapten conjugates but not to the carrier protein alone were selected for further analysis. A second ELISA was carried out in the presence or absence of free antigen (1μM - 100μM). Those clones which showed a significant reduction in binding to the conjugate in the presence of free antigen were then sequenced. The genes encoding the antibody variable regions of unique clones were excised and cloned into the soluble expression vector pIMS147, a development of the vectors pIMS100 and pHELP.[9] Soluble antibody was produced by expression of *E. coli* cultures using IsoPropyl-β-D-Thiogalactoside (IPTG) induction, and protein purified by Immobilised Metal Affinity Chromatography (IMAC) via a hexa-histidine tail, using published protocols.[10] The diagnostic potential of purified antibodies was determined using an indirect competitive inhibition ELISA.[11,12] A constant sub-saturating concentration of antibody was bound to antigen conjugates immobilised onto 96 well plates in the presence of a range of concentrations of free antigen, and binding compared to that observed with no free antigen present. Naïve synthetic and immune libraries, each containing >10^8 clones, were screened.

Panning of both libraries yielded several different antibodies able to recognise the target chemical phthalic acid. Competition ELISA data for the most sensitive from each library (Figure 3) indicates the extent of the differences observed. With an IC_{50} of 850 nM, the best antibody from the immune library is approximately 300 times more sensitive to antigen than that achieved using a naïve library (250 μM). The limits of detection, taken as a 20% signal reduction (IC_{20}), are 250 nM and 5 μM respectively, representing a 20 fold difference. It should be noted however, that the antibodies from the naïve library were isolated and characterised within one month, as immunisation and construction of a target specific library were not necessary.

2.4   Antibody stability

Preliminary, experiments have been performed to confirm the viability of the proposed approach (Figure 4). In produced water samples, immunoassay performance of an existing engineered anti-atrazine antibody[11] showed 100% activity in a 1:20 dilution of oil field waters. The limit of detection of this antibody increased to around 80 (parts per trillion) ppt of free antigen. This is still >1200 times more sensitive than the required level of sensitivity needed for the equivalent scale-inhibitor assay (100 ppb).

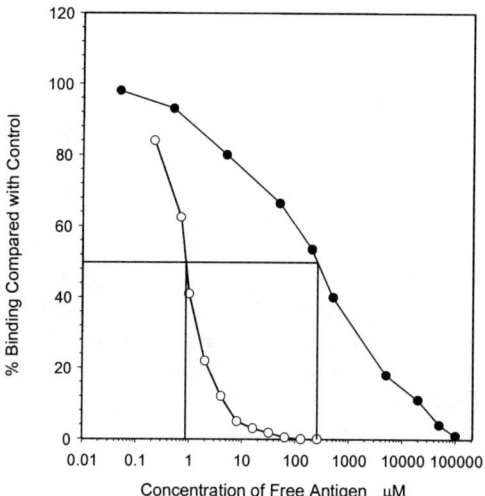

Figure 3 *Competitive inhibition ELISA comparing the binding of anti-phthalic acid antibodies isolated from (●) naïve synthetic and (○) immune libraries. IC_{50} values are indicated by drop lines. Data points represent means of 3 replicates*

To "toughen-up" the anti-chemical antibodies and allow them to function in the extreme conditions associated with produced waters, it may be necessary to modify them further using protein engineering techniques. We have shown this to be a very successful approach for a number of our antibody structures.[13,14] Stabilisation is achieved by the modification to cysteines of framework residues pairs, conserved in most antibodies, and predicted by computer modelling to allow disulphide bond formation (stabilisation) without affecting antigen binding.[15] The resulting stAbs (stabilised antibodies) show increased function in a range of non-physiological conditions including the presence of organic solvents, protease, elevated temperature and denaturants (Figure 5).[13]

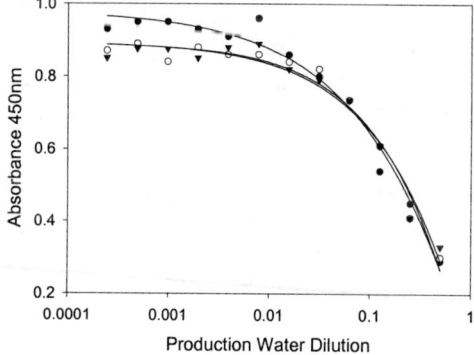

Figure 4 *The ability of an anti-atrazine antibody to bind to antigen in the presence of oil well production water. Data for three replicate analyses are shown*

2.5   Animal Immunisation

In order to generate high affinity antibodies to the scale inhibitor CM40-026, two BALB/c mice were immunised with CM40-026 conjugated to BSA. Immunisation comprised an initial injection (interperitoneal) of 200 µg protein in 200 µl Freund's complete adjuvent, followed by three successive boosts with 200 µg protein in 200 µl Freund's incomplete adjuvent at two week intervals. Blood samples were taken prior to immunisation and two days post-injection for analysis by ELISA for binding to CM40-026 conjugates (Figure 6). Both mice showed a typical strong response to the immunising antigen, the dramatic increase following boost 2 corresponding to affinity maturation and class switching. Significantly, both sera also show strong binding to CM40-026-TG (Thyroglobulin), indicating that a substantial part of the observed response is due to antibodies recognising the hapten component of the conjugate. An immune phage display library is currently being constructed from B-cells derived from these mice.

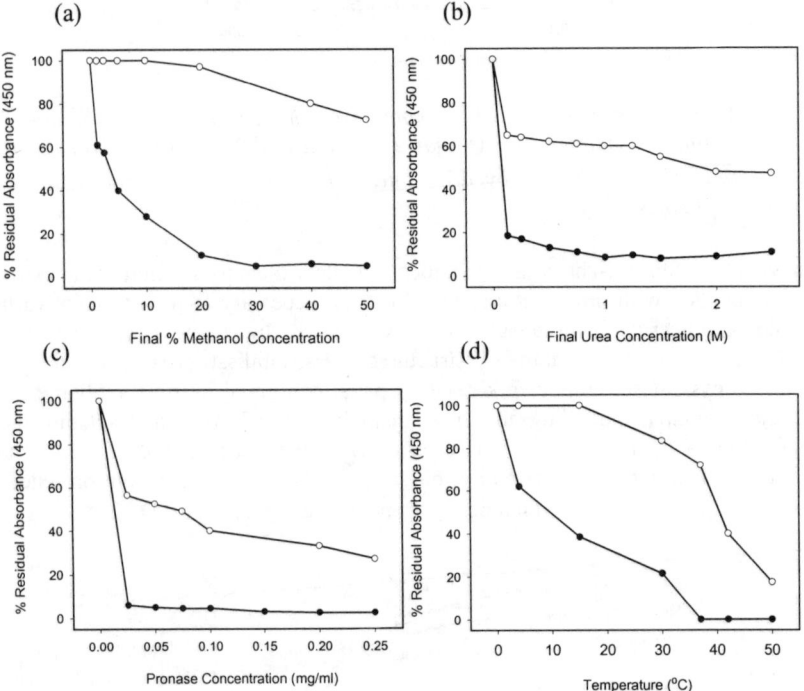

Figure 5   *A Comparison of the functionality of 'natural' (scAb) (●), and stabilised (stab) (○) anti-atrazine antibodies in non-physiological environments. Antibody fragments were incubated with denaturant for 3 h at 4ºC* prior to being applied to a 96 well plate coated with atrazine-BSA conjugate. *incubation at temperature indicated. Data taken from ref[13]*

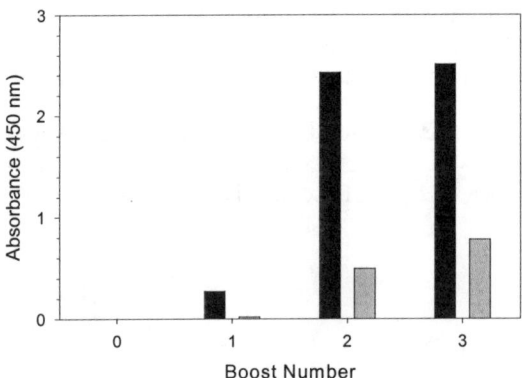

Figure 6   *ELISA analysis of the sera from mice immunised with CM40-026-BSA binding to CM40-026-BSA* (black bars) *and to CM40-026-TG* (grey bars). *Data represents the mean response from two mice. Sera diluted 1:5000 prior to analysis*

3 CONCLUSIONS

We are becoming more familiar with the concept of genetic engineering applications impacting on our daily lives. Here we show that antibody engineering approaches can provide powerful new diagnostic tools for the rapid, reliable and robust analysis of chemical targets. The advantages of this technology are the ease and speed of use and the ability to design assay formats for point of need applications. In this case, tailor-made diagnostics suitable for use offshore. Assay formats will tolerate sample extremes allowing valuable information on down-hole chemical concentrations to be produced quickly and simply.

References

1   G.P.Smith, *Science*, 1985, **228**, 1315
2   J.McCafferty, A.D.Griffiths, G.Winter and D.J.Chiswell, *Nature*, 1990, **348**, 552
3   C,Sawyer, J.Embleton and C.Dean, *J.Immunol.Methods* 1997, **204(2)**, 193
4   T.J.Vaughan, A.J.Williams, K.Pritchard, J.K.Osbourn, A.R.Pope, J.C.Earnshaw, J.McCafferty, R.A.Hodits, J.Wilton, and K.S.Johnson, *Nature Biotechnology* 1996, **14**, :309
5   J.D.Marks, H.R.Hoogenboom, T.P.Bonnert, J.McCafferty, A.D.Griffiths and G.Winter, *J. Mol. Biol.* 1991, **222**, 581
6   J.Schlaeppi, W.Fory and K.Ramsteiner, *J. Agric. Food Chem.*, 1989, **37**, 1532
7   A.E.Karu, M.H.Goodrow, D.J.Schmidt, B.D.Hammock and M.W.Bigelow, *J. Agric. Food Chem.*, 1994, **42**, 301
8   J.McElhiney, L.A.Lawton and A.J.R.Porter, *FEMS Microbiol. Letts.*, 2000, **193**, 83
9   A.Hayhurst and W.J.Harris, *Prot. Exp. Purif.*, 1999, **15**, 336
10 G.Strachan, S.D.Grant, D.Learmonth, M.Longstaff, A.J.Porter and W.J.Harris, *Biosen. Bioelectron.*, **13**, 665

11   S.D.Grant, A.J.R.Porter and W.J.Harris, *J. Agric. Food Chem.*, 1999, **47**, 340
12   K.A.Charlton, W.J.Harris and A.J.R.Porter, *Biosen. Bioelectron.*, 2001 (In Press)
13   H.Dooley, S.D.Grant, W.J.Harris and A.J.Porter, *Biotechnol. Applied Biochem.*, 1998, **28**, 77
14   G.Strachan, J.A.Whyte, P.M.Molloy, G.I.Paton and A.J.Porter, *Environ. Sci. Technol.* 2000, **34**, 1603
15   U.Brinkmann, Y.Reiter, S-H.Jung, B.Lee and I.Pastan, *Proc. Natl. Acad. Sci. USA*, 1993, **90**, 7538

USING ELECTROCHEMICAL PRE-TREATMENTS FOR THE PROTECTION OF METAL SURFACES FROM THE FORMATION AND GROWTH OF CALCIUM CARBONATE SCALE

A. P. Morizot,[1] S Labille,[2] A Neville[1] and G. M. Graham[2]

[1]Corrosion and Surface Engineering Research Group, Department of Chemical and Mechanical Engineering
[2]Oilfield Scale Research Group, Department of Petroleum Engineering, Heriot-Watt University, Edinburgh

ABSTRACT

This study examines the potential of adsorption of scale inhibitor and indeed other cations such as magnesium and calcium, promoted by electrochemical pre-treatment, to effectively protect metallic surfaces from the adhesion and growth of calcium carbonate scale. Tests have been conducted which examine the surface of stainless steel rotating disk electrodes (RDE) under ambient conditions. The involvement of divalent cations such as Mg^{2+} in the inhibition of scale is clearly demonstrated. Visualisation of the amount of scale deposition, with and without electrochemical pre-treatment, has been conducted using scanning electron microscopy (SEM).
   In summary, this paper describes the beneficial effects of using an electrochemical pre-treatment to inhibit scale deposition on metal surfaces and assess the cation/inhibitor interactions and their effect on inhibitor efficiency.

1 INTRODUCTION

The nucleation and growth of scale (i.e. insoluble mineral salts) on surfaces is one of the main aspects of crystal formation which causes operational problems in industrial plant and facilities. Formation of scale in the pores of rock can cause plugging of wells and deposition on production equipment (e.g. pipework) can lead to increased turbulence in flow systems and can eventually block flow lines. Notwithstanding this fact, the main effort in scale research has been to develop an understanding of scale formation (precipitation) in the bulk solution and several models have been developed to assess the scaling tendency of particular waters based on thermodynamic data [e.g. 1]. Information from these models is often used in well-management programmes to control scale formation and indicate inhibitor dosing rates. The methodology commonly adopted for assessing the efficiency of inhibitor chemicals is based on NACE standard TM0197 [2] in which the scale-forming ion concentration is measured (by Inductively Coupled Plasma (ICP) for instance) when two brines are mixed and scaling occurs. The effectiveness of inhibition is evaluated by comparing the ion concentration in presence and in absence of inhibitor after bulk precipitation has occurred. This method has been used to rank the efficiency of inhibitors in a wide range of environments [e.g. 3]. However, there are several limitations of this method in relation to the inability to assess the effectiveness of inhibitor treatments in preventing deposition of scale on surfaces. Hasson et al. [4] also expressed their opinion that that although the large bank of work carried out on bulk precipitation is valuable there is a real need to understand the kinetics of scale formation at a solid surface and this requires alternative test procedures.

In recent work it has been shown that surface deposition can be monitored using an electrochemical method to assess the rate of oxygen reduction reaction at the electrode. This has been used by the current authors to compare the efficiency of inhibitors in preventing precipitation in the bulk solution and deposition at metal surfaces [5,6]. Other techniques which have been used and show promise for monitoring surface deposition include in-situ microscopy [7], the quartz crystal microbalance [8] and electrochemical impedance spectroscopy [9].

In studies of surface deposition and scale inhibition it is important to consider the inhibitor action at the metal surface. Many of the polymeric scale inhibitors have also been shown to reduce corrosion rates [10]. Their efficiency with regard to corrosion has often been attributed to their ability to adsorb on metal surfaces and their action is therefore one where active corrosion sites are blocked [11]. In relation to scale control one of the likely mechanisms stated for control of growth is adsorption onto growth sites [12]. In previous communications [13, 14] the formation of an inhibitor film on metal surfaces has been reported and it has been demonstrated that conditions at the surface (e.g. cation concentration and species, inhibitor concentration, applied electrode potential) can all affect the level of film coverage.

In this paper the efficiency of several pre-treatment conditions, in which the Mg^{2+}, Ca^{2+} and inhibitor combinations are varied, in reducing deposition of $CaCO_3$ on metal surfaces is assessed.

2 EXPERIMENTAL TECHNIQUES

Stainless steel rotating disk electrodes (RDE), as shown schematically in Fig. 1a, were used as the surface onto which deposition occurred. In this study the two main experimental phases were: 1) pre-treatment of the RDE surface and 2) scale deposition tests to assess the efficiency of the pre-treatment.

2.1 Pre-treatment

The RDE was rotated, in a solution of 5g/l NaCl containing inhibitor at pH=10, at 600 rpm with a potential of –1V/SCE (Saturated Calomel Electrode) applied for 2 minutes using the three-electrode cell as shown in Fig. 1b. The electrode was then rinsed with distilled water prior to scale deposition tests.

The environmental conditions used in the pretreatment (inhibitor, Ca^{2+} and Mg^{2+} concentration) are given in Table 1. The inhibitor used in this study was Polyphosphino Carboxylic Acid (PPCA), with mean molecular weight of 3,600 g/mol. The molecular structure of PPCA is shown in Fig 2.

2.2 Scale deposition

Two synthetic brines were used in this study. They were prepared in such a way that when mixed in a 50%:50% ratio the resulting solution reproduced the composition of a 100% formation water typical of the Banff field situated in Block 29/2a of the UK sector of the

(a)

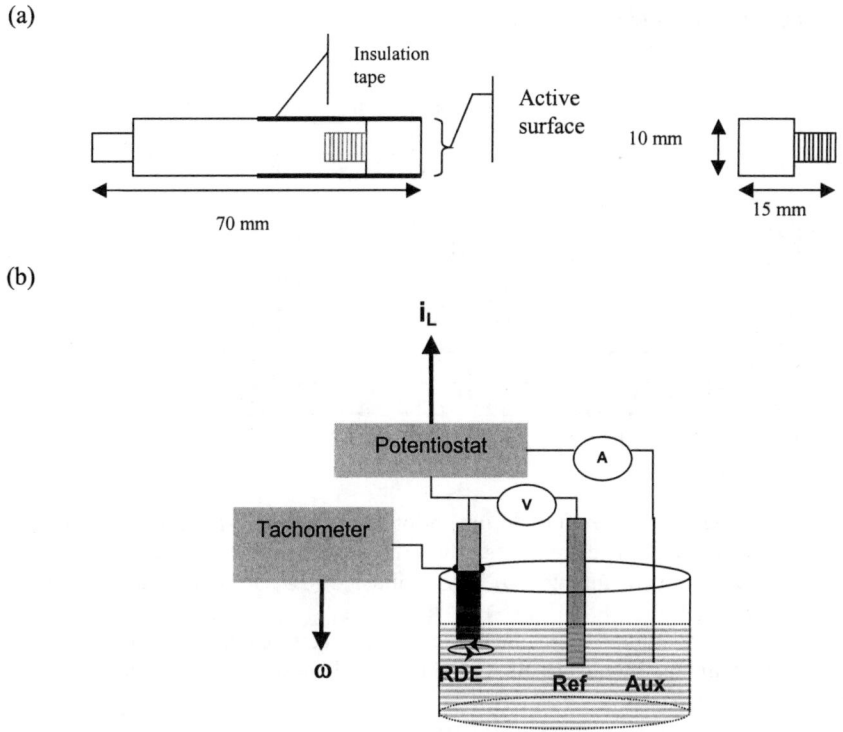

(b)

Figure 1 *(a) RDE sample and (b) 3 electrode cell set up for pre-treatment of RDE surfaces*

Table 1 *Parameters for the pre-treatment of RDE samples prior to scaling tests*

Inhibitor PPCA	Magnesium (C_{MgCl_2})	Calcium (C_{CaCl_2})	Images showing scale deposition
250ppm	500ppm	0ppm	Figure 6
250ppm	0ppm	500ppm	Figure 7
0ppm	0ppm	500ppm	Figure 8
0ppm	500ppm	0ppm	Figure 9

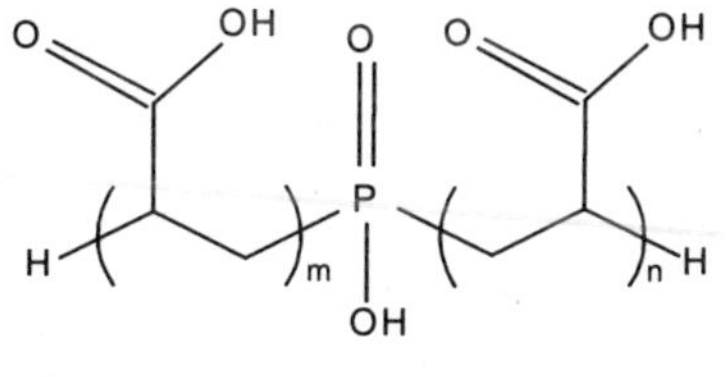

Phosphino polycarboxylic acid - PPCA

Figure 2 *Molecular structure of the PPCA inhibitor species used in the study*

Table 2 *Compositions of the brines used in this study*

	Concentration (ppm)	
	Brine 1 - FW	*Brine 2 - SW*
Na^+	25,210	25,210
Ca^{2+}	5,200	0
Mg^{2+}	690	0
K^+	1,170	0
Ba^{2+}	0	0
Sr^{2+}	270	0
SO_4^{2-}	0	0
HCO_3^-	0	1,120
Actual Cl^- (ppm)	51,368	38,225
Total dissolved ions (ppm)	*83,908*	*64,555*

North Sea. The brines were filtered (0.45 μm) prior to use, in order to remove impurities, which might provide some nucleation sites. The brine compositions are given in Table 2. The pH of brine 2 (Brine containing CO_3^{2-}) was adjusted to 9, in order to accelerate the scaling procedure.

The electrodes were immersed in the brine mixture at room temperature using the experimental set up in Fig. 3 with a rotation of 600rpm. The tests were two hour duration.

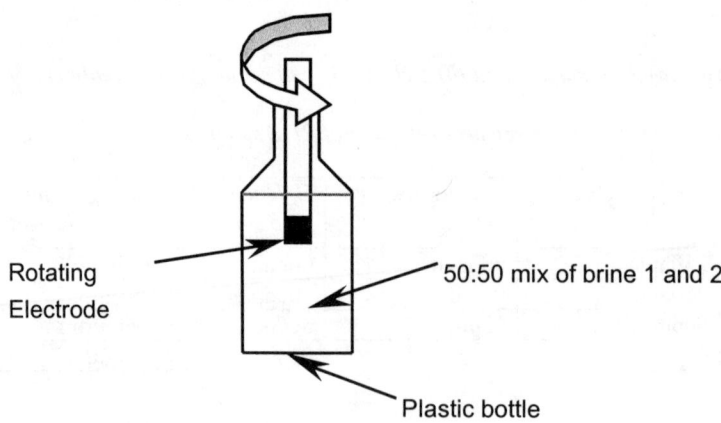

Figure 3 *Experimental set-up for scale deposition on RDE samples*

Following deposition tests using the RDE a thorough examination of the extent of scale formation on the surface was conducted using the SEM.

3 RESULTS

Without electrochemical pre-treatment extensive deposition occurred on the stainless steel electrode from the supersaturated formation water in the one hour immersion period as shown in Fig. 4a. The crystals were of a cubic form and from Fig. 4b the size is typically 10-20μm (maximum length dimension). In Figs 5-8 the SEM images corresponding to the four different pretreatment conditions are shown. The (a) figure represents a lower magnification view from which the general scaling extent can be seen and (b) shows a higher magnification view to enable the crystal characteristics to be seen. Some interesting observations can be made from these as reported in the next paragraphs.

Firstly it is clear that in comparison to the untreated reference sample there is a significant reduction in scale deposition when pre-treatment has been carried out in the presence of Mg^{2+} ions. The beneficial effect of pretreatment is greatest when both inhibitor and Mg^{2+} ions are present (compare Figs. 5 and 8). Although visibly less scale is produced when Mg^{2+} ions are present during the pretreatment there is no obvious change in the crystal size or morphology.

Where the pre-treatment was performed in a solution containing Ca^{2+} ions and no inhibitor there was little visible reduction in the amount of scale deposited as can be seen comparing Figs. 4a and 7a. Addition of inhibitor during the pretreatment reduces the scale deposition compared with the reference sample but the pretreatment is less effective than when carried out in the presence of Mg^{2+} ions. There is no change in the crystal morphology – both cases produce cubic crystals of similar size.

Figure 4 *Scale deposition of CaCO$_3$ from the 50:50 brine mix on a metal RDE sample without pre-treatment*

4 DISCUSSION

From previous studies reported in the literature it has been confirmed that polymeric inhibitors can effectively adsorb on metal surfaces to form a film which is effective in reducing corrosion rates [10]. In previous work by the authors [13,14] the extent of film formation by PAA and PPCA inhibitors has been studied using an electrochemical technique and the film formation kinetics have been shown to be dependent on Ca and Mg ion content, inhibitor concentration, hydrodynamic regime and applied electrode potential.

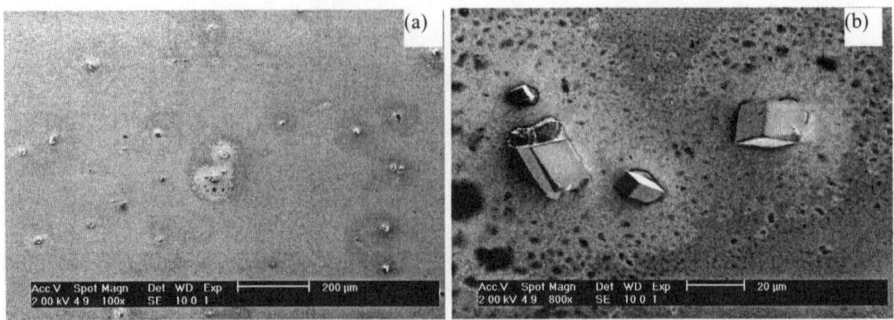

Figure 5 *Scale deposition with electrochemical pre-treatment in a solution containing 250ppm PPCA and 500ppm MgCl$_2$*

Figure 6 *Scale deposition with electrochemical pre-treatment in a solution containing 250ppm PPCA and 500ppm CaCl$_2$*

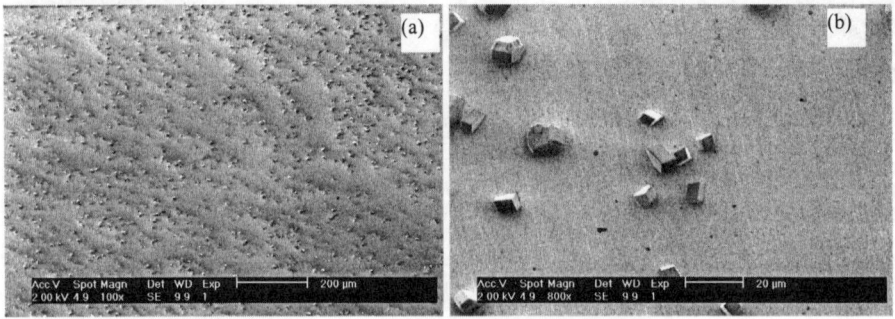

Figure 7 *Scale deposition with electrochemical pre-treatment in a solution containing no inhibitor and 500ppm CaCl$_2$*

In the current study it has been shown that the film formed during electrochemical pretreatment can be effective in reducing the extent of CaCO$_3$ deposition from a supersaturated solution. An adsorbed film can be formed without electrochemical pretreatment and in a study by Mueller et al. [15] the reduction of crystal (CaCO$_3$) formation rate on stainless steel when pretreated by immersion in a solution containing polyaspartate was reported. They also observed a reduction in average crystal size which was not the case in the present study.

Figure 8 *Scale deposition with electrochemical pre-treatment in a solution containing no inhibitor and 500ppm MgCl₂*

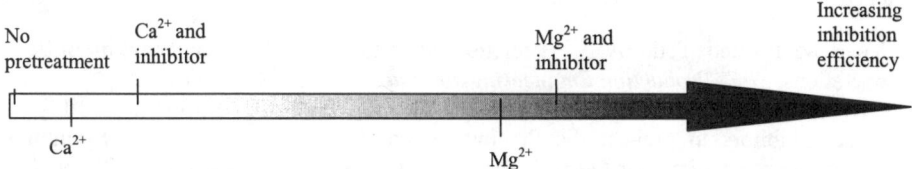

Figure 9 *Qualitative summary of inhibitive effects of pre-treatment in presence of inhibitor and absence and presence of Ca²⁺ and Mg²⁺ ions*

Figure 9 is a schematic summary of the efficiency of each of the pretreatments applied in this study. In the presence of inhibitor, addition of Ca^{2+} and Mg^{2+} ions during pretreatment enabled better inhibition to be achieved. This indicates that the Ca^{2+} ions and more effectively Mg^{2+} ions promote the ability of the inhibitor to bind with the surface and develop an efficient inhibitor film to retard deposition. Two binding mechanisms are proposed. The first involves an electrostatic cation bridge between the dissociated acrylate functional groups on the PPCA and adsorbed divalent cations. Adsorption of divalent cations leads to a more positive surface charge through which the negatively charged dissociated acid units to bind [16, 17]. Alternatively, in the presence of magnesium cations, a magnesium hydroxide film can form at the electrode surface. This may lead to strong hydrogen bonding mechanisms between the carboxylate acid groups and hydroxyl groups in an analogous manner to that described for adsorption at the silica surface [18]. The mechanism involved in the electrochemical adsorption of PPCA at the electrode surface is currently under examination.

Interestingly pretreatment in a solution containing Mg^{2+} without inhibitor produced a significant inhibitive effect. It has been widely reported in the literature that the presence of Mg in solution can affect the formation of $CaCO_3$ and in particular it can promote formation of aragonite rather than calcite [19]. In the current study there is an obvious inhibition of calcite formation on pretreatment in the presence of Mg ions. In cathodic protection it is known that the initial layer of calcareous deposit which forms is typically Mg-rich [14] and so in the absence of inhibitor it is feasible that a precursor Mg-rich layer has formed during the pretreatment. This Mg-rich layer forms rapidly and has been detected by XPS [14] although the nature of it is still not fully clear. However this layer has a significant inhibiting effect on $CaCO_3$ deposition possibly through blocking initiation sites at the metal surface through formation of a thin Mg-containing layer.

5 CONCLUSIONS

- Electrochemical pre-treatment of metal RDE coupons can lead to effective surface inhibition of $CaCO_3$
- The presence of Mg^{2+} ions during the pretreatment enables a significant reduction in scale to be obtained and this is most effective when PPCA is present
- The pretreatment of surfaces to enhance film formation and hence reduce scale deposition may have practical implications for oilfield scale control

References

1. Yuan M D. and Todd A.C., Prediction of sulphate scaling tendency in oilfield operations, *SPE- Production Engineering Journal*, Feb., 63-72 (1991)
2. NACE Standard TM 0197-97, Laboratory Screening Test to Determine the Ability of Scale Inhibitors to prevent the Precipitation of barium Suphate and/or Strontium Sulfate from Solution (for Oil and Gas Production Systems), Item no. 21228, NACE International, 1997,
3. Graham, G.M., Boak, L.S. and Sorbie K.S.: "The Influence of Formation Calcium on the Effectiveness of Generically Different Barium Sulphate Oilfield Scale Inhibitors" **SPE 37273** presented at the SPE Oilfield Chemistry Sym., held in Houston, 18-21 Feb. 1997. Accepted for publication SPE Production & Facilities, in press.
4. Hasson D. *et al.*, Influence of the flow system on the inhibitory action of $CaCO_3$ scale prevention additives, *Desalination*, **108**, 67-79 (1996)
5. Neville A. *et al.*, Electrochemical assessment of calcium carbonate deposition using a rotating disk electrode (RDE), *Journal of Applied Electrochemistry*, **29** (4), 455-462 (1999)
6. Morizot A. P. *et al.*, Studies of the deposition of CaCO3 on a stainless steel surface by a novel electrochemical technique, *Journal of Crystal Growth*, **198/199**, 738-743 (1999)
7. Davis R. V. *et al.*, The use of modern methods in the development of calcium carbonate inhibitors for cooling water systems, Mineral Scale Formation and Inhibition, edited by Zahid Amjad, Plenum Press, New-York, 33-46 (1995)
8. Noik C. *et al.*, Development of electrochemical quartz study microbalance to control carbonate scale deposit, CORROSION/99, paper N°114, NACE, Houston (1999)
9. Gabrielli *et al.*, Study of calcium carbonate scales by electrochemical impedance spectroscopy, *Electrochimica Acta*, **42**(8), 1207-1218 (1997)
10. Fivizzani K. P. et al, Manganese stabilisation by polymers for cooling water systems, CORROSION/89, Paper No. 433 (Houston, TX : NACE) 1989
11. Chen Y. et al, EIS studies of a corrosion inhibitor behaviour under multiphase flow conditions, *Corrosion Science*, **42** (2000) pp979-990
12. Verraest D. L. et al., Carboxymethyl Inulin : A new inhibitor for calcium carbonate precipitation, *JAOCS*, Vol. 73, No. 1, 1996, pp55-62

13. Morizot A. P. and Neville A., A study of inhibitor film formation using an electrochemical technique, CORROSION/2000, Paper N°183, Orlando, March 2000
14. Morizot A.P., Electrochemically based technique study of mineral scale formation and inhibition, PhD thesis, Heriot-Watt University, November 1999
15. Mueller E. et al, Peptide interactions with steel surfaces : inhibition of corrosion and calcium carbonate precipitation, *Corrosion*, Vol. 49, No. 10, 1993, pp829-835
16. Sorbie, K. S. et al, The effect of pH, calcium and temperature on the adsorption of inhibitor onto consolidated and crushed sandstone, , *SPE 68th Annual Technical Conference, Houston*, 1993, Paper No. SPE 26605
17. El Attar, Y et al, Influence of calcium and phosphate ions on the adsorption of partially hydrolysed polyacrylamides on TiO_2 and $CaCO_3$, *Progr Colloid Polym Sci*,82, 1990, 43-51
18. Iller, R. K., The chemistry of silica, J Wiley and Sons, New York, Chapter 6, 1979
19. Jaouhari R et al, Influence of water composition and substrate on electrochemical scaling, *Journal of Electrochemical Society*, 147 (6), June 2000, pp2151-2161.

Applications

THE CHALLENGES FACING CHEMICAL MANAGEMENT: A BP PERSPECTIVE

S. Webster and D. West

BP, Burnside Road, Dyce, Aberdeen, UK AB21

1 INTRODUCTION

To be a successful oil and gas company in the 21st century is extremely challenging, not only in meeting the world's increasing demand for hydrocarbons but also in meeting the ever-increasing expectations of the consumer, such as reduced unit costs, improved health and safety aspects and environmental performance.

Conservative estimates indicate that the world demand for oil will grow by around 15% over the next 9 years and the demand for gas will be even greater. This increase in demand and the underlying decline in production from traditional areas are forcing oil companies to find then produce oil in more remote and hostile environments. This challenge brings with it unique technological and commercial demands, plus the requirement to minimise any negative impact it has on the environment and the communities in which we operate.

Turning specifically to the North Sea region. It has often been referred to as a mature province, and yet over the last decade it has grown by almost 50%. However, if it is to continue to grow rather than enter decline, it will need a great deal of innovation to make it happen. Within BP we see that this growth will be centered on small pool development. It will be less about stand-alone developments and more about utilising existing infrastructure. It will be about investment in mature fields, infill drilling, satellites and subsea developments.

It is quite clear that the challenges facing BP as a corporation that is active in the North Sea oil and gas industry can be applied to the area of chemical management.

This paper discusses the challenges facing chemical management focusing on:

- Technology: What are the key technical challenges facing chemical management and how are we addressing them?
- HS &E: What are BP's aspirations and how do they impact chemical management?
- Commercial: How can commercial relationships be developed and implemented to drive performance, and reward both operators and contractors, whilst still accessing new technology?

2 TECHNOLOGY CHALLENGES

Some of the key technological challenges facing the oil industry are:

Deep water. In the 1980s BP's Magnus field, at a water depth of 186m, was seen as a technical challenge and as pushing the known engineering design envelope. Today the industry is designing and building facilities to operate at water depths of 2000m in areas such as the Gulf of Mexico and West Africa. These water depths bring unique challenges to the size and operation of the facilities and impose significantly different operating environments on what have traditionally been manageable fluids.

Subsea developments. The current trend for deep-water developments, and also for the exploitation of satellite fields tied back to existing facilities, is to minimise the infrastructure deployed on the platform and to install the facilities subsea. At present this is mainly focused on wellheads, manifolds and flow lines, but the technical challenge is to extend the equipment to include subsea water separation, water injection and metering.

Again these designs bring with them unique challenges which are related not only to the operating conditions but also access to wells and flow lines for both surveillance and chemical deployment. In 1997 BP had 20 subsea wells, by 2001 this number had risen to 150.

Complex wells/intelligent wells. Drilling and completion technology has rapidly advanced over the last few years yielding reduced rig times and increased well productivity. The current trend is to move from vertical to horizontal wells, single bore to multilateral and to install sophisticated equipment downhole not only for data acquisition but also for zonal control. These advances again import significant challenges to managing production chemistry risk such as monitoring both well and fluid performances, deployment of chemicals and ensuring the integrity of the well.

Having set the scene, we will now refer to some specific key technical challenges facing chemical management in the North Sea:

2.1 Hydrate management

Given the long flow line distances and low seabed temperatures, hydrate formation is a significant risk. The control of hydrates with chemicals is often the preferred option especially when capital expenditure is constrained. The use of methanol is often restricted due to HS&E concerns, specifically the volumes required and the management of downstream impacts. Therefore the focus at present is on the development of low-dosage hydrate inhibitors. These chemical technologies will either slow down the formation of hydrates or prevent them from forming a plug. The challenge now is to extend the current range of chemicals, which operate from 10–14°C sub-cooling to 25+°C sub-cooling.

2.2 Wax and asphaltene management

The key challenge at present is to improve our capability to predict the occurrence and rate of deposition of wax and asphaltenes, both downhole and in our subsea facilities. This will allow us to optimise our current designs and also gain a better understanding of the impact of the operating conditions on the rheology of the fluids. At present there is also considerable activity in developing both chemical and engineering solutions for wax and asphaltene control.

2.3 Scale management

The challenges facing scale management fall into several categories. One key area is to improve our prediction capability so that we can accurately predict not only the type of scale but also where it will form and at what rate. This capability would have a significant impact on our exposure to increased capital expenditure. The other area is in scale inhibitor deployment technology rather than just improving chemical efficiency. Focus within the industry is towards oil based technology which will allow pre-emptive treatment of wells before water production has started, deployment within our sand control completions and solid scale inhibitors which can be deployed in the well to provide scale control on water breakthrough.

2.4 On-line fluid and risk monitoring

At present most production chemistry risks are managed by taking samples and analysing them either at the operational site or sending them to a remote laboratory for analysis. This is not only inefficient but also means that the risks are not proactively managed. The technology of on-line analysis and data management needs to be developed if we are to successfully operate complex subsea facilities in remote parts of the world. At present technology development is focusing on on-line water analysis using conventional electrochemical techniques. The challenge is being able to tap into the thriving electronics and medical monitoring technology to promote the transfer of this technology to the oil industry. This technology is one area that has the potential to create significant value to the operators and chemical management providers.

2.5 Deployment

The one area that is lagging behind both engineering design and chemical development is deployment technology. If we are to successfully operate these innovative developments then we must be able to deploy the required chemicals to where they are required, without reliance on expensive well intervention equipment. A particular area of interest is diversion technology which can be "bull-headed" down flowlines, ensuring accurate chemical placement with minimal risk to well performance on clean up.

3   HEALTH, SAFETY AND ENVIRONMENTAL PERFORMANCE

Within BP HS&E performance is one of our core principles for doing business and our goals are simply stated as *no accidents, no harm to people, and no damage to the environment.* Within BP we aim not only to meet the current legislation of the regions in which we work, but where appropriate, set internal aspirations that will drive performance beyond that set by legislation. Within the UK BP has not only set aspirations in the area of atmospheric emissions, which is driving technology in the area of gas recovery, but also in the area of water discharges. Our aspirations are to eliminate all routine water discharges from our offshore installations by 2005, and to show a year-on-year improvement in performance throughout that period. In this area we are focusing on:

- Reducing water volumes discharged. This involves not only optimising water flood management but also utilising water control technologies such as chemical water shut-off. In addition we have an active programme of introducing produced water re-injection for our existing fields with this being the base case for all new developments.

- Reducing the environmental impact of the water discharged. This focuses not only on reducing the amount of oil discharged with the water, which we have reduced by over 45% during the last 5 years, but also on the toxicity of the separated water. This is achieved by carefully monitoring the quantities of chemicals used and the selection of more environmentally friendly chemicals where appropriate. As a result we have seen our chemical usage increase with increasing water production whilst our environmental impact has decreased

4 COMMERCIAL CHALLENGES IN THE CONTEXT OF CHEMICAL MANAGEMENT

In addition to the many technical challenges faced in chemical management, we are also presented with a number of commercially related challenges, and if our business is to be a success we must overcome or manage these issues. Commercial challenges can be identified in three distinct sections; these are not necessarily easily managed and it requires much involvement and collaboration between the operator and the contractor personnel to overcome several difficulties. The discussion that follows concentrates on the three areas:

- Supply chain management philosophy.
- Scope and purpose of the contracting relationship.
- Successful management of the relationship.

4.1 Philosophy

It is perhaps appropriate at the outset to communicate what BP's supply chain management strategy is for the UK Continental Shelf (UKCS).

In order to be successful in growing our business we need to understand contracting relationships and the risk and uncertainty profiles that exist in these relationships. For example it is imperative that the operator and contractor are aligned and buy in to the challenges, targets and goals. We need to continuously demonstrate appropriate behaviour and we must ensure that a robust assurance process exists. In an environment where 80% of BP's expenditure is with third parties (which equates to almost $2billion in UKCS and more than $30 million on chem-ical management alone), it is necessary to review our supply chain management performance by benchmarking against other key industries in the market place. Having completed such an exercise, we concluded that our supply chain management strategy needed to recognise four key principles:

- Operate regional contracts (streamline into single federal contracts),
- Manage the supply chain (correct level of control and influence on what we get and how we get it),
- Performance transparency (manage performance quarterly with joint target setting),
- Access to technology (a clear focus on technology, an explicit part of the formal review process).

These will all contribute to success in developing the goal of "Performance-based Relationships to Increase Value and Deliver Innovation".

Specifically in the context of chemical management additional key elements were reviewed. Firstly there was a need to see a clear shift from the traditional supply arrangement where profits were directly linked to the amount of chemicals sold or pumped down-hole. This needed to be strategically managed by considering a more innovative approach, whereby the cost/price model contained agreed overheads and profit margins in relation to the total manufactured cost of a product. Secondly, in order to attain agreement in this respect, it was necessary to operate an "open book" cost structure where full transparency is allowed from raw materials, through blending costs, to logistics. Finally, the revised cost model was fundamentally different from the traditional model, and it returned lower profits to the contractor. Thus it was agreed that an appropriate mechanism would be applied to offer reward and recognition to the contractor in return for innovative approaches to chemical problems, specifically where the Total Cost of Operations (TCO) could be reduced. It has generally been recognized that TCO (including replacement pipework, remedial work, etc.) can equate to as much as five times the actual chemical treatment costs. Hence a value-added mind-set provides scope for significant benefits to all parties concerned in the venture.

4.2 Scope and purpose of the contracting relationship

Whilst this document says little about the selection of the contractor, it is imperative that the operator identifies clearly the nature of the service to be provided. Whenever possible capable contractors should be chosen from a contested market place to provide that service. This may be done on a local, regional or even global basis (when it is appropriate to do so).

What is even more important however, is the challenge of setting the right expectations in respect of appropriate deliverables under the contract. For the operators this might be reducing chemical costs, and at a more strategic level, identifying reduced treatment costs or even "non-chemical" solutions. Conversely, the contractor will have certain profit aspirations; perhaps continuation of contract duration, but more importantly the key challenge will be the question of how to achieve an acceptable "return-to-shareholders" which is particularly problematical if the operator aspires to innovative "non-chemical" solutions.

It is obvious that alignment, although essential for the successful delivery of results and future growth of both parties, may become problematical when attempting to achieve the desires or corporate aims of everyone involved. One thing is quite clear, challenges of this nature must be closely monitored and both parties must make a real effort in understanding not only their own aims, but working in a collaborative manner to identify some key common goals.

One example of this relates to treatment of down-hole scale. This can severely affect production flow rates to the extent that if treatment is unsuccessful then production can be lost entirely. The challenge for the contractor is to find an effective, innovative, scale treatment, this may result in higher product costs but if each subsequent treatment is effective for a longer duration, then overall treatment costs will be reduced – a "win-win" scenario.

A second example focuses on the very important topic of safety and the environment. BP's stated corporate philosophy is clear, *no accidents, no harm to people, and no damage to the environment.* We are committed to working with partners, contractors, competitors and regulators to raise the standards of our industry. The challenge in the context of this document is to ensure alignment with our contractors. For chemicals management this means obvious efforts like reducing the effect on the environment caused by certain chemical products and minimising chemical discharges to the environment. However more importantly is the visible commitment of the contractor's management personnel in "walking the talk" and encouraging safe practices amongst all its staff.

4.3 Successful management of the relationship

When considering the practical implementation of the contracting relationship, each contract 'sector', in this case chemicals, is managed by a sector specialist who is the informed buyer with technical specialists who know the market and BP's needs. These sectors dictate the shape of our federal contracts, spanning across all our business units in the UK. Within each of these sectors there are typically two or three federal contractors, as we fundamentally believe in healthy market competition. For example in the UKCS, BP has two chemical "managers". They 'manage' our requirements by providing chemical products; but of more importance, their technical staff are 'embedded' into our business units, providing real opportunities for innovation and provision of technical solutions that are by no means only limited to chemicals.

By engaging our contractors, both BP and suppliers are kept mutually aware of the needs and opportunities that usefully create the recognition and desire within the business units for performance improvement. Each of our contracts are founded on a robust performance management process utilising continuous performance scorecards covering:

- HSE.
- People and Competence.
- Operational Performance and Cost.
- Technology and Innovation. The key to improved performance is the implementation of quarterly performance reviews with each session attended by senior BP and contractor management.

Key to this process is the enrolment of the personnel within the operating units or business units to the extent that real success is achieved when they themselves run the process. They are the individuals who really understand the true performance and they are therefore best positioned to discuss this with the contractors. The key challenge for success is to make this performance management process part of the normal business activity, *not just a procurement/supply chain management initiative.*

In summary, we used the expression "Performance-based relationships to increase value and deliver innovation". This does not always come easily; experience has shown that it must be carefully managed; and certainly on a regular basis. It has to be a formal process, and you must get "buy-in" from your operational staff. The challenges are many; and often easily brushed aside due to time constraints and other operational priorities. However, if managed properly, it will reap significant rewards for all parties concerned.

THE DEVELOPMENT AND APPLICATION OF DITHIOCARBAMATE (DTC) CHEMISTRIES FOR USE AS FLOCCULANTS BY NORTH SEA OPERATORS

Paul R Hart

Baker Petrolite, 12645 West Airport Blvd, Sugar Land, Texas 77478, USA. E-mail: Paul.Hart@BakerPetrolite.com

1 INTRODUCTION

Underground reservoirs of oil and gas also contain water. This water comes up along with the hydrocarbons. The unwanted water must then be reinjected underground or added to surface waters, such as rivers, streams, or oceans. Such is the case in the North Sea, where it is discharged overboard from offshore platforms.

1.1 Value of Water Clarification

After the initial separation of the bulk produced fluids, the produced water still contains finely dispersed solids and oil. Where the water is reinjected, residual solids can blind off the reservoir, reducing its production, or plug filters, raising back pressures, which wastes energy, damages equipment or can even shut down production. The energy needed to push the water back downhole pollutes the air to some extent. Where the water is discharged, excessive residual oil, in addition to being lost production, can damage human health, local eco-systems or the broader environment. In the North Sea, strict overboard discharge limits are set by corporate commitments and government regulations.

Baker Petrolite has developed a proprietary line of chemical additives commonly referred to as water clarifiers. Water clarifiers, also called deoilers, reverse breakers, coagulants, flocculants, and flotation aids, when applied as recommended by the local Baker Petrolite experts, assist in purifying the produced water to meet or exceed effluent water specifications. Water clarifiers consist of special blends of polymers, surfactants, and inorganic coagulants. These enable the process systems to recover oil and even water-soluble organics from the water. They reduce the turbidity, or cloudiness, of the water, and remove particulate matter that could plug up downhole producing or disposal formations. Among the most powerful and generally useful of these clarifiers is a unique and patented class of flocculants based on dithiocarbamate (DTC) chemistry. In many cases, it has simply not been possible to meet discharge limits without the use of these compounds.

1.2 Physical Chemistry of Water Clarification

1.2.1 Types of Emulsion. Petroleum emulsions can be either water-in-oil or oil-in-water. The water-in-oil type, called "inverse" in colloid chemistry, is considered "normal", "obverse" or "forward" in petroleum chemistry. Conversely, a "normal", oil-in-water emulsion in colloid chemistry is referred to as a "reverse" emulsion in the oilfield. Emulsions in which the discontinuous phase is undispersed but unresolved, called "condensed" in colloid chemistry, accumulate in the middle of separation vessels where they are referred to as "interface", "cuff", "rag" or "pad". These might be settled water, floc'd oil, or co-continuous, sponge-like layers. Condensed oil-in-water is called "floc", especially if it floats. A layer on the bottom of the water is generally called "mud" or "sludge". Solids, both oil-wet and water-wet, both organic and inorganic, are typically entrained and concentrated in these emulsions. Gas bubbles are also intentionally entrained in floc'd emulsions to enable or enhance their separation.

Even when the proportion of water is small relative to the oil, when they flow together in a line, the lower viscosity of the water causes it to flow much faster and more turbulently past the oil. This causes the oil to emulsify into the water, forming a reverse emulsion. The water also becomes emulsified into the oil. When that obverse emulsion breaks, the oil between the water droplets becomes a reverse emulsion.

Dispersions of oil, solids and gasses in water are stabilised by a range of forces. The longest-range force is coulombic, charge repulsion. Water molecules at a hydrophobic surface turn their relatively cationic (positively charged) hydrogens away, toward the hydrogen bond accepting, relatively anionic (negatively charged) oxygens in the bulk water. This orientation bias imparts an anionic surface potential to a hydrophobic particle in water that repels similar anionic surfaces on other particles. In addition, the majority of native surfactants, derived from the phospholipid membranes of bacterial decomposition, even 500 million years ago, are acidic, as are the surface groups formed from subsequent oxidation. At neutral pH, these impart an additional anionic charge as they deprotonate and their counter cations drift away. Any attempt to join these particles must overcome this charge repulsion.

The distance over which this force operates depends on the ionic strength, or salinity, of the water. The fresher the water, the greater and more far reaching the repulsion. The saltier the water, the weaker and shorter ranging. Water produced in the North Sea typically contains about 5% salt, similar to seawater (Table 1). Compared to the clarifiers developed for fresh water industrial and municipal applications, those developed for the more brackish and briny waters in the oilfield must rely more on shorter range forces for their effects.

At shorter range, the nature and distribution of the surface groups become important. Cationic, long chain or polynuclear aromatic amines, of proteinaceous origin, and associated with the asphaltene fraction in the crude, are present along side the more numerous anionic groups. Alcohol, phenol, ether, amide, ester, carbonylic, heterocyclic and porphyritic species can be found. Synthetic sulfonate and phosphenate surfactants, and various acrylic, maleic, succinic and cellulosic polymer additives may be present. Partly hydrophilic metal silicates, carbonates and hydroxides—silts, clays and salts from the formation, scales, rusts and mud

Ion	Conc. (mg/L)
Na^+	20,480
K^+	735
Mg^{+2}	135
Ca^{+2}	830
Ba^{+2}	490
Sr^{+2}	87
Cl^-	34,330
SO_4^{-2}	50
HCO_3^-	1,840
pH	7.5

Table 1   *Chemistry of typical North Sea produced water*

from the production process—adsorb at these interfaces too. These polar sites form a structured hydration layer in the water that prevents the surrounding hydrocarbons from contacting and sticking to each other.

Moreover, even after the hydrocarbons contact, the more hydrophobic surfactants and solids on the oil side of the interface—the tarry asphaltenes and slimy sulfides—must move out of the way for the floc to be resolved into separate oil and solid phases.

Appropriate clarifiers are selected by evaluating each emulsion using a scientifically chosen basis set of model compounds at actual process temperature, interfacial age, and surface to volume ratio. Each type of clarifier in the test kit has a unique set of characteristics, such as charge, size and lipophilicity, important to the resolution of emulsions. Table 2 lists the characteristics of one such basis set of 30 clarifiers. The significance of each characteristic is discussed below.

1.2.2 Charge Mobility. Ionic surfactants and polymers are salts in which one ion stays put (on a surface or in solution) as the other diffuses away. The charge of the surfactant or polymer derives from that on the less mobile ion. The "charge density" expresses the type and amount of this charge per mass of active compound. This value (in mole equivalents per kg) is listed for each clarifier in Table 2.

Charges are "neutralised" by introducing counterions as immobile as the ions that are staying put. Ions might be less mobile because they are big, binding or both. Less mobile cations include polyvalent metal salts, polymeric ammonium salts, micellar ammonium surfactants and even covalently bonding protons (from mobile acids). Polymeric or hydrophobic acids create less mobile anions. The source of the ions characterising each clarifier in Table 2 is listed as the Clarifier Type.

Though all of these types can be effective, they each have their own characteristics. Protons are universal but react with water (to form hydronium) and metal. This renders them inefficient and corrosive. Metallic hydrates are also inefficient; loosely associated, they are only marginally less mobile than their monovalent counterparts. At least they are predictable and stable. Surfactants are efficient but associate in non-linear ways; they can stabilise just as easily as destabilise emulsions. Polymers can be as big and immobile as needed. But can be so big and immobile, they impede coalescence. So viscous, they can be hard to feed. So extended, they can be torn apart by turbulence. So much charge per molecule, they can deliver too much at once and restabilise the particle with the opposite charge.

There are several ways to achieve large size. The direct route is high molecular weight (MW). The logarithm of the average MW of each clarifier is listed in Table 2 under Size

Factors as Covalent Bonding. The MW of clarifiers is limited by the form—emulsion or solution—in which it is delivered. The form of each clarifier is noted in Table 2.

Hydrophilic monomers can be polymerised inside micron sized water droplets suspended in mineral oil. These inverse emulsion polymers, or "inverts", can achieve MWs in the 4-40 million dalton (MDa) range. They are the biggest and most efficient molecules used as water clarifiers. Those with high charge density on the polymer backbone exhibit an internal charge repulsion that causes them to stretch to their maximum length. Even those with low charge density backbones will form an extended random coil. This greater extension of the polymer allows better bridging between particles but also makes them fragile to shear degradation. To prevent this, they are most efficiently used toward the end of the clarification process.

Another downside to inverts is that the water-in-oil emulsion must be "made down" or (re)inverted into at least a 100-fold excess of fresh water before being fed. The inversion surfactants, or "breakers" put into the emulsion are not strong enough to allow dilution directly into salt water, yet are too strong to permit long term storage of the emulsion without stratification.

More hydrophobic monomers can be polymerised as dispersions in water or brine. These dispersion polymers, or "latexes", are typically in the 1-10 MDa range. They are charge stabilised and so can have good long-term stability. They can be added directly to brine (though the ones made in brine can't be added to fresh water without congealing). Although these can be as large as the invert polymers, the nature of their hydrophobicity makes their conformation globular rather than extended. This conformation is more shear stable, but not as able to bridge between particles as the extended invert conformation.

Solution polymers are limited by viscosity considerations to the 1-100 kDa range, the higher end being more dilute. Size comparable to the emulsion polymers can be achieved, however, if more tenuously, via self-association. Self-association allows the polymer complex to survive shear forces and reassemble to bridge particles in quiescent zones. The effective size of the complex and its speed of formation *in situ* then are limited only by the strength of that association. The type of self-association exhibited by each clarifier is listed in Table 2, in order of decreasing strength. Some associate in more than one way. Hydrogen bonds form the weakest link. Colloidal metal salts and highly hydroxylated polymers both form hydrogen bonded networks. Amphoteric polymers (those with both cationic and anionic sites) can form ion pair crosslinks. Surfactants and hydrophobic regions on polymers can form crosslinking micelles. The DTC group can form bridging organometallic complexes with native polyvalent metal ions. This last is the strongest type of associative link.

1.2.3 Lipophilicity. In addition to their charge mobility characteristics, water clarifiers differ in their attraction to oil, or lipophilicity. Sticking to the surface of the particles, whether oil, solid or gaseous, further immobilises the clarifier and allows changes in its conformation to pull particles together. The best adhesion to the surface occurs when the clarifier contains groups that complement the surface characteristics of the particle. Cationic sticks to anionic, anionic to cationic, hydrophilic to hydrophilic and lipophilic to lipophilic. This is where the choice of clarifier becomes specific to the emulsion, or emulsion component. The bulk fluid makes a difference too. As noted, the more saline the brine, the more critical these short-range adhesive forces are relative to the long-range charge repulsion.

The specific lipophilicity, or lipophilicity density, of each clarifier overall, excluding the extremely hydrophilic effect of being charged, is listed in Table 2. This is the theoretical lipophilicity per mass of the molecule after immobilisation with a tightly bound counterion of neutral philicity. These are calculated from the logarithm of the partition coefficient between aliphatic hydrocarbon and water [Log $P_{(h/w)}$], which is proportional to the free energy of phase transfer.[1] (The more commonly employed octanol/water coefficients are not appropriate for predicting performance on crude oil.) The Log $P_{(h/w)}$ contribution of each molecular fragment is summed then divided by the MW of the whole molecule. The contribution of any non-ionic co-monomer block is also broken out and listed separately.

In general, the heavier, less refined and more residual an oil source, the more polar and hydrophilic its surface is and the better it will bind with a more hydrophilic clarifier. In contrast, light crude in primary production, such as that in the North Sea, tends to have a less polar surface and can be expected to bind better with more lipophilic molecules.

In addition to particle adhesion, conformational changes in the bulk fluid also depend on the lipophilicity of the polymer. The type and degree of change depends on its distribution in the polymer and the nature of the water. Upon neutralisation of their internal repulsive charge (by adhesion to the particles being removed), polymers whose ionic and non-ionic monomers (if any) are both hydrophilic transform from stretched linear to random coil in fresh water. In highly saline brines, they start random coiled but coil a bit tighter. Co-polymers with lipophilic ionic monomers and hydrophilic nonionic monomers (most inverts) go from stretched to micellar globules when neutralised in fresh water, coiled to globular in brine. Co-polymers with hydrophilic ionic monomers and lipophilic nonionic monomers (the latexes) stay globular but become tighter globules upon neutralisation in fresh water or brine. Polymers whose ionic and non-ionic monomers (if any) are both lipophilic (such as the DTCs) go from globule to a completely collapsed oil ball when neutralised in fresh water or brine.

A clarifier's effect on coalescence also depends on its lipophilicity. Overcoming long- and short-range repulsions sufficient to stick particles together may be all that is necessary to clarify water *per se*. Many industrial applications can simply discard or indefinitely store or reprocess material that has been removed from water. The clarifiers used there generally do not waste material promoting coalescence of the flocculated oil. In the oilfield, however, and especially offshore, this is not desirable or even allowable. Recovering the oil and minimising the discharge of oily solids is required. To do this, the immobile, barrier surfactants impeding coalescence must now be mobilised. They can be pulled into the water by hydrophilic clarifiers, pushed into the oil by lipophilic clarifiers and/or made more laterally mobile by liquefying clarifiers of neutral philicity. All of these might be done—there are layers of barriers and each layer can be desorbed differently. The choice depends on the nature of the surfactants and their environment. As a general rule, the lighter, less polar crudes that bind better to lipophilic clarifiers also have more lipophilic surfactants that are earlier to push into the oil than pull into the water.

Clarifier		Charge Density (Eq/kg)	Covalent Bonding (Log MW)	Size Factors Organo-Metallic Association	Micellisation	Ion Pairing	H-Bonding	Lipophilicity Excluding Charge (LogP/kg)	Non-ionic Co-Monomer Only
Form	Type								
Invert	Acrylic # 1	-2.8	6.7				x	-62.5	-66.9
Invert	Acrylic # 2	1.1	6.5				x	-52.8	-66.9
Invert	Acrylic # 3	2.8	7.0				x	-30.4	
Invert	Acrylic # 4	3.6	6.6				x	-6.0	-66.9
Invert	Acrylic # 5	3.6	6.6				x	-18.9	-66.9
Invert	Acrylic # 6	4.5	6.6				x	-1.5	-66.9
Latex	Acrylic # 7	-4.7	6.0		x		x	-22.1	0.2
Solution	Acid	13.2	1.9					-77.6	
Solution	Acrylic # 8	2.0	5.0			x	x	-20.0	-19.9
Solution	Acrylic # 9	4.7	5.0				x	-0.5	
Solution	DTC	4.0	2.9	x	x			3.2	0.3
Solution	DTC-aminated	6.8	2.6	x			x	9.4	
Solution	DTC-hydroxylated # 1	4.4	3.9	x	x		x	2.8	6.6
Solution	DTC-hydroxylated # 2	5.1	3.5	x	x		x	5.1	14.0
Solution	Polyamine-hydroxylated # 1	5.1	2.8				x	-7.0	
Solution	Polyamine-hydroxylated # 2	5.6	2.8				x	-15.8	
Solution	Polyamine-hydroxylated # 3	6.9	2.7				x	-27.1	
Solution	Polyamine-hydroxylated # 4	9.9	2.9				x	-41.2	
Solution	Polyamine-hydroxylated # 5	10.2	3.8				x	-11.2	
Solution	Polyamine-hydroxylated # 6	10.4	3.2				x	-19.4	
Solution	Polyarylamine # 1	3.2	4.1		x			14.3	28.7
Solution	Polyarylamine # 2	6.5	2.8				x	-6.4	
Solution	Polyarylamine # 3	17.3	3.2				x	-45.9	
Solution	Polyvalent metal # 1	3.0	2.7				x	-37.4	
Solution	Polyvalent metal # 2	4.6	2.3				x	-24.9	
Solution	Polyvalent metal # 3	5.7	2.5				x	-51.6	
Solution	Surfactant # 1	0.0	3.4		x			0.3	
Solution	Surfactant # 2	0.0	3.0		x			11.0	
Solution	Surfactant # 3	3.5	2.8		x			28.7	
Solution	Surfactant # 4	5.2	2.3		x			16.3	

Table 2   *Structural properties of a non-redundant set of water clarifier bases*

The obverse breaker, added to coalesce the water in the oil, is one of the surfactants present at the interface that also helps coalesce the oil-in-water. The direction the clarifier attempts to move a given barrier surfactant should reinforce, or at least not fight, the direction the demulsifier is attempting to move it; and *vice versa*—the clarifier should help, not hurt, the dehydration of the oil. Clarifier lipophilicity is thus a guide to demulsifier compatibility.

1.2.4 Treatment Strategy. For best results, clarifiers should be added early and often. A shear-stable clarifier, generally referred to at this point as a reverse breaker, should be added as far upstream as free water flows. This allows bulk oil to wash the reverse emulsion and helps prevent more oil from being entrained in the water phase during the extraction process. Offshore, this is generally just ahead of or just after the primary separator on the platform. Since the native emulsion is generally anionic, the primary clarifier will generally be cationic. In cases of low charge or high brine strength, a lipophilic anionic might be added to intensify the charge, followed by a cationic to break it. In rare cases, a very low pH or the addition or recycling of synthetic surfactants will create a cationic primary emulsion, which will require an anionic primary breaker.

The reverse emulsion leaving the primary separator will be different from the original. The easy emulsion will be gone, and the rest will have been treated. It may also now come from the settled obverse emulsion as that breaks in the separator or oil coalescer. It may even come from unresolved, or re-emulsified, floc, skimmed and recycled from the secondary clarification system. As it passes to the secondary clarification process, it can be treated again, this time with a smaller amount of cationic, nonionic or anionic clarifier, depending on the residual, post-treatment charge. The secondary process might be a setting tank or drum, plate or filament coalescer, hydrocyclone, centrifuge, gas flotation cell or any series of these. A final filtration or adsorption is sometimes employed as a tertiary treatment. As the emulsion continues to change through each unit, it can continue to be retreated, each time with a smaller, adjusting dose, often with different, complementary chemicals. A cationic might be followed by an anionic, or a lipophilic by a hydrophilic, or a low MW by a high MW, for instance. It this way the water gets progressively clearer and cleaner until it meets the discharge specification.

In the ideal case, all the oil emulsified alone or entrained on solids is returned to production and only perfectly clean, invisible solids remain. In reality, flocculated oily solids skimmed from the secondary clarification are recycled back to primary separation. There they accumulate until they are fine enough and few enough to leave with the produced oil or clean enough to exit with the water. This accumulation equilibrates only when the rate of floc resolution equals its rate of production. Excessive accumulation can produce bad oil as well as bad water. Accelerating the final resolution of the floc to minimise its equilibrium accumulation may require an adjustment to the clarifier and/or the demulsifier treatment. A more appropriate oil demulsifier can thus produce better water, a more appropriate water clarifier, better oil.

2 CHEMISTRY OF DITHIOCARBAMATE CLARIFIERS

2.1 Synthesis

Dithiocarbamate based water clarifiers are produced by the reaction of polymeric or oligomeric primary or secondary amines with carbon disulphide and caustic in aqueous or alcoholic solution (Scheme 1).

$$[-RR'NH-]_n + nCS_2 + nKOH \rightarrow [-RR'(NC(=S)S^-K^+)-]_n + nH_2O$$

R and R' are organic radicals, one of which can be H,
n = 2-200, typically 3-30.[2-7]

Scheme 1

Primary dithiocarbamates are in equilibrium with some isothiocyanate and bisulphide (Scheme 2). Excess caustic is added to prevent any formation of hydrogen sulphide gas (Scheme 3), the loss of which would be hazardous and shift the equilibrium away from the dithiocarbamate.

$$[-R(HNC(=S)S^-K^+)-]_n \leftrightarrow [-R(N=C=S)-]_n + nHS^- K^+$$

Scheme 2

$$HS^- + H^+ \leftrightarrow H_2S \uparrow$$

Scheme 3

2.2 Structure

2.2.1 Structural Variations. DTCs can be synthesised from a variety of base polyamines. The first generation employed simple, relatively low MW (200-800 Da) polyalkyleneamines or polyetheramines.[2-3] The DTC group was essentially the only functional group. Increasingly complex substrates were then developed which added extra functionality. For example, by copolymerizing the polyamines with epichlorohydrin and diepoxides, free amine, hydroxyl and aromatic groups were incorporated into larger (up to10 kDa) and more highly branched structures.[4-7] These featured scores of DTC groups, mostly of the more stable secondary amines. This allowed greater and faster crosslinking to a larger effective size. Furthermore, the hydroxylation softened their extreme lipophilicity and created hydrophilic associations that prevented them from gelling the oil into which they partitioned. The overall effect of these larger but gentler bases was retention of the ability to flocculate the oil with improvement in the ability to completely resolve the resultant floc.[4-6]

2.2.2 Property Comparison. The DTCs' remarkable effectiveness results from their unique combination of properties. One difference is the nature of their charge. Although anionic *in vitro*, they are cationic *in situ*. This is because of the tenacity with which they bind polyvalent transition metal cations. Iron has a particularly strong binding constant. A few parts per million ferrous ion (Fe^{+2}) is all that is needed. Produced water generally is has no shortage of these (though if needed, they are easily added).[7] The Fe^{+2} rapidly binds, converting the DTC⁻ anion into the DTC-Fe⁺ cation. Double binding the ferrous ion creates organo-metallic crosslinkages: DTC-Fe-DTC. At least three DTC sites per molecule produces a large polymer backbone with pendant metallic cations (Figure 1). These metal cations in turn bind well to the anionic carboxylate sites on the surface of the oil particles. In contrast, the polyvalent metal type of clarifier, though it binds in the same way, has little size to back it up. Adding hydroxylated polyamines to the colloidal metals (a common practice) contributes size but not organo-metallic strength. The inverts have similar size and even greater strength, but are hard to feed, especially to brine. Furthermore, they are hydrophilic.

The DTCs are lipophilic. The sulphurs love oil so much, they completely cancel the effect of the nitrogen's hydrophilicity. On a charge-neutralised basis, only the surfactant class and the non-hydrogen-bonding polyarylamine share with the DTCs the driving force to partition to the oil. Moreover, neither of the other lipophiles has any size increasing mechanism other than hydrophobic micellisation; and micellisation, though a moderately strong linkage in water, breaks up on contact with the oil and stops pulling particles together. The latex has an effective size and micellises (because of the size of its hydrophobic region) but its hydrophilic monomers still keep it on the water side of the interface. The DTCs penetration of the interface forms a much stronger attachment than mere surface adhesion. Where the others are merely glued, the DTCs are glued and screwed to the surface. This penetration also pokes destabilising holes in the surfactant barrier to coalescence.

Among the DTCs, the main structural different is whether they have hydrogen bonding sites. As noted, one drawback to the simple, purely hydrophobic DTCs, is that once on the oil side, they can gel the oil just like any other high MW oil soluble polymer. Adding just the right number of hydrophilic interactions still allows them into to the oil, but has them coil up out of the way once they get there.

2.3 Mechanism

2.3.1 Transformation. DTCs are delivered to the system as low viscosity, water soluble, anionic salts. On contact with the produced water, they are converted to high viscosity, partly oil soluble (the charge neutralised part), cationic polymers. They do this via association with di- or polyvalent heavy metal ions, such as ferrous iron (Figure 1).

Primary DTCs *in situ* also undergo a slow decomposition to isothiocyanate and iron sulphide (Scheme 4).

2.3.2 Floc Formation and Resolution. The backbone of the organo-metallic polymer formed *in situ* becomes hydrophobic upon neutralisation of its charge. It would collapse into a micellar ball. The pendant cationic salts, however, are still hydrophilic, charge repelled from each other, and attracted to the anionic sites on the particles dispersed in the water. They reach out to those sites, stretching the polymer chain from its lowest free energy conformation. The combination of the hydrophobic effect and the polymer uncoiling leaves it both enthalpically and entropically strained. On contact with a dispersed particle, the pendant DTC group looses both its hydrophilicity and its charge repulsion, as does the site of the particle it contacts. The extended polymer collapses catastrophically, bringing with it the dispersed particles to which it is now inextricably coupled. This coagulation forms a micro-floc, which, as the process continues, becomes increasingly macro.

The surfactants complexed by the polymer at the interface are pushed into the oil. It is so big, however, that it doesn't completely dissolve in oil. Its lack of unassisted mobility in the oil can eventually create a barrier to total resolution of the floc into bulk oil and clean solids. Paradoxically perhaps, adding more DTC groups as well as hydroxyl groups to the molecule cause it to eventually collapse into a tighter, less gelatinous ball. This is more easily dispersed in oil and produces less resistance to final floc resolution. The obverse demulsifier employed to dehydrate the oil (by mobilising oil side barriers to coalescence) can also greatly assist in mobilising this barrier as well.

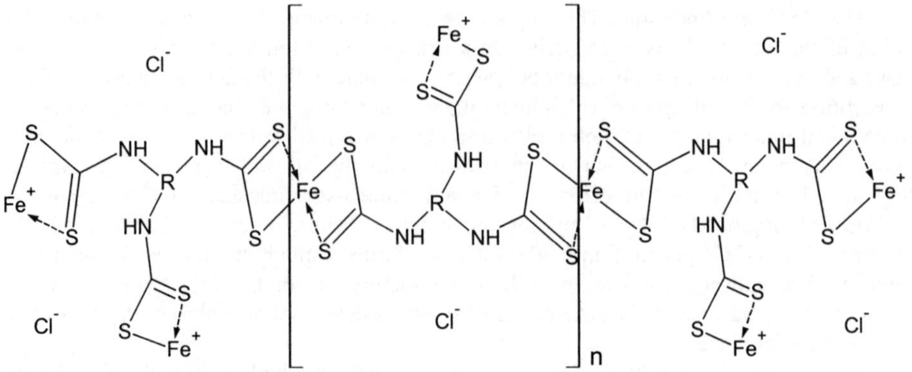

Figure 1 *Simple DTC in situ, crosslinked and cationized with ferrous iron*

$$[-R(HNC(=S)S-FeOH)-]]_n \rightarrow [-R(N=C=S)-]_n + nFeS + nH_2O$$

Scheme 4

2.3.3 Formulatory Improvements. The latest development to improve DTC based clarifiers is the explicit incorporation of oil side mobility aids into the formulation. These are non-ionic lipophilic surfactants similar to those used in obverse demulsifiers but which have been specifically selected for their compatibility and positive interaction with specific DTC components.[4-5]

2.3.4 Systemic Fate. DTCs are effective at extremely low rates of usage: 1-5 ppm active (4-50 ppm product) based on produced water. In addition, unlike most water clarifiers, whose collapsed, spent forms are still hydrophilic and exit with the water, spent DTCs exit with the oil. Together this minimises, if not eliminates, the effect of their use on the aquatic environment.

2.4 Environmental Issues

The two most effective DTC clarifiers for North Sea production are Magnaclear W-243 and W-285. With respect to the new OSPAR Harmonised Mandatory Control Scheme (HMCS) Regulations, their overall environmental profile is favourable. Table 3 summarises the environmental data, which are discussed below.

Aquatic Toxicity	*Criterion*	*W-243*	*W-285*
Acartia Tonsa	LD50, 48 hrs	<10 ppm	>10 ppm
Skeletonema Costatum	ED50, 72 hrs	<10 ppm	<10 ppm
Corophium Volutator	LD50, 10 days	>500 ppm	>1000 ppm
Bioaccumulation Potential			
Octanol/Water Partition	Log P_{ow} (range)	0 to 3.9	-1.5 to 1.7
Coefficient (OECD 117)	Log P_{ow} (wt. avg.)	2.6	0.6
Biodegradation			
Saltwater (OECD 306)	28 days	29%	47%
CHARM Hazard Quotient	@ 20 ppm	0.5	0.04

Table 3 *Ecotoxicological profile of W-243 and W-285*

2.4.1 Aquatic Toxicity. Aquatic toxicity only becomes a relevant parameter if the chemical under assessment is likely to enter the receiving environment during use. This is unlikely for DTCs, since, as discussed in earlier sections of the paper (2.2.2 and 2.3.4), they are very lipophilic and so partition entirely into the oil phase during normal chemical usage. If they were exposed to the marine environment, they would, like most water clarifiers, exhibit some aquatic toxicity. However, this toxicity tends to be less than that of alternative clarifiers, such as polymeric quaternary ammonium salts, which are noted for their high acute aquatic toxicity.

2.4.2 Bioaccumulation Potential. DTCs should not bioaccumulate in fatty tissues in the food chain due to the polymeric nature of the molecules. DTCs *in situ* have a molecular weight in excess of 10,000 Da, which makes them too large to pass through or be incorporated into the membranes of cells or liposomes in marine organisms.

An estimate of a compound's partition between aqueous and polar organic phases is often used as an indicator of bioaccumulation. In the OSPAR HMCS protocol deionized water is used as a proxy for blood plasma and octanol is used as a proxy phospholipid membranes. Even so, the partition between these two proxy phases is not measured directly but estimated from liquid chromatographic elution times. This method is generally inappropriate for predicting the partition tendencies of high MW surface active polymers, which tend to form their own competitive colloidal phases.

DTCs are particularly problematic because the measurement is of the more water soluble form in which they are delivered, not the radically less soluble form in which they work prior to discharge (as discussed in section 2.3). Thus the results from the Log P_{ow} studies performed for W-243 and W-285 as per HMCS protocols show more, and more variable, water partitioning than would occur in actual use. Even in the case of an accidental spill, except perhaps into a chromatograph, the dilution into seawater would instantly convert them into insoluble, non-bioavailable solids.

2.4.3 Biodegradation. DTCs are inherently biodegradable. Both W-243 and W-285 achieve biodegradation rates of > 25% in a 28 day saltwater test.

2.4.4 CHARM Hazard Quotient. The overall Chemical Hazard Assessment and Risk Management (CHARM) Hazard Quotients for W-243 and W-285, at typical platform dosages of 20 ppm, are 0.5 and 0.04, respectively. These figures, especially that for W-285, are considerably less than 1.0, the value at which the probability of environmental harm is considered significant.

3 APPLICATION OF DTCs

3.1 The Separation Process

3.1.1 Flow Schematic. A typical treatment scheme for treating a North Sea platform is shown in Figure 2. Treatment includes adding 50-60 ppm scale inhibitor and 10-20 ppm obverse demulsifier ahead of the primary 2- or 3-phase separators. The water leaving the separator is treated with 10-50 ppm primary water clarifier and then optionally with 5-30 ppm of a secondary clarifier prior to a gas fluxing flotation unit. A corrosion inhibitor is added to the produced oil. The clarified water is discharged overboard.

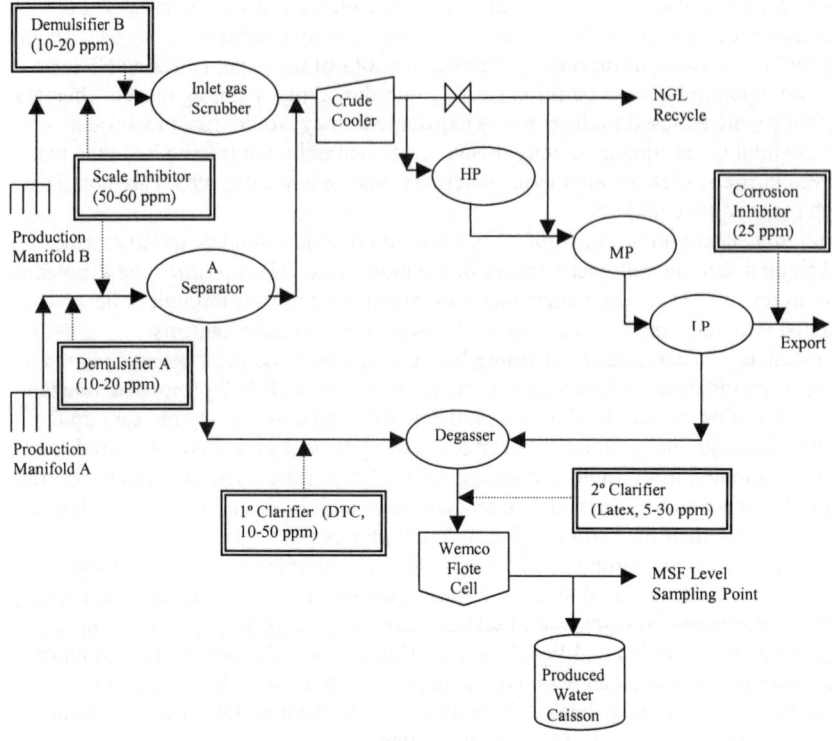

Figure 2   *Simplified flow and chemical treatment scheme for one North Sea platform*

3.2 DTC Effectiveness

On one such platform, the replacement of a conventional (i.e. non-DTC) clarifier with 50 ppm of the DTC based Magnaclear W-243, resulted in a dramatic reduction in oil counts from the 1000-1200 ppm range to the 5-50 range (Figure 3). After a few months of equilibration, however, excursions of high oil-in-water started to recur. The unprecedented retention of oil was perhaps causing a build-up of cationic charge. The clarifier was switched to 45 ppm of a milder and more dilute DTC, Magnaclear W-285-50 (half strength W-285). In addition, 30 ppm of a secondary clarifier, an anionic latex, Magnaclear 2317W, was added to hydrophilise the spent DTC and sop up any excess charge. This stopped the oily excursions (except for one brief upset) and maintained even lower average levels than before.

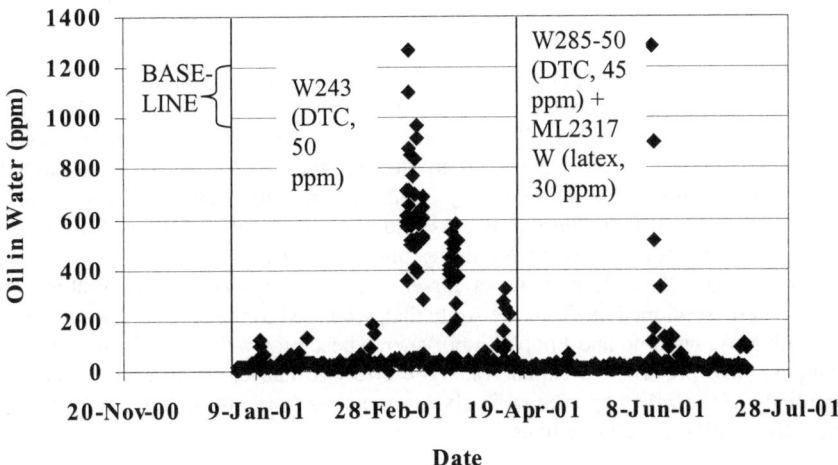

Figure 3   *Initial reductions of oil-in-water are maintained by vigilant re-optimisation*

On another platform, an even smaller amount of DTC, 11 ppm of Magnaclear 285-50, was able to reduce the oil in the discharged water to the 40-ppm range (Figure 4). The subsequent addition of only 5 ppm of the same secondary clarifier, the anionic latex, Magnaclear ML2317W, dropped the oil counts to the 23-ppm range. Finally, switching the demulsifier to one that worked better with this new water clarifier treatment resulted in a further reduction in the oil-in-water to the 15-ppm range.

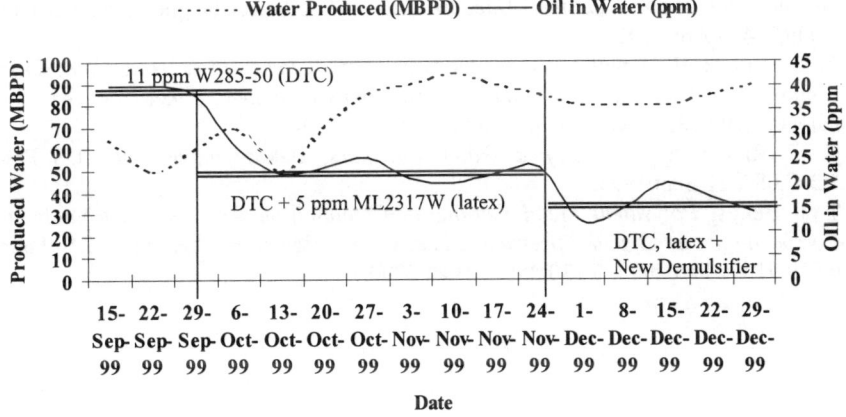

Figure 4   *Successive reductions in residual oil-in-water resulting from successive application and adaptation of chemical*

4 CONCLUSION

Produced water in the North Sea contains dispersions of anionically charged oil droplet and solids. To cleanup the water, these dispersions must be destabilised with high MW cationic water-based polymers. Conventional cationic monomers, generally quaternary ammonium compounds, are expensive, ineffective and toxic. Conventional high MW polymers are viscous solutions or messy emulsions that are difficult to use. The water solubility needed to diffuse these polymers into water limits their affinity for the oil and thus their ultimate effectiveness on North Sea produced waters.

To overcome these limitations, a novel class of low MW, water-soluble anionic polymers were developed by Baker Petrolite that, when fed to the produced waters, form *in situ* a high MW, cationic and lipophilic polymer. These high polymers, formed from DTC precursors, have proven less expensive, more effective and easier to use. Moreover, the lipophilicity of the active species radically limits its discharge to the environment relative to the water partitioning alternatives

References

1   R. F. Rekker and R. Mannhold, *Calculation of Drug Lipophilicity*, VCH Verlagsgesellschaft mbH, 1992.
2   N. E. S. Thompson and R. G. Asperger, *Dithiocarbamates for Treating Hydrocarbon Recovery Operations and Industrial Waters*, Petrolite Corp., US Patent 4,894,075, 5 Sep 1989.
3   N. E. S. Thompson and R. G. Asperger, *Methods for Treating Hydrocarbon Recovery Operations and Industrial Waters*, Petrolite Corp., US Patents 4,956,099, 11 Sept 1990; 5,013,451, 7 May 1991; 5,019,274, 28 May 1991; 5,026,483, 25 Jun 1991; 5,089,227, 18 Feb 1992; 5,089,619, 18 Feb 1992.
4   D. K. Durham, U. C. Conkle and H. H. Downs, *Additive for Clarifying Aqueous Systems without Production of Uncontrollable Floc*, Baker Hughes Inc., US Patent 5,006,274, 9 Apr 1991.
5   E. J. Evain, H. H. Downs and D. K. Durham, *Water Clarification Using Compositions Containing a Water Clarifier and a Floc Modifier Component*, Baker Hughes Inc., US Patents 5,190,683, 2 Mar 1993; 5,302,296, 12 Apr1994.
6   G. T. Rivers, *Epoxy Modified Water Clarifiers,* Baker Hughes Inc., US Patent 5,247,087, 21 Sept 1993.
7   T. J. Bellos, *Polyvalent Metal Cations in Combination with Dithiocarbamic Acid Compositions as Broad Spectrum Demulsifiers*, Baker Hughes Inc., US Patents 6,019,912, 1 Feb 2000; 6,130,258, 10 Oct 2000.

OPTIMISING OILFIELD OXYGEN SCAVENGERS

Andrew J McMahon, Alison Chalmers and Heather Macdonald

TR Oil Services Limited, Howe Moss Avenue, Kirkhill Industrial Estate, Dyce, Aberdeen, AB21 OGP, UK

1 INTRODUCTION

Oxygen scavenger chemicals are widely deployed within the offshore oil industry to remove dissolved oxygen from sea water streams,
- injected into reservoirs for pressure maintenance
- used in pipelines for hydrotesting

The oxygen is removed to a level < 20 ppb or lower in order to protect the carbon steel or alloy steel in the pipelines, topsides facilities, or downhole tubing from corrosion. The removal also inhibits the growth of general anaerobic bacteria (GABs).

Ammonium bisulphite oxygen scavenger (NH_4HSO_3, ie "ABS") is one of the most commonly used oxygen scavengers in the oil industry. It has the advantage over sodium bisulphite ($NaHSO_3$, ie "SBS"), an alternative scavenger, that it is soluble in concentrated solution (ie 65% w/w) at low ambient temperatures around 5°C. Under similar conditions SBS would produce a precipitate. This makes ABS preferable for low temperature environments such as the North Sea.

In locations where higher ambient temperatures are the norm some operators prefer SBS to ABS on the grounds that the ammonium ions in ABS will provide a food source for bacteria. However, any such effect will be minimised if biocide deployment is carried out in a proper fashion.

This paper presents laboratory and field results on the scavenging performance and corrosion effects of ABS. The work recommends a series of best practise guidelines for,
- assessing scavenger in the laboratory
- deploying and monitoring scavenger performance in the field
- applying other chemical treatments to sea water injection streams

2 ASSESSING THE SCAVENGER IN THE LABORATORY

2.1 Apparatus

The aim of this work was to devise a simple laboratory test method for routine assessment of oxygen scavenger chemicals. Early work showed that it was often difficult to completely exclude air from the various designs of prototype apparatus and this prevented accurate work at the extremely low oxygen concentrations (< 20 ppb) which were desired. Ultimately it proved necessary to build a compact apparatus comprising a single glass cell with lid, incorporating the Orbisphere membrane measuring probe hanging down inside the vessel (Figure 1). This set-up minimised sources of potential leaks such as external tubing and seals. Sparging of air saturated test brines with nitrogen showed that the oxygen levels <5 ppb could be achieved successfully.

The apparatus was thoroughly cleaned and dried prior to each test. All ports were sealed and oxygen-free nitrogen was sparged into the cell for 10 minutes to remove air. Sparging was stopped and then 600 ml of air saturated brine was added and the oxygen measurement with the Orbisphere probe was started. The magnetic stirrer was switched on at around 800 rpm to produce a significant vortex in the liquid so as to simulate, as far as possible, the turbulent conditions expected after scavenger addition to an oilfield sea water injection system. After monitoring the baseline oxygen concentration for a few minutes the scavenger chemical was dosed into the port at the bottom of the cell through the rubber septum. The fall of the oxygen concentration with time was then monitored.

Tests were carried out in both synthetic sea water and real sea water (obtained from Cove Bay, Aberdeen). The compositions of these sea waters are discussed in Section 2.2. Most of the tests were carried out at ambient temperature (22°C) and also some at 10°C.

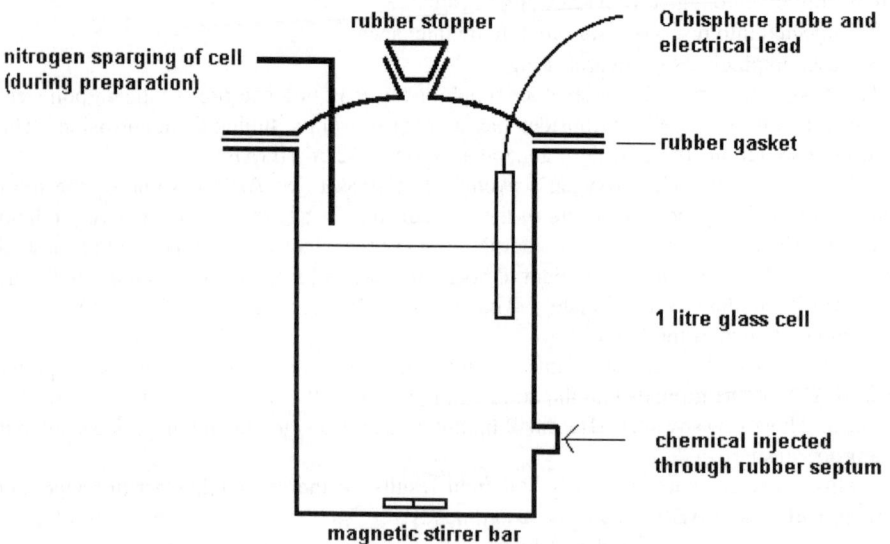

Figure 1   *Oxygen scavenger test cell*

2.2   Scavenger Performance

2.2.1 Synthetic Sea Water. A simple synthetic sea water composition was used in the initial work. This composition contained all the major dissolved ions and is one of the standard North Sea compositions used for routine work on oilfield sulphate scaling (Table 1).

Species	Concentration (ppm)
Na^+	10890
K^+	460
Ca^+	438
Mg^{2+}	1368
Ba^{2+}	0
Sr^{2+}	7
Cl^-	19766
sulphate	2960
bicarbonate	140
total dissolved solids	3.60%

Table 1   *Composition of Synthetic Sea Water*

Figure 2 shows results for blank, ammonium bisulphite scavenger (ABS), and catalysed ABS tests. Transition metal catalysts are sometimes added to ABS to improve performance as it is widely accepted that they provide a kinetic benefit [1]. In the present test work 1 ppm of catalyst was added separately to the test cell after the ABS scavenger had been added. The chemicals were added separately in order to avoid any incompatibility effects between neat ABS and neat catalyst which can sometimes occur.

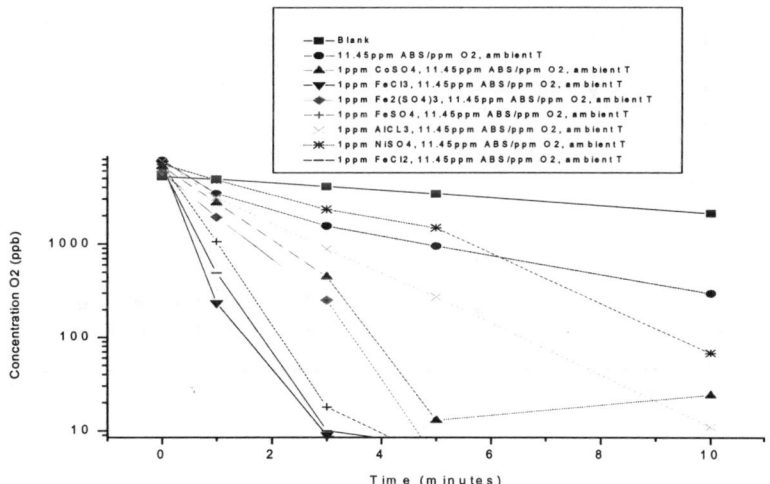

Figure 2   *Scavenger tests in synthetic sea water at 22°C*

The blank result is for no chemical added to the cell. The slow decrease in oxygen concentration from the starting value of 5-8 ppm reflects the operation of the Orbisphere probe which measures oxygen by means of an amperometric oxygen reduction method across a permeable membrane. Hence the probe will gradually consume oxygen over time.

The blank consumption rate is insignificant compared to the scavenger tests, which are discussed next, and so can be ignored.

Addition of ABS alone provides only a moderately faster reaction than the blank. The ABS was added at 11.45 ppm ABS solution per 1 ppm of dissolved oxygen, with respect to the dissolved oxygen concentration measured at the start of the test. This approach ensured that the ABS / oxygen ratio was the same at the start of all the tests. This 11.45 : 1 ratio is about a 25% excess with respect to the 9.4 : 1 stoichiometric ratio calculated for 65% w/w ABS solution and oxygen (see Section 3.3).

The performance of ABS improves markedly in the presence of a variety of catalysts such as Co, Fe, Al and Ni salts. The best catalyst is 1 ppm $FeCl_3$ which achieves <20 ppb oxygen after only 2 minutes.

2.2.2 Real Sea Water. The synthetic sea water composition does not contain the multitude of minor components which have been measured in sensitive analysis of real sea water (Table 2). The Table shows most of the components which are present at concentrations >10 ppt (ie parts per trillion, 10^{-9}). Co is also shown (3 ppt). There are hundreds of other components present at even lower concentrations.

When the scavenger tests were repeated in real sea water the result for ABS alone was significantly faster compared to synthetic sea water (Figure 3).

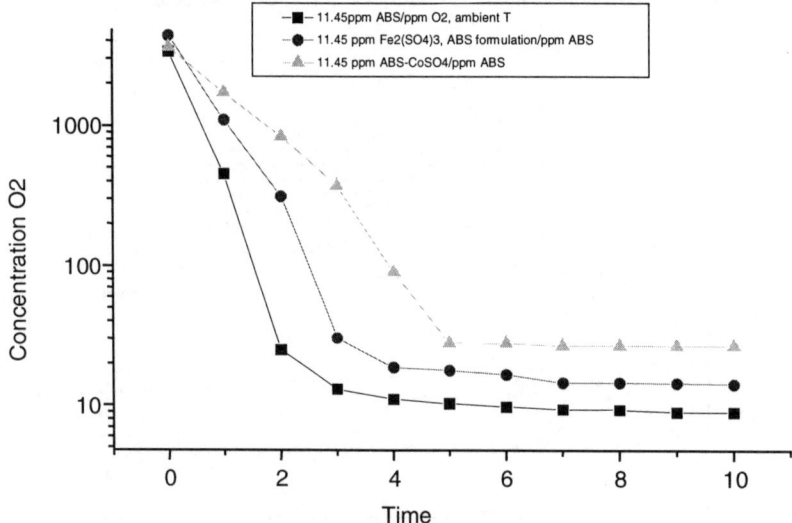

Figure 3   *Scavenger tests in real sea water at 22°C*

Species	Concentration (ppm)
Na^+	10770
K^+	380
Ca^+	412
Mg^{2+}	1290
Ba^{2+}	0.02
Sr^{2+}	8
Cl^-	19500
sulphate	2715
bicarbonate	142
Br	67
N	11.5
B	4.4
Si	2
F	1.3
Li	0.18
Rb	0.12
P	0.06
I	0.06
Mo	0.01
As	0.0037
U	0.0032
V	0.0025
Ti	0.001
Zn	0.0005
Ni	0.00048
Al	0.0004
Cs	0.0004
Cr	0.0003
Sb	0.00024
Mn	0.0001
Cd	0.0001
Cu	0.0001
W	0.0001
Fe	0.000055
Zr	0.00003
Bi	0.00002
Nb	0.00001
Th	0.00001
Tl	0.00001
Co	0.000003
total dissolved solids	3.53%

Table 2   *Composition of Real Sea Water*
Note: this composition is for Hawaiian reef sea water (see reference 2), but it will also be similar to typical sea water from the North Sea

This change is probably due to the presence of the many minor components in real sea water which act collectively as catalysts and improve the performance of ABS. ABS alone in real sea water achieves <20 ppb after about 2 minutes. It appears that added catalyst is not necessary in real sea water and it may even retard the kinetics. However, there may still be circumstances when sea water is mixed with other water streams in the field, possibly deactivating the natural sea water catalysts, and so added catalyst may still be useful. Overall, these effects show the importance of using real sea water in laboratory work with oilfield oxygen scavengers.

The catalysed tests in real sea water give similar results to synthetic water for the Co^{2+} and Fe^{3+} catalysts (compare Figures 3 and 2). The Co^{2+} catalyst was added separately from the ABS in both waters. The Fe^{3+} catalyst in Figures 3 and 4 was mixed with the ABS first and then the mixture injected into the cell

The good repeatability in the test work is demonstrated in Figure 4 for three different runs under identical conditions.

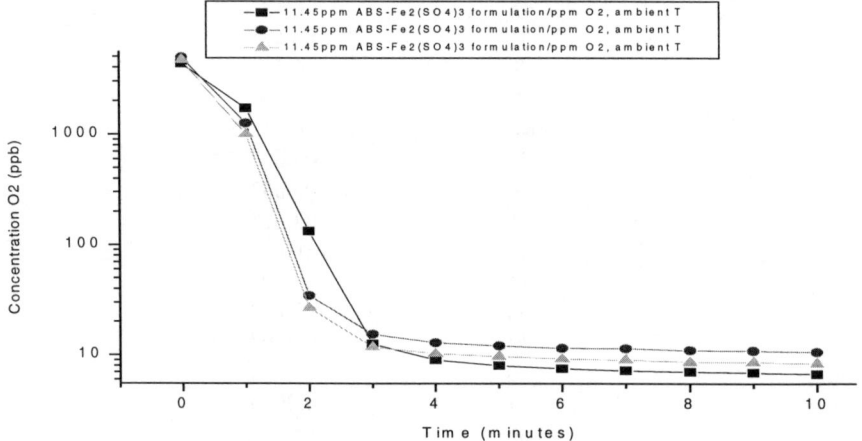

Figure 4   *Repeatability of scavenger tests in real sea water at 22°C*

2.2.3 Effect of Temperature. All of the results presented so far are at the ambient temperature of 22°C. When the temperature is reduced to 10°C the performance of ABS in real sea water is significantly slower, as would be expected (Figure 5). It was found that uncatalysed and catalysed ABS both required about 40-60 minutes to reach <20 ppb at 10°C.

This duration does not give any problems for pipeline hydrotesting since the sea water is often present for several days or longer. In sea water injection systems it is important that sufficient residence time is provided to allow the scavenging reaction take place, or alternatively, the sea water should be heated, ideally by using it as the cold fluid in a heat exchanger system, which is the normal practise in the process engineering for offshore oil platforms.

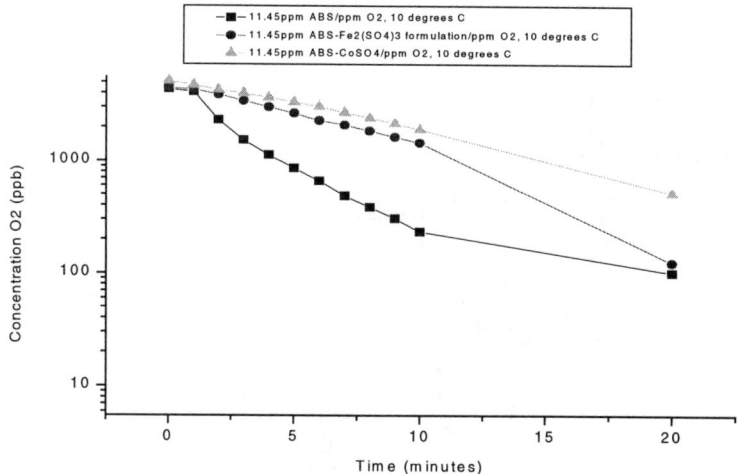

Figure 5 *Scavenger tests in real sea water at 10°C*

3 ASSESSING THE SCAVENGER OFFSHORE

3.1 Description of the Offshore System

The use of oxygen scavenger to remove all the dissolved oxygen from sea water (as carried out in Section 2) is normal practise during pipeline hydrotesting. However, for a sea water injection system on an offshore oil platform it is normal to pass the raw sea water through a deaeration tower first, lowering the oxygen concentration to a few hundred ppm, and then use the oxygen scavenger to "polish" the water and further reduce the oxygen to <20 ppb. The deaeration tower can operate by using gas stripping (ie produced natural gas) or, more commonly, by vacuum stripping.

This Section presents results from an offshore field trial on uncatalysed ABS oxygen scavenger on a North Sea platform. The platform used a gas stripping deaeration tower to reach <20 ppb oxygen and did not normally require any oxygen scavenger. However, they wished to use more of their produced gas as a fuel rather than for sea water deaeration duty. The trial was carried out to assess whether ABS could be used to compensate for a reduction in the stripping gas flow rate in the deaeration towers.

Figure 6 shows a schematic diagram of the sea water injection system and the location of the sample points used in the trial. The sample point descriptions CF, BP, and IP - indicated on the Figure - will be used to describe the work. Sodium hypochlorite solution (ie NaHOCl) is injected downstream of the supply pumps as a biocide. The sea water temperature at CF was 2°C but it rose to 13°C before entry into the deaeration tower due to passage through a heat exchanger. The water injection system is constructed in Cunifer (a corrosion resistant Cu / Ni / Fe alloy) up to the dearation tower. The tower itself and all material downstream is carbon steel. The sea water flow rate through the deaeration tower was constant at 12500 m^3/day throughout the trial.

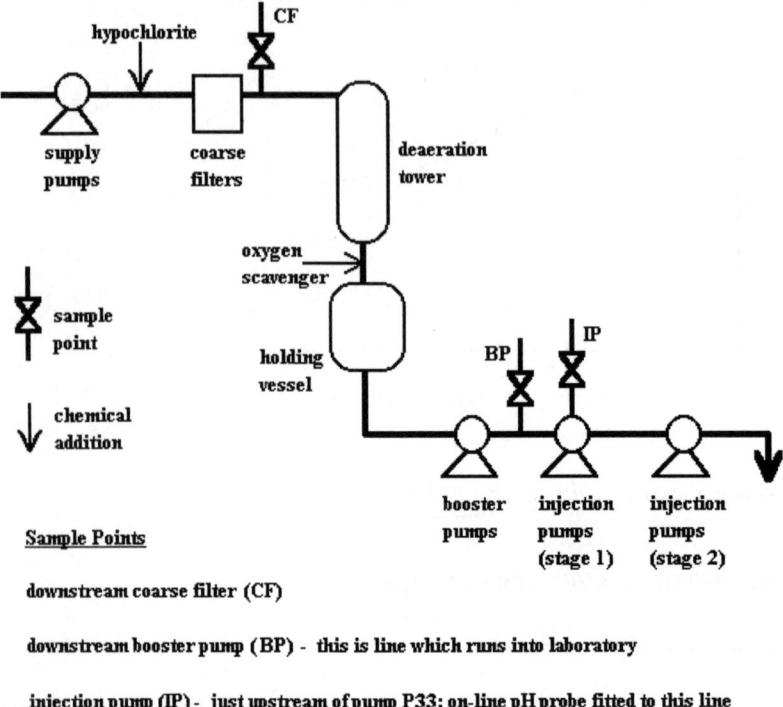

Sample Points

downstream coarse filter (CF)

downstream booster pump (BP) - this is line which runs into laboratory

injection pump (IP) - just upstream of pump P33; on-line pH probe fitted to this line

Figure 6 *Schematic diagram of offshore sea water injection system*

3.2 Effective Oxygen Measurement

A portable Orbisphere oxygen meter, as used in Section 2, proved to be the most reliable oxygen measurement technique during the offshore trial. It gave steady, consistent values in the sea water and was not affected by either chlorine (up to 1 ppm) or oxygen scavenger (ie up to 4 ppm ABS). "Chemet" rhodamine colorimetric ampoules for oxygen measurement agreed with the Orbisphere oxygen values when chlorine was switched off, but **overestimated** the oxygen concentration when the chlorine was on (Figure 7). The details along the top of the Figure show the prevailing chlorine concentration and gas stripping rate in the deaeration tower during the oxygen measurements by both Orbisphere and Chemet. When the chlorine is switched on and rises to 0.3 ppm, the Chemet reading increases but the Orbisphere reading remains steady. An increase in the stripping gas flow rate after about 6 hours has the expected effect of improving the oxygen removal to around 15 ppb (Orbisphere measurement).

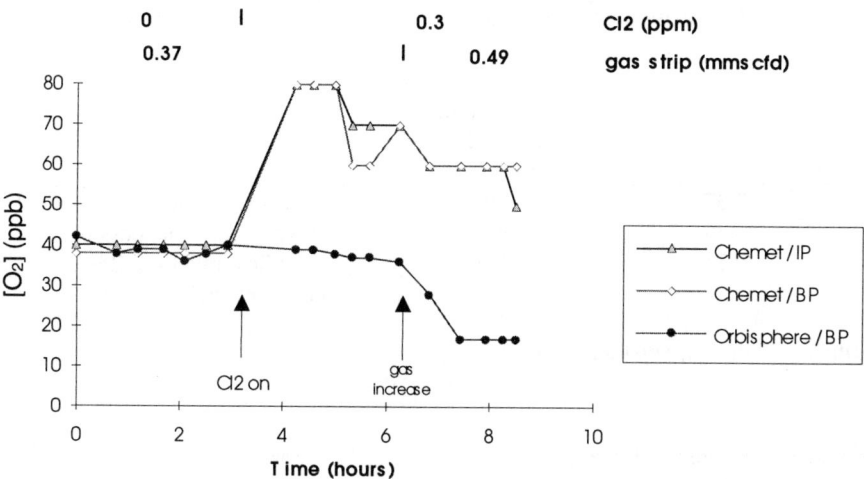

Figure 7 *Effect of chlorine on oxygen measurements*

Clearly the Chemets were sensitive to chlorine. Every 0.1 ppm chlorine (determined using Lovibond DPD no 1 tablets) contributed about 10 ppb to the observed oxygen measurement. The Chemets were not affected by up to 4 ppm oxygen scavenger when that was used later in the trial.

Orbisphere Ltd agree that their probe will be unaffected by chlorine for these conditions. This is because at pH 7 - the prevailing pH of the sea water - virtually all the chlorine will remain in the hypochlorite form (OCl⁻) rather than the free chlorine form (Cl_2). The membrane on the Orbisphere probe is ionophobic (ie it will restrict transport of ions) and therefore will not be affected by the OCl⁻ ions.

$$Cl_2(g) \leftrightarrow Cl_2(aq)$$
$$Cl_2(aq) + H_2O \leftrightarrow H^+ + Cl^- + HClO$$
$$HClO \leftrightarrow H^+ + OCl^-$$

(the equilibrium constants are $K_1 = 0.062$, $K_2 = 0.00042$, $K_3 = 3.4E\text{-}8$ respectively)

3.3 Oxygen Scavenger Performance

The system was first run over a range of stripping gas flow rates in order to establish baseline values for oxygen, pH and corrosion. The stripping gas was then kept at the low value and this was used for the oxygen scavenger trial (see Table 3).

Figure 8 shows the oxygen concentration at sample point BP during both the baseline phase and the oxygen scavenger phase.

stage	Stripping Gas Flow Rate		Comments
	MM SCFD	*SCF per 6 minutes (this is the value shown on flow meters in MOL Control Room)*	
1.	**0.37**	**1500**	..."standard" flow rate on arrival
2.	**0.49**	**2000**	...highest flow rate possible
3.	**0.22**	**860**	...low flow rate for ABS scavenger trial

Table 3   *Stages in Offshore Trial*

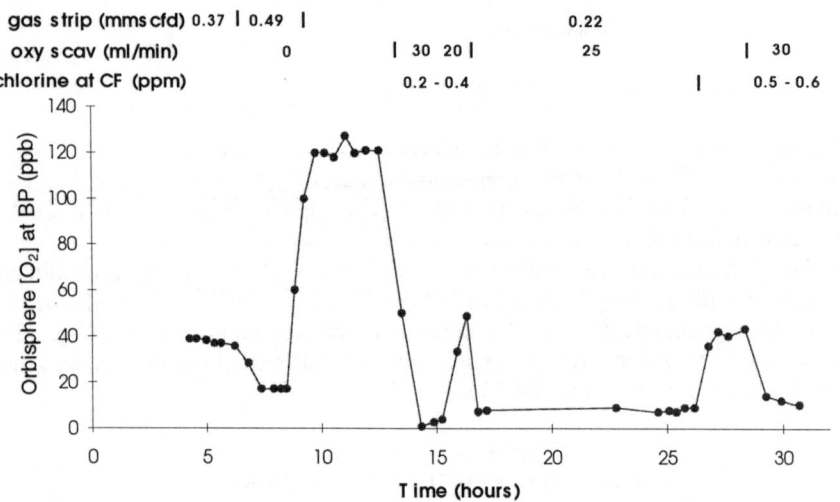

Figure 8   *Oxygen scavenger performance during trial*

The baseline period, from 0 - 13 hours, shows the effect of changing the stripping gas flow rate. The normal oxygen level for 0.37 MMSCFD stripping was 40 ppb, and the level varied inversely with the gas flow as expected. ABS oxygen scavenger was turned on at 13 hours and the oxygen concentration quickly fell to under 5 ppb. Section 3.3 gives more detail about the ABS composition and the calculation of the required dose rate. The dose rate was adjusted during 15-20 hours and showed that 25 ml/hour ABS solution was the minimum dose which gave < 20 ppb oxygen. Therefore, the ABS successfully compensated for the 40% reduction in stripping gas throughput.

The chlorine level was 0.2-0.4 ppm for all the data up to 13 hours. The values at CF and BP were always similar which showed little or no loss across the system. When ABS was started at 13 hours, the chlorine level at BP fell immediately to zero and remained at zero even when the oxygen level subsequently rose to 10-50 ppb. This shows that the ABS reacts more quickly with chlorine than with oxygen. Therefore, use of ABS leaves **no**

chlorine in the water injection header. Regular batch biocide treatment would be needed to compensate.

The ABS + Cl$_2$ reaction means that the required dose of ABS depends on the prevailing chlorine level as well as on the oxygen level. This is demonstrated at 27 hours when the chlorine was increased from 0.4 to 0.6 ppm (the chlorine was increased because the dosing pump gave frequent airlocking problems when delivering the lower concentration). The higher chlorine dose consumed more of the ABS and, therefore, the oxygen level increased to ca 40 ppb. When the ABS dose was also increased (from 25 to 30 ml/hour) at 29 hours, the oxygen fell back to 10 ppb, as expected.

The optimum process parameters at the end of the trial (ie the end of Figure 8) are listed in Table 4.

Parameter	Value	Comments
water temperature	13°C at BP	...was only 2°C at CF
liquid flow rate in deaeration tower	12500 m^3/day	
chlorine	0.6 ppm at CF; 0 at BP	
oxygen	5 ppb at BP	
pH	7.2 at IP	...pH only varied between 6.8 and 7.2 for all the conditions in Figures 7 and 8
oxygen scavenger	30 ml/min (ie 3.5 ppm)	...see Section 3.3
bisulphite residual	1.3 ppm at BP	...1.3 ppm bisulphite = 2.0 ppm sodium sulphite; see Section 3.3
corrosion rate	0.11 mm/yr (ie 4.3 mpy) at WI header	...see Section 3.4

Table 4   *Optimum Process Parameters at the End of the Trial*

Note that there is a 1.3 ppm residual bisulphite at BP despite the presence of 5 ppb oxygen. The bisulphite residual was measured using a Hach test kit. The residual is present because the ABS + O$_2$ reaction is slow at 13°C (see Figure 4) and there is not enough residence time in the holding tank for the reaction to go to completion. All the oxygen would eventually be removed at some point further downstream in the system, possibly in the injection well. Decreasing the ABS dose rate from 3.5 ppm in an attempt to reduce the bisulphite residual would not be successful. Oxygen removal would simply become even slower.

3.4   Analysis of ABS Dose Rate Required in Trial

The following reaction scheme calculates the theoretical demand for ABS and compares it against the actual optimised demand during the field trial. This scheme is intended as a detailed reference for the field deployment of ABS.

Ammonium bisulphite (NH$_4$HSO$_3$) was supplied as a 65%w/w solution in water with a specific gravity of 1.35 g/cm^3.

Reaction with oxygen

$$2H_2O + 2HSO_3^- \rightarrow 2SO_4^{2-} + 6H^+ + 4e^-$$

$$O_2 + 4H^+ + 4e^- \rightarrow 2H_2O$$

$$\text{overall:} \quad O_2 + 2HSO_3^- \rightarrow 2SO_4^{2-} + 2H^+$$

Hence, 1 ppm w/v NH_4HSO_3 reacts with 0.16 ppm w/v oxygen (ie 163 ppb),

    1 ppm w/v 65%w/w NH_4HSO_3 solution reacts with 0.106 ppm w/v oxygen

Reaction with chlorine

$$H_2O + HSO_3^- \rightarrow SO_4^{2-} + 3H^+ + 2e^-$$

$$Cl_2 + 2e^- \rightarrow 2Cl^-$$

$$\text{overall:} \quad Cl_2 + HSO_3^- + H_2O \rightarrow SO_4^{2-} + 3H^+ + 2Cl^-$$

Hence, 1 ppm w/v NH_4HSO_3 reacts with 0.71 ppm w/v chlorine,

    1 ppm w/v 65%w/w NH_4HSO_3 solution reacts with 0.47 ppm w/v chlorine

Comparing theory with practise

What dose rate of 65% NH_4HSO_3 is required to remove the 120 ppb O_2 and 0.4 ppm Cl_2 in the field trial at 12 hours in Figure 8 and leave 1.3 ppm residual bisulphite?

120 ppb O_2	requires	1.13 ppm w/v 65%w/w NH_4HSO_3 solution
0.4 ppm Cl_2	requires	0.85 ppm w/v 65%w/w NH_4HSO_3 solution
1.3 ppm HSO_3^-	requires	2.48 ppm w/v 65%w/w NH_4HSO_3 solution

               total =   **4.46 ppm w/v 65%w/w NH_4HSO_3 solution**

                   =   3.30 ppm v/v 65%w/w NH_4HSO_3 solution

                   =   29.8 ml/min 65%w/w NH_4HSO_3 solution into

                      12500 m^3/day water

    The observed dose rate was 3.5 ppm (25 ml/min). Therefore, the theory agrees reasonably well with practise (to within 20%).

3.5   Corrosion Monitoring Data During the Trial

Corrosion rates were measured using a linear polarisation resistance probe (LPR probe) fitted with 3 projecting electrodes (7.86 cm^2 per electrode) which was located at the 10 o'clock position on the 10 inch sea water injection header downstream of stage 1 injection pump (see Figure 6). The probe was installed 2 weeks prior to the trial. Cormon and Rohrback LPR meters were used on the same probe for comparison. The corrosion data appeared to be satisfactory because,

- the Cormon meter did not give any error messages during intensive manual use over 3 days
- the Cormon and Rohrback meters gave similar absolute and relative corrosion rates (Figure 9)
- the corrosion rates changed over the range 2.5 - 6.5 mpy depending on solution conditions (Figure 10)

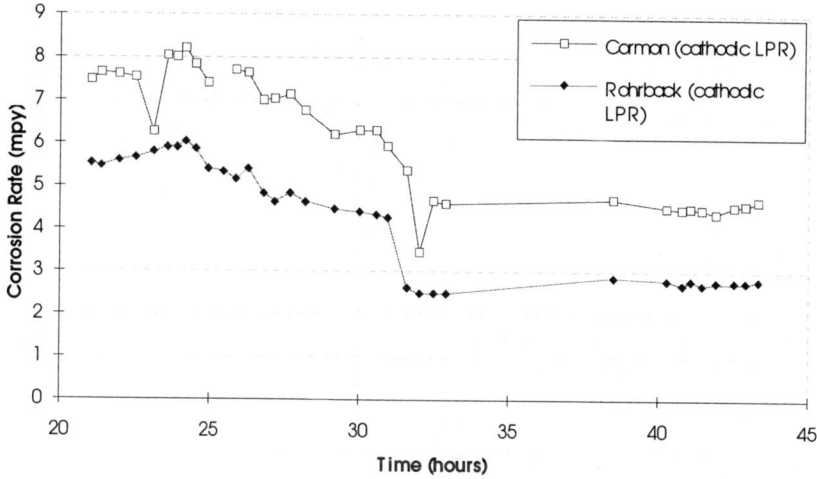

Figure 9 *Comparison of corrosion data from Cormon and Rohrback LPR meters*

Figure 10 shows that the corrosion rate was <6 mpy (ie <0.15 mm/yr; mpy is thousandths of an inch per year, 100 mpy = 2.54 mm/yr) for 0 - 120 ppb O_2, 0 - 0.7 ppm Cl_2, or 0 - 3.8 ppm bisulphite, during the trial. This is a low corrosion rate. The corrosion rate was <3 mpy under the optimum conditions at the end of the trial.

The corrosion rate showed a small step change (5- 5.5 mpy) at ca 17 hours when chlorine was increased from 0 to 0.4 ppm (Figure 10). The highest rate in the trial was 6.5 mpy recorded at 0 hours for 0.7 ppm chlorine. Hence, the corrosion rate was moderately dependant on chlorine level, as would be expected.

The largest step change (4.3 - 2.6 mpy) was at ca 32 hours when the residual bisulphite concentration steadied at 1.3 ppm. At low concentrations bisulphite does have some corrosion inhibiting properties on carbon steel probably through the formation of FeS surface films.

gas strip (mmscfd)   0.37   0.49   0.22

Cl2 (ppm)   0.7 0.5 | 0 | 0.2 - 0.4 | 0.5 - 0.6
at header

HSO3 at header (ppm)   0 | 3.8 2 1.3   1.3

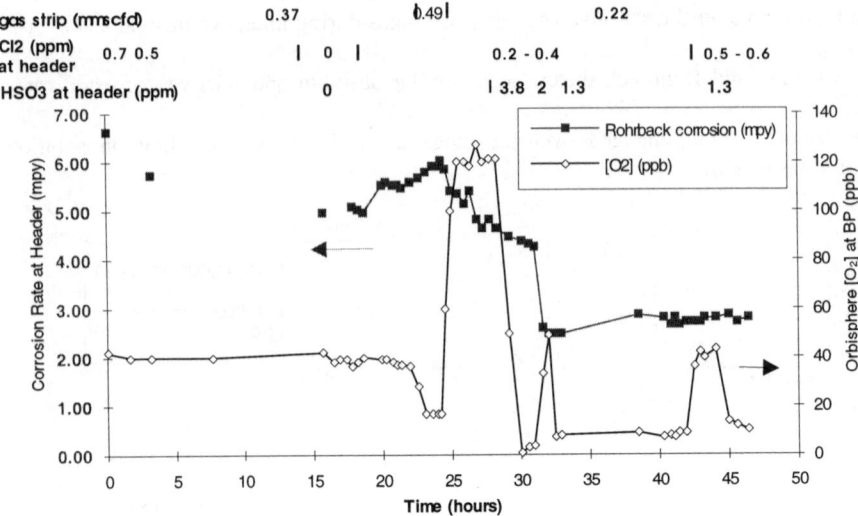

Figure 10 *Corrosion rate and process conditions during the trial (NB "header" means the sample point BP in Figure 6)*

4 EFFECT OF RESIDUAL BISULPHITE IN INJECTION WATER

A small residual concentration of bisulphite in the injection water is desirable in order to provide some capacity to mop up any small oxygen leaks and thereby ensure continuous chemically reducing conditions : 3.2 ppm bisulphite can consume about 320 ppb of dissolved oxygen. It is important to avoid swinging between excess oxygen and excess bisulphite because regular switching between oxidising and reducing conditions could create a patchwork of oxide and sulphide regions on a carbon steel pipewall and thereby increase the risk of galvanic corrosion.

Many operators specify an upper limit on the concentration of residual bisulphite which is permitted in injection sea water; typically this is in the range 0.64 - 3.2 ppm. These values arise from the Hach bisulphite test kit in which each drop of analyte represents 0.64 ppm bisulphite (ie equals 1 ppm sodium sulphite) and therefore 3.2 ppm bisulphite is equivalent to 5 drops of analyte.

The various limits are often based on accumulated operating experience and inherited perceptions of best practise with no detailed compilations of test or field data which unequivocally define the acceptable levels. There is experience within several operators that > 3 ppm bisulphite can produce enough FeS solids on pipeline and tubing walls, and loose in solution to eventually cause problems opening and closing downhole valves, and frequent sticking of wirelines and tools. Corrosion coupons with black sulphide films have also been retrieved from systems where there has been inadvertent gross overdosing of bisulphite.

A further factor influencing the adoption of the limits is the desire to avoid the temptation - in design or operation - of over-reliance on large quantities of oxygen scavenger at the expense of efficient deaeration by vacuum or gas stripping. Such over-reliance can lead to gross overdosing during plant upsets or variation.

4.1　Corrosion Measurements on Bisulphite

Figure 11 shows dynamic corrosion data for bisulphite in contact with carbon steel using a rotating cylinder electrode (RCE) running at 10 Pa wall shear stress, a level representative of a typical sea water injection line.

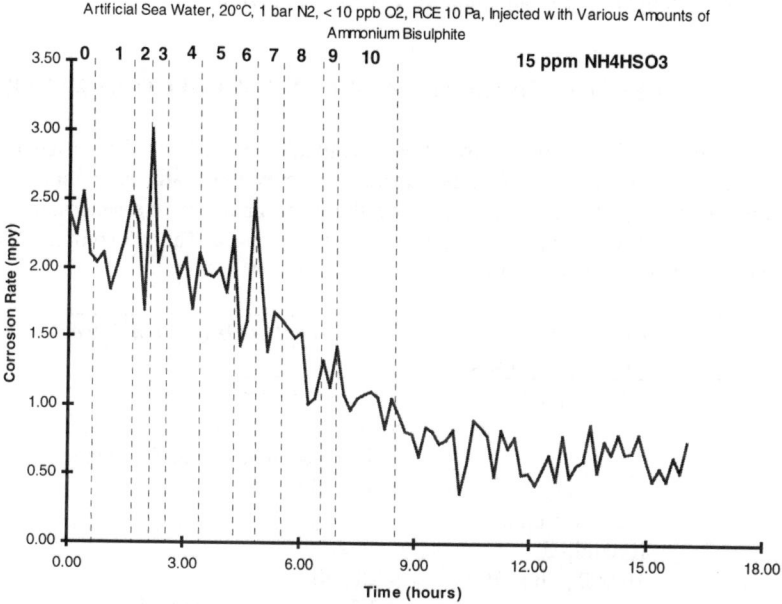

Figure 11 *Bisulphite corrosion of carbon steel in synthetic sea water using the RCE at 20°C*

Residual bisulphite caused no short term corrosion damage up to 12 ppm bisulphite (ie 15 ppm NH_4HSO_3) at 20°C. The corrosion rates were all <2.5 mpy. This measurement is consistent with the <3 mpy observed at the end of the field trial with 1.3 ppm residual bisulphite (Figure 10).

At low temperature, however, the real corrosion concern with bisulphite is not short term general corrosion but pitting corrosion arising from long term formation and then damage of sulphide films. There is a field example of < 4 ppm bisulphite in sea water showing only 4 mpy general corrosion but 40 mpy pitting corrosion [3]. However, if a sulphide film is damaged it will arguably lead to pitting corrosion only if some oxygen is present. Oxygen levels will obviously increase if the performance of the deaeration tower drops periodically. Overall, provided that the deaeration tower performs consistently, and that chemically reducing conditions are always maintained by ensuring a residual level of bisulphite, then pitting corrosion will be minimised.

4.2　Further Comments on FeS Solids Build-Up

Note that FeS solids can be produced by the action of sulphate reducing bacteria (SRBs) on the sulphate in the sea water, as well as by the reaction of excess bisulphite. Such bacteria require continuous monitoring and control using biocides. Bacterial and FeS solids monitoring can be carried out using carbon steel studs or weight loss coupons fitted to the main process line or, more conveniently, fitted to a sidestream on the sea water system. If any significant FeS films are observed then there are reports that an isotopic ratio experiment (ie $^{32}S/^{34}S$) can help to determine whether the FeS is of biological or inorganic origin. The presence of significant levels of SRBs would be also be telling. If of biological origin then the hypochlorite and biociding regimes will need review. If of inorganic origin then the residual bisulphite level will need review.

5 BEST PRACTISE FOR CHEMICAL TREATMENT OF INJECTION WATER

Oxygen scavenger is only one of the treatment chemicals normally used in offshore sea water injection systems. Hypochlorite is injected into raw sea water as a biocide and filtration aid. Antifoam can be used to improve the performance of the deaerator tower. Batch biocide is used downstream of oxygen scavenger injection. Table 5 summarises the suggested best practise for chemical treatment of injection water.

Parameter	Upstream of Deaeration Tower	Downstream of Deaeration Tower
hypochlorite	0.5 - 1 ppm Cl_2	• 0.2 - 0.3 ppm Cl_2 maximum • will be zero if oxygen scavenger is in use, hence need for batch biocide
antifoam	0.25 - 2.5 ppm if required	
oxygen	10,000 ppb at 10°C	< 20 ppb
oxygen scavenger	not present	< 3.2 ppm bisulphite residual
batch biocide	not present	• 300 ppm, 50% glutaraldehyde for 3 hours weekly (NB at least 200 ppm is required for adequate performance) • review concentration or duration depending on regular microbiological monitoring data, but always retain weekly frequency

Table 5   *Summary of Chemical Treatment Recommendations for Sea Water Injection*

Notes:
1   Hypochlorite is known to improve filtration efficiency in coarse and fine filters, as well as being a biocide.
2   A concentration of 3.2 ppm residual bisulphite is capable of removing about 320 ppb O2.
3   A suitable microbiological monitoring system is essential, ideally a slip stream with bioprobes which are pulled and analysed frequently (2-6 weeks) and also bioprobes in the main flow stream which are pulled at 6-12 monthly intervals. The bioprobes should also be used to monitor any excessive build up of FeS deposits. **4.** Flush with biocide before shutdowns so that the residual stagnant water is treated.
4   Reduce water rates during biociding. This is often a cost effective way of reducing chemical costs but the decision depends on the practicability of changing flow on the reservoir water-flood requirements.

5   Traditional oxygen scavengers and biocides can react with each other and so cancel each other out. Hence, some operators switch off the oxygen scavenger during biociding and thereby sustain a high oxygen peak for a short period.

6 CONCLUSIONS

- Transition metal catalysts can significantly improve the performance of ammonium bisulphite oxygen scavenger (ABS) in synthetic sea water in laboratory tests.
- The effect of added catalyst disappears in real sea water probably because real sea water already contains small amounts of many transition metals which act as natural catalysts.
- Added catalyst may still improve the field performance of ABS in circumstances where the natural catalysts have been deactivated.
- Real sea water should always be used for ABS performance tests in the laboratory.
- In an offshore field trial an Orbisphere oxygen monitor gave reliable results. It was unaffected by chlorine biocide in the sea water.
- The "Chemets" colorimetric method is only accurate when there is zero chlorine. Every ca 0.1 ppm chlorine contributes ca 10 ppb to the observed oxygen concentration in the "Chemets" test.
- In the offshore trial ABS was successfully used to compensate for a reduction in stripping gas flow rate in the deaerator towers. The optimum dose rate also agreed reasonably well with theory.
- ABS performance is critically dependent on temperature. Ideally the sea water stream should be heated in a heat exchanger prior to the deaerator tower and ABS injection.
- General corrosion rate measurements showed a low corrosion rate (<5 mpy, ie < 0.13 mm/yr) under optimum conditions at the end of the trial (5 ppb O_2, 0.6 ppm Cl_2, and 1.3 ppm bisulphite).
- Provided that the deaeration tower performs consistently, and that chemically reducing conditions are always maintained by ensuring an residual level of bisulphite (up to about 3.2 ppm), then pitting corrosion will be minimised.
- Oxygen scavenger removes chlorine from the sea water before it removes oxygen. Therefore, use of oxygen scavenger leaves no chlorine downstream of the deaerator. Frequent batch biociding is needed to compensate.
- Corrosion probes in the main line and / or stud-probes in a side stream can be used to any excessive build up of FeS solids build-up resulting from either residual bisulphite or bacterial action.
- A set of best practise guidelines have been compiled for oxygen scavenger, hypochlorite, antifoam, and biocide treatment of sea water injection streams.

References

1   H.H. Ulig, R.W. Revie, *Corrosion and Corrosion Control*, John Wiley, New York, 1985, **279**.

2   G. Bearman, *Ocean chemistry and deep-sea sediments*, Pergamon, Sydney, 1989 (see also web-site http://www.personal.psu.edu/dept/aquarium/Chem.htm).

3   J.T.A. Smith, *Minimising Corrosion of Carbon Steel in Sea Water Systems - Guidelines for Water Quality*, BP Sunbury Report No ESR.94.ER.005, 1994 (available on the *BP Corrosion and Materials Guidelines 2001*, public domain CD-ROM).

ENHANCING RELIABILITY, PERFORMANCE AND ENVIRONMENTAL ACCEPTABILITY OF SUBSEA HYDRAULIC PRODUCTION CONTROL FLUIDS, A TRUE CHEMISTRY CHALLENGE.

R. Rowntree, R. Dixon

Castrol Subsea Technology Group, Castrol Technology Centre, Whitchurch Hill, Pangbourne, Reading, Berks, RG8 7QR

1   ABSTRACT

Subsea production control fluids have been widely used as the hydraulic media for the last twenty years to operate subsea production control systems. Both closed and open systems are widely used throughout the world. However a combination of tightening legislation, environmental constraints and increasing performance demands have led to significant changes in the formulations of these products. The resolution of these demands requires both innovative chemistry and novel engineering to work together as a system. The overall objective is to provide a reliable subsea control system.

2   INTRODUCTION

The objective of this paper is to outline:

- The environments in which subsea production control fluids operate.
- The legislation that these products are subject to.
- The performance demands of the end user's applications.

There is increasing focus on the environmental impact of the subsea hydraulic control fluids, including the application design and the raw material components used in formulating such products. By considering the control fluid as part of the complete subsea production control system, which interacts with materials, contaminants and changing environmental conditions, the product development process needs to be adapted to address this change in focus. The challenge for the product development chemist is to manage this changing environment.

3 SUBSEA PRODUCTION CONTROL SYSTEMS

3.1 General Overview

Subsea Production Control Systems have become an economic way of recovering hydrocarbons from offshore reservoirs, especially those that are situated in deepwater, remote locations or where hydrocarbons exist in smaller quantities. After drilling operations, production tubing is installed to bring the hydrocarbon to the surface. In simple terms, a series of valve actuators (X-Mas Tree) are placed on the seabed at the top of the production tubing to control hydrocarbon flow. The Xmas trees are functioned hydraulically by a control pod that is supplied with electronic and hydraulic power, via a control umbilical, from a surface installation or the land. Offset distances (the length between the surface installation and the X-Mas Tree) range from a couple of kilometres up to, more recently, tiebacks of 90 km to the land.

The control pod also provides hydraulic power to function the sub surface safety valve, (a fail safe, well shut in device) which is normally situated approx 80 - 100 metres below the seabed, in the hydrocarbon stream, in the production tubing. The significance of this is that the hydraulic fluid will have to maintain performance and function at reservoir flowing well temperatures, which can be up to 200°C.

Subsea Production System configurations are defined by a number of factors such as; the offset distance from the surface installation, number of X-Mas Trees, water depth from 10metres to >1000metres) and the pressure and temperature of the hydrocarbon reservoir.

The response of a system is dependent on certain critical factors such as fluid viscosity, hydraulic line diameter and offset distance. Over shorter offset distances (1 to 5km) it is possible to use direct hydraulics and achieve acceptable response performance, for example, minimal pressure drops from the surface installation to the X-Mas Tree. However, over longer distances (>5km) it is necessary to employ other techniques to control seabed equipment. The Electro hydraulic Multiplexed system is such a system and it uses a combination of electronic and hydraulic power. The hydraulic fluid is stored in accumulators close to position of use on the seabed and released by electronically controlled solenoid valves. The critical aspect is the recovery time of the accumulators to full pressure, to allow the next valve actuation to occur.

There are two main types of Subsea Production Control System.

- An **open** system, where the hydraulic control fluid is used to actuate a series of valves and after actuation the fluid is discharged into the marine environment. A water glycol based fluid is used in this application.

- A **closed** system, where the hydraulic control fluid is stored in seabed accumulators and after use is returned to the platform using a return line in the control umbilical. In general synthetic oil based fluids are used in this application.

Both systems have their respective benefits, capital and operational costs. A study is underway with major operators to determine the through field costs, typically over 15 to 20 year period, for both types of system.

3.2 Performance Demands

The technical demands that the product are expected to meet are increasing.   Subsea production control systems are being designed to meet more severe external environments such as higher pressures (up to 1200 bar) and higher temperatures (up to 200°C) in the well.   Exploration and production operations in increasingly deepwater mean that reliability is essential to reduce maintenance and work over costs.

Internal changes to the subsea production system are occurring too.   New system materials are continuously being introduced into the hydraulic system.   For example, metals which are more resistant to seawater, new grades of elastomer, thermoplastic or ceramics.

The hydraulic control fluid is expected to be compatible with all these materials. There is also the expectation that the subsea control fluid will be tolerant to a variety of potential contaminants (see list below) that can migrate into the hydraulic system during installation or operation activities.

Seawater
Completion fluids - Brines NaCl, CaBr , formates, polymer muds
Chemical Injection fluids - Methanol
Acid Fracturing Chemicals – Hydrochloric acid 1M to 3M

4 CHEMISTRY OF SUBSEA PRODUCTION CONTROL FLUIDS

4.1 Control Fluid Function

The same product development approach is used for both water glycol and synthetic oil based control fluids. This follows the scheme shown below.

Control Fluid Function = Control Fluid Properties required = Component Selection

Firstly, consideration must be given to the key function characteristics of the control fluid. Some of the main functions are listed below:

- To efficiently transmit power
- To lubricate pumps, motors, and valves
- To protect against corrosion
- To remove debris and contamination
- To maintain thermal stability
- To show compatibility with elastomers, plastics and metals
- To show compatibility with solvents and completion fluids

From these fey functions, a list of fluid properties are drawn up, as below:

- Viscosity
- Suitable anti-wear
- Suitable anti-corrosion
- High temperature stability

Then finally, appropriate raw materials are selected and tested to meet these criteria, and conform to specific legislation requirements. For example the hydraulic response of a system will be specified, leading to a product viscosity requirement which will govern the type of base fluid selected.

4.2 Water Glycol Based Control Fluids

These are the most frequently used products. The main constituents are water and monoethylene glycol (the ratio of these can be adjusted to change the density of the final product). Additives include:

Anti-wear – usually provided by phosphorous or sulphur based chemistry.
Anti-corrosion – usually provided by amine-neutralised salts of organic acids
Biocides
Dyes
Anti-foam

Typical properties are in the Appendix, Table 1.

4.3 Synthetic Control Fluids

These products are also used, but generally in much lower volumes than the water glycol based fluids and for closed systems only. The products are based on the synthetic base oil polyalphaolefin (1-decene). Typical additives include

Anti-oxidants
Anti-wear
Anti-corrosion
Viscosity modifiers

Typical properties are in the Appendix, Table 2.

5 LEGISLATION

5.1 Offshore Legislation

Legislation exists in many countries around the world, to control the use of chemicals in Offshore Oil and Gas production. The following section discusses these requirements.

The Oslo and Paris Commissions
    The Oslo Commission was initially established to regulate and control the dumping at sea of industrial wastes such as sewage sludge. The Paris Commission was initially established to regulate and control materials discharged by the power generation industry, to the sea from land-based sources. The Oslo and Paris Commissions came together in September 1992. Ministers responsible for the marine environment of each of the 14 signatory states agreed to a new convention for the Protection of the Marine Environment of the North-East Atlantic, known as the "OSPAR Convention".

Like many industries the Offshore oil and gas industry is subject to Governmental environment legislation. One of the guidelines laid down by the OSPAR Commission was the voluntary Harmonised Offshore Chemical Notification Scheme (HOCNS). The scheme sets out a framework for testing oilfield chemicals that are used on the NE Atlantic sector of the North Sea. The HOCNS is administered by a department in each of the Governments of the 14 signatory states. The HOCNS applies to all chemicals, which are used in the exploration, exploitation and associated offshore processing of petroleum in NE Atlantic sector. The HOCNS applies to "operational" chemicals that through their mode of use are expected in some proportion to be discharged. Hydraulic fluids that are used to control subsea production systems fall into this category.

In January 2001, the HOCNS became mandatory for the Use and Reduction of the Discharge of Offshore chemicals.

The HOCNS guidelines require a level of ecotoxicological information on each product, so that a complete HOCNS document will contain the following information:

- Product details - i.e. estimated product usage, physical properties, marine toxicity data
- Raw material details - i.e. concentration, biodegradability / bioaccumulation data. This data must be generated regardless of concentration in the product.

Each country that has signed up to the mandatory scheme may in fact assess the data as they see fit. The HOCNS forms the basis of any assessment, but each European country is at liberty to interpret the data differently.

5.2   UK Offshore

The HOCNS information is assessed and the product is assigned a classification A to E, where E contains products that are least damaging to the marine environment. This system is not entirely satisfactory since the bands A to E increase logarithmically and testing has shown that two products from the same band (category E) tested against the marine toxicity show eight fold difference in performance. This is now being addressed by the introduction of the CHARM model, which is a computer based modeling programme. CHARM uses the HOCNS data set but it determines the potential risk associated with discharge of chemicals in the North Sea. This will result in a calculated hazard quotient being awarded for the product. CHARM operates on a single continuum and so products can be compared directly with one another for risk.

In the UK the scheme will become the Offshore Chemical Regulations (OCR). Once classified, a product remains on the HOCNS for 3 years when it becomes due for re-certification. The data set that is used for the HOCNS will be carried over to the new OCR scheme.

5.3 Norway Offshore

In Norway, the key focus is on the additive components in a product, assessed against biodegradability and bioaccumulation. Acceptance limits are set that determine what additives are acceptable (green), must be phased out (yellow) or prohibited (red), see Appendix, Table 3. This focus ensures that by minimising the use of harmful additives, the finished products should have less impact on the marine environment.

This has a significant impact on the method of product development and the speed with which a product can be introduced to market. Traditionally, product development has focused on meeting performance requirements, overall product environmental performance, and HSE requirements in conjunction with European legislation such as COSHH, CHIP, and DPD etc. Increasingly, product development must first consider the environmental performance of the raw materials and also proposed formulations against the clear specifications laid out by the HOCNS. Raw materials that deliver the performance characteristics required, maybe prohibited for use under certain local legislation. This raises a number of issues, such as

- The provision of different formulations for different countries
- The time to market for new product developments
- The product life cycle has shortened considerably

5.4 Material Legislation

In parallel with the offshore chemicals legislation there exists the general guidelines that control the use of chemicals, such regulations as COSHH, Chemicals (Hazard Information & Packaging for Supply) Regulations 1994 CHIP2, Dangerous Preparations Directive (DPD) are essential to control the use of hazardous substances and notify users of potential hazards. However, the increasing information available in this area and the continuous re-classification of substances means that raw materials are continuously being re-assessed for their potential hazards. Many chemical suppliers have their own internal guidelines that further restrict the use of substances. Overall, these factors result in a gradual reduction of raw materials that are available for use in product development.

5.5 Impact of the Legislation

Often the customer expectation is for an enhancement in product performance. This presents a difficult challenge to the chemist to meet these customer and application demands, whilst being mindful of the environmental and raw materials classification constraints. The impact of all this means product formulations need to be continually assessed and kept up to date. It is noticeable that in recent times, product development times have reduced significantly to attempt to match the speed with which products are expected to conform to new guidelines.

The product development approach considers the product as an integral part of the system. As such the raw materials used often perform more than one function. Therefore, should such raw materials fail the environmental criteria outlined earlier, the development programme becomes further complicated by identifying alternative materials that can perform multi functions.

6 DATA GENERATION

The required data set for completion of the HOCNS is as follows:

6.1 Marine Toxicity

On the finished product:

Organism
Skeletonema Costatum
Acartia Tonsa
Corophium Volutator
Scopthalmus maximus

6.2 Biodegradation

On all raw materials using method OECD 306 or BODIS (for water insoluble materials)

6.3 Bioaccumulation

On all raw materials using method OECD 117 or OECD 107

Drawbacks of this method are that surface-active raw materials cannot be measured using the OECD 107 / 117 methods because they do not elute down the HPLC column. In these instances a reasonable attempt must be made to estimate or calculate a value based on the chemical structure of the component. This can present further confidentiality problems in obtaining permission for the raw material supplier to release this information to a test house for them to carry out the assessment.

7   TESTING RESULTS

To illustrate some of the challenges the following additive functions have been selected for discussion, dyes, surface-active chemistry.

7.1 Dyes

Water-soluble dyes are of particular importance in the water-based products to distinguish the product from water, a vital Health and Safety need.

A number of dyes were tested according to the Biodegradation method OECD 306 (HOCNS guidelines)

4,5-dihydro-5-oxo-1-(4-sulfophenyl)-4-[(4-sulfophenyl)azo]-1H-Pyrazole-3-
carboxylic acid, trisodium salt, also known as Tartrazine, has FDA approval status (FD&C Yellow No.5) for use in foods. A biodegradation of < 1% in seawater at 28 days using OECD 306 was obtained. Current Norwegian legislation labels this chemical for phase out, also irrespective of treat rate. In addition, natural dyes such as Curcumin I, Curcumim III the active ingredient in Turmeric and Indigo all contain an aromatic ring and phenolic functionality. This is associated with poor biodegradation properties.

Overall, this presents an interesting challenge for selecting new dye additives. Firstly they must perform to the HOCNS standards, secondly as a colour for HSE purposes and thirdly, they must not compromise the other performance characteristics of the fluid.

7.2 Surface Active Chemistries

Hydraulic fluids need to provide anti-corrosion and anti-wear performance in a hydraulic system. Therefore, surface-active chemicals are essential to provide a coating on the metal surfaces to deliver performance in these areas. Under the HOCNS testing guidelines the bioaccumulation test OECD 107 or 117 should be carried out. However, surface-active chemicals cannot be tested using these methods because they are too surface active and do not elute from the HPLC column that is used in the test. It may be possible to estimate a value based on the chemistry of the material but this requires release of this information from the raw material supplier under confidentiality agreements. If a value cannot be measured or estimated then a default value of Log Pow >6 is assigned to the material. The only way to gain acceptance for use of such chemicals is a biodegradation result of > 60% at 28 days.

7.3 Yellow Metal Passivators

A small group of chemicals exist that provide protection to copper. Typically they are based on triazole chemistry. However, testing carried out according to the HOCNS guidelines indicate that all the chemicals in this group perform poorly and this type of chemistry must be phased out under the Norwegian guidelines. As yet there are no known equivalent chemistries that will provide similar performance. Currently, the only solution is to design out copper based metals from the hydraulic system.

8   SOLUTIONS

This paper has focussed on a number of external factors that are driving change in the development of subsea production control fluids. What solutions are available to our industry today?

8.1 Chemistry Based Solutions to Identify New Raw Materials for use in this Application

As well as identifying new raw materials for use, a fast product development process must be in place to reduce the time to market of new formulations to maximise the product lifetime. In developing new formulations there are a number of aspects to consider.

1. Do the new materials meet and perform to the required environmental standards?
2. Do raw materials deliver the functional performance required of them?
3. Do raw materials enhance and not compromise the overall performance of the finished product?
4. The marine toxicity of the finished product must be re-tested to ensure there is no significant change from the original formulation.

What does this mean for the reliability of the production control system? A new formulation must be thoroughly evaluated and qualified in equipment to demonstrate acceptable performance. In addition, if possible formulations should be backwards compatible with existing products and systems materials to allow retrofit to an improved environmental product if available.

8.2 Engineering Based Solutions to Identify Different Methods of Production Control

- To make all systems closed loop and minimise opportunity for fluid discharge.
- Material selection must become flexible to be compatible with control fluid not the other way around.
- Use electric actuators, or seawater, to operate the subsea production system. This removes the need for hydraulics.

9 CONCLUSION

It is likely that a pragmatic combination of both chemistry and engineering will be required to allow subsea production control fluids to be used in such applications in the near future.

10 APPENDIX

Table 1   *Water Glycol Based Control Fluid - key properties*

Property	Units	Typical Value
pH @ 20°C		8.8 to 9.4
Kinematic Viscosity @ 40°C	cSt	2.2 to 2.4
Density	g/ml	1.07

Table 2   *Synthetic Control Fluids - key properties*

Property	Units	Typical Values
Kinematic Viscosity @40°C	CSt	10
Density @ 20°C	g/ml	0.82

Table 3   *Norwegian Classification of Components for discharge into the marine environment*

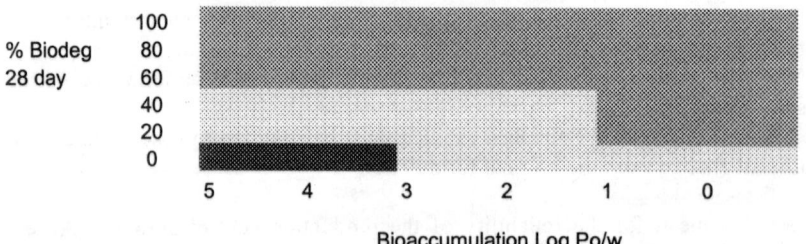

RESERVOIR DRILLING FLUIDS: AN OVERVIEW OF CURRENT TECHNOLOGY AND NEW POTENTIAL DEVELOPMENTS

D. A. Ballard and C. A. Sawdon

M-I Drilling Fluids UK Ltd; Research and Technology Centre, Ashleigh House, West Tullos, Aberdeen AB12 3AD

1 INTRODUCTION

The chemicals utilised in a reservoir drilling fluid (RDF) play a critical part in the success of drilling and completion operations and ultimately the productivity of the well. Modern well-completion techniques, such as a horizontal open hole in the reservoir, place even greater demands on the chemicals used whether it be in terms of improved rheological profile, fluid loss control, or chemical treatments to ameliorate formation damage.

The paper will review the chemical technology that has been applied to RDF systems and to the treatments that are used to minimise the impact of a RDF on the productivity of the well. Examples of RDF systems, such as calcium carbonate and sized salt based fluids will be discussed along with clean up treatments like breakers, acids, and solvents. The paper will then go on to discuss new high potential developments in this area, such as "smart" additives (filter cakes that do not need clean up treatments) and fluids that work synergistically with treatments and completion techniques.

2 EVOLUTION OF RDFS AND CLEAN-UP TREATMENTS

2.1 Reservoir Drilling Fluids

RDFs using sized calcium carbonate bridging or weighting agents have been used for many decades to minimise formation damage due to solids or filtrate invasion of the producing formation. These fluids are often based upon a brine to provide density and to inhibit the mobilisation of the clays often present in sandstone reservoirs. Viscosity and suspension stability are provided by a soluble polymer such as xanthan gum (more recently schleroglucan gum). Cross-linked and functionalised starches, for example hydroxypropyl starch, are used as a colloidal dispersion to reduce filtration.

The solid particles utilised in a RDF typically have a particle size distribution designed to promote rapid bridging of the pore or micro fracture openings of the rock formation, thereupon building a compact filter cake. The "removability" of the calcium carbonate solids by dissolution in hydrochloric acid clean-up treatments is an asset, but acidisation is frequently not necessary. This is particularly so for the cased and perforated completions in near vertical wells more commonplace prior to the mid-1980s. Here the minimal solids invasion depth into the permeable rock could easily be penetrated by perforation, especially when the larger tubing-conveyed perforating guns were introduced.

In fact, for this reason, specialised drill-in fluids were often deemed unnecessary, regular drilling fluids weighted with barite being used to drill the reservoir.

With the advent of horizontal-hole completions came a re-emphasis on non-damaging drilling fluid design. The use of steel casing or liners followed by perforation became less frequent. Open-hole completions are now often used where the strength of the rock allows. Where more support is needed slotted liners can provide hole stability while maintaining connection to the reservoir. Gravel packing the producing interval is used to avoid producing sand along with (say) the oil when a sandstone is poorly consolidated. Wire-wound cylindrical screens or expandable screens are also employed to avoid sand production.

In each of these cases solid residues such as mud filter cake could cause damage to the near wellbore and plugging of screens or gravel packs, greatly reducing well productivity. This prompted increased attention into finding ways to avoid or remove solids plugging, whether by mechanical, physical or chemical means. Although carefully sized solids have been developed which might flow back through the apertures of screens or gravel packs, a preferred route was to use bridging solids that are soluble in a clean-up fluid pumped into the well during the final stages of completion.

Alternative (to calcium carbonate) removable bridging and weighting agents have been introduced. One system uses water-soluble salt particles, suspended in a saturated solution of that salt, to provide temporary bridging and sealing[1]. The clean-up treatment is simply an under-saturated aqueous solution. Sodium chloride is effective where the higher density imposed by the brine is needed or tolerated. A more sparingly soluble salt, ulexite (sodium calcium borate), has been used to avoid the higher densities and considerable material requirements of the sodium chloride system[2].

Ideally the solid residues in the well should disappear spontaneously when the well is brought onto production, avoiding the need to pump a chemical clean-up treatment. An early example of such "smart" solids is the oil-soluble resin introduced in the 1960s[3] for use in aqueous drilling, under-reaming or completion fluids. The resin particles form a filter cake on the rock surface, minimising fluid losses into the formation. When oil production starts, the residual solids and filter-cake simply dissolve in the crude oil. The applications of these resins have been limited, however, by their temperature limits (softening point), and the fluids' sensitivity to contamination by oil, lubricants and so on.

There remains a need for improved "smart" solids that automatically disappear at the onset of production. Frequently clean-up treatment fluids such as acids do not effectively contact all the residues. If the residues in a small section of the producing interval are dissolved, the acid can rapidly leak-off into the formation instead of reaching the remaining plugging solids.

A disappearance of the solids triggered by a natural stimulus appearing with the produced fluids would be much more effective at maximising productivity. The "clean-up solution" is the produced fluid itself, and the "treatment time" is long. Expensive rig-time is saved by the avoidance of pumping a treatment fluid and its reaction time.

The development of a novel bridging material exhibiting automatic clean-up is described in Sections 3 and 4.

2.2 Clean-up Treatments

Acid treatments, typically with 15% hydrochloric acid, have traditionally been utilised to dissolve calcium carbonate residues, and to degrade polymers such as starch derivatives. Handling dangers and corrosion problems are obvious drawbacks, and specialised pumps are required. Even treatment of (especially) a long horizontal interval is difficult to engineer because the acid can leak-off into part of the formation, leaving part of the wellbore untouched.

Mildly acidic sequestering agents such as partially neutralised EDTA can be effective replacements for hydrochloric acid. Although the reaction with the calcium carbonate is slower, plugging by the precipitation of materials such as iron hydroxides is avoided. However they have little ability to degrade the polymeric additives such as starch derivatives, hydroxyethyl cellulose or xanthan gum.

Enzyme treatments are now available for "breaking" all of these polymers. Although there is no effect on the solids in a filter-cake, the polymers concentrated in the pores of the cake can be degraded, causing the cake to become permeable and dispersible. Their use is however restricted to down-hole temperatures of less than about 90 °C. Again the evenness of the treatment is difficult to control.

Oxidative polymer-breakers such as calcium hypochlorite, persulphate salts and peroxides have been used to degrade viscous or colloidal polymer residues. Although useful on occasion after drilling a reservoir, they are more commonly applied as a component of fracturing fluids. Problems associated with oxidisers include handling, corrosion, and evenness of treatment.

One development aimed at improving the completeness of treatment utilises insoluble particles of magnesium peroxide as an evenly dispersed component of the drilling fluid[4]. At the normal high pH condition of a drilling fluid the reaction of the peroxide with the drilling fluid polymers is very slow. After installing the completion hardware, clean-up is initiated by pumping an acidic solution to activate the peroxide. The high mobility of hydrogen ions allows their rapid penetration into the cake. The intimate admixture of the magnesium peroxide with the polymers in the cake promotes a more complete degradation of the polymers. A drawback is that some premature polymer breaking during drilling can occur if the downhole temperature exceeds about 120 °C.

Treatment fluids have been developed[5] to reverse the effects on completion efficiency of the various oil-based drilling fluids (OBF). OBFs normally consist of an invert (water-in-oil) emulsion of brine droplets in (usually) a low toxicity oleaginous liquid such as mineral oil, liquid polyolefin, an ester or an n-alkane mixture. The OBF must carry a significant excess concentration of emulsifiers and oil-wetting agents in order to ensure emulsion stability. The presence of these in the filtrate that invades the rock formation can change the wettability of the mineral surfaces in the pores of the rock from water-wet to oil-wet. This will cause a significant reduction in the relative permeability to oil, and a drop in productivity.

Often during the well completion process the OBF is displaced from the well by a brine-based completion fluid. Unless special procedures are adopted, the fluids will mix near the interface resulting in a high viscosity, very high ratio brine-in-oil emulsion sludge which can cause plugging.

Treatment fluids can be applied to an affected zone or used in a "spacer" fluid between the OBF and the brine. Generally they are cocktails of aggressively water-wetting surfactants, solvents and co-solvents such as ethylene glycol mono-butyl ether (EGMBE). The idea is to return the rock wettability to water-wet and to avoid emulsion blocks.

Their efficiency is not always optimal, especially in wettability reversal. Often a residue of oil-wetting agents remains persistently adsorbed on the rock surfaces. Where the oil-wet film is removed by the treatment, most water-wetting treatment surfactants will adsorb on to the mineral grain surfaces. With multi-layer surfactant adsorption the surfaces remain water-wet, but during production of oil and formation water the adsorbed surfactant can elute down to a monolayer with exposed hydrophobic tails. The formation can again become more oil-wet.

However, Patel and Growcock (SPE 54764 1999) have shown that the design of the OBF emulsifier/wetting-agent package can be altered to greatly improve the reversibility of the fluid. The switchable surfactants allow avoidance of the viscous brine/OBF interface problems and improvement of the water-wetness of the formation. This is described in the next section.

3 RECENT AND NEW DEVELOPMENTS

3.1 Reversible Invert Emulsion Drilling Fluid

A W/O emulsion drilling fluid has been developed (Patent Publication Number WO 98/05733) which remains stable while drilling by maintaining an alkaline environment. The fluid itself or the filter-cake from the fluid can be readily switched and dispersed to a water-wetting O/W emulsion by contact with a mild acid solution. The process is illustrated in figure 1. The acid treatment avoids the problems noted in the previous section of high-viscosity sludge formation. Perhaps more importantly, the wettability of the formation contacted by the filtrate can be switched to a water-wet condition without generating an emulsion block.

For oil wells the water-wet condition provides optimal oil mobility and production rate. In the case of a seawater-injection well, an acidic flush will allow the oil filtrate saturation in the near-wellbore area to be swept away as emulsified droplets, increasing the water-saturation and the relative permeability to water. Similarly for gas wells, the removal of the foreign third phase (oil filtrate) from the normal gas-water system will enhance the relative permeability to gas. Recently, the reversible invert emulsion fluid has been successfully used in 16 horizontal sections in Cabinda, West Africa for both oil producing wells and seawater injection wells.

The key to the reversible behaviour is the surfactant package. The proprietary surfactants chosen exhibit a marked increase in hydrophilic character upon protonation, promoting direct O/W emulsions and the removal of oil from mineral surfaces (see figure 1). During drilling the fluid is maintained in an alkaline condition by the addition of about 1.0 to 5.0 lbs of lime per barrel of fluid (2.86 to 14.3 Kg/cubic meter). This causes the surfactants to assume a more hydrophobic character as the water affinity of the hydrophilic group is suppressed, resulting in the stable W/O emulsion required for the excellent drilling performance associated with oil-based fluids.

Until recently, it was thought that a treatment with a clean-up solution such as citric or acetic acid (at about 10-50% concentration) was required to achieve the benefits from the reversible OBF. However recent laboratory results indicate that where a well's produced fluids contain significant amounts of carbon dioxide[6] there is sufficient acidity[7] (see figure 2) to flip residues of the reversible invert emulsion to a water-continuous condition, and to render the filter-cake solids water-wet. In our tests the filter cake became much more permeable and friable, and more likely to disperse in any water produced along with hydrocarbons. This is of particular importance in situations where the pumping of clean-up

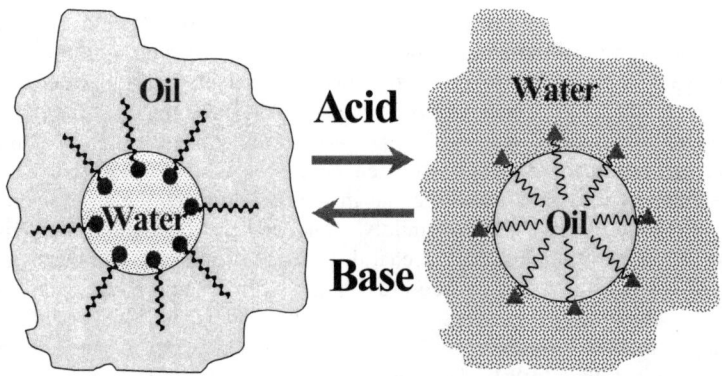

Figure 1 *Representation of the reversible invert fluid mechanism; invert emulsion under alkaline conditions, direct emulsion under acid*[15]

Effect of CO2 partial press. on pH of water

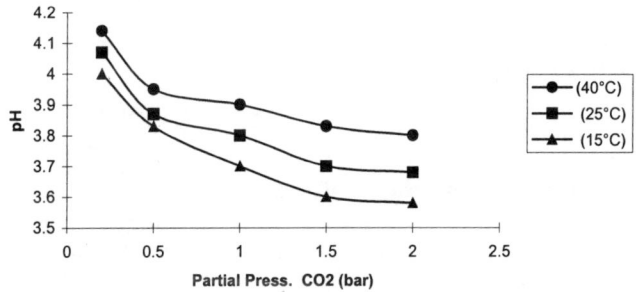

Figure 2 *Effect of carbon dioxide partial pressure on pH of water*[16]

solutions is impractical, and offers promise of improved automatic "smart" clean up over a whole producing interval.

As before, pumped clean-up solutions may not contact all the residues, and the treatment times are short. In contrast, for the reversible OBF, the flowing formation fluids or gaseous carbon dioxide are far more likely to contact the residues efficiently, and the "treatment time" is long.

One advantage of carbon dioxide is that it is very soluble in oil (in comparison to citric acid, which is not oil soluble). Hence the "acidity" as carbon dioxide can diffuse through the continuous oil phase of the invert emulsion to the brine droplets where the "switchable" surfactants and the lime/ alkalinity are located. It is thought likely that the carbon dioxide will ultimately diffuse to all fluid and filtercake residues left in the well, probably allowing a more complete reversal than a non-oil-soluble short term acid wash.

The laboratory tests illustrating the effects of carbon dioxide on a filter cake from a reversible invert emulsion fluid are detailed in Section 4.

3.2 Degradable Bridging Polysaccharide

As implied earlier, the ideal bridging solid to incorporate into a RDF would be a material that is stable during drilling and completion operations but which breaks down naturally when the well is brought into production. This would eliminate the need to pump a remedial treatment solution, and promote more complete clean up.

Based upon a low molecular weight polysaccharide, a novel water-insoluble bridging solid that exhibits this behaviour has been developed. The polysaccharide is rendered water-insoluble by cross-linking with bonds that are stable under the alkaline conditions of a RDF, but readily hydrolysed under mildly acidic well-flowing conditions. This contrasts with conventional cross-linked polysaccharides used in drilling fluids. For instance starch-based fluid loss control additives are cross-linked with epichlorohydrin or phosphorous oxychloride, which means that they are highly stable under both mildly acidic or alkaline conditions. Another difference is that these starch derivatives disperse in water to a colloidal "solution" with little bridging ability for pores or micro-fractures.

Elsewhere, colloidal starches for use in the paper, adhesives and textile industries have been developed which contain acetal cross-links. Acetal bonds are stable under neutral or alkaline conditions, but are readily broken at an acidic pH. This is represented in Figure 3.

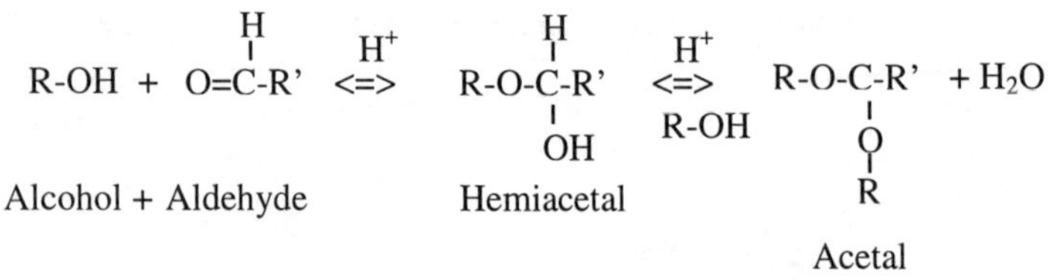

$$R\text{-OH} + \underset{\text{Aldehyde}}{\overset{\overset{\displaystyle H}{|}}{O=C\text{-R'}}} \overset{H^+}{<=>} \underset{\text{Hemiacetal}}{\overset{\overset{\displaystyle H}{|}}{\underset{\overset{|}{OH}}{R\text{-O-C-R'}}}} \overset{H^+}{\underset{R\text{-OH}}{<=>}} \underset{\overset{|}{\underset{R}{O}}}{R\text{-O-C-R'}} + H_2O$$

Alcohol + Aldehyde Hemiacetal Acetal

Figure 3   *Formation and breakdown of acetal bonds*

By analogy, a bridging solid formed from a low molecular weight polysaccharide highly cross-linked with labile bonds can be hydrolysed to soluble products under low pH well-flowing conditions. A commercially feasible material of this type has recently been developed[8], here referred to as a "bridging polysaccharide". Compared to rigid bridging solids like calcium carbonate, it displays an enhanced sealing action because the particles swell in the water-based fluid to become more pliable, promoting close packing in the filter cake. Another advantage of the bridging polysaccharide is that its degradation is catalysed by acid, compared to the conventional stoichiometric reaction of hydrochloric acid with calcium carbonate. The acid is not neutralised by the polysaccharide so the carbonic acid is still available to degrade more material. Coarser-ground grades of the material hold promise as a lost-circulation material to temporarily seal fractures in a reservoir into which large volumes of drilling fluid were being lost.

The bridging polysaccharide has a low density, enabling a wider range of applications to be covered. It will not contribute to the fluid's density compared to calcium carbonate (specific gravity 2.7) which, when added at the desired concentrations for bridging, imparts a fluid density increase that is sometimes unacceptable. Another benefit from the low density is the improved removal of undesirable, non-acidisable, contaminating solids. Contaminating solids from the formation have roughly the same density as calcium carbonate, which makes solids separation difficult in a conventional fluid. However, the bridging polysaccharide has a density of approximately 1.0, enabling the use of high efficiency centrifugation to remove the drilled solids and not the bridging polysaccharide.

The filter cake therefore contains less non-degradable material resulting in improved clean up. The results of testing a bridging polysaccharide produced on a pilot plant by a commercial manufacturer are discussed in section 4.

4 METHODS AND RESULTS

4.1 Assessment of Relative Filter Cake Integrity after Exposure to CO_2

Typically, leak off tests are carried out to determine the effectiveness of remedial clean up treatments. The test usually involves exposing the filter cakes to various types of clean up treatments for a set period at high temperatures and pressures and then measuring the flow of the treatment fluid through the filter cake. This gives a comparative assessment of how each treatment has degraded or weakened the filter cake. This is pertinent as the yield strength of the filter cake is an important factor in filter cake removal during well-flow initiation[9]. Moreover, it has been suggested that the more robust filter cakes containing bridging solids bound together by polymers give higher flow initiation pressures[10] than filter cakes where the polymer has been degraded and only contain bridging material[11].

However, the objective of this work was not to determine the effectiveness of clean up treatments, but to identify the types of filter cakes that are closest to being self-cleaning under the natural reservoir conditions (or to discover which types of filter cakes are weakened the most by these conditions). The performance of the Reversible OBF was compared to that of conventional oil-based and water-based fluids. The formulations are shown in Tables 1&2. Each test used a modified version of the leak-off test simulating the mildly acidic conditions that are found in the reservoir due to the presence of gases such as carbon dioxide and hydrogen sulphide.

Table 1 *Fluid formulations for typical reservoir drill in fluids*

Conventional Oil Base RDF (10.6ppg, 60/40 OWR, 0.75 Aw)		Conventional Water Base RDF (10.6ppg)	
Product	g per lab barrel (350 ml)	Product	g per lab barrel (350 ml)
Base Oil	127.5	NaBr/NaCl brine*	382.0
Primary Emulsifier	8.0	Potassium Chloride	15.0
Oil Wetting Agent	2.0	Glycol	11.5
Clay Based Viscosifier	2.0	Biopolymer	1.3
Polymeric Fluid Loss Additive	1.0	Starch Fluid Loss Additive	5.0
Lime	6.0	Sized Calcium Carbonate	5.0
CaCl2 (83.5%)	45.4	Sized Calcium Carbonate	45.0
Water	106.1	Simulated Clay contamination	10.0
Dolomite	96.7	Biocide	0.2
Sized Calcium Carbonate	5.0	*Brine formulation for 350 ml	14.5g NaBr
Sized Calcium Carbonate	45.0		102.5g NaCl
Simulated Clay contamination	10.0		306.4g Water

Table 2   *Fluid formulations for reversible oil base drill in fluids*

Reversible Oil Base RDF (10.6ppg, 70/30 OWR, 0.75 Aw)	
Product	g per lab barrel (350 ml)
Base Oil	141.6
Reversible Emulsifier	12.0
Oligomeric Fluid Loss Additive	1.0
Reversible Oil Wetting Agent	4.0
Clay Based Viscosifier	5.5
Polymeric Fluid Loss Additive	1.0
Lime	6.0
Calcium Chloride (83%)	33.5
Water	77.8
Dolomite	112.3
Sized Calcium Carbonate	45.0
Sized Calcium Carbonate	5.0
Simulated Clay contamination	10.0

The test involves growing filter cakes from heat aged fluids on 10-micron pore diameter ceramic discs, at 80°C and 500-psi nitrogen differential pressure, for 2 hours. After the fluid loss test, the fluid is removed from the cell and it is refilled with 3% potassium chloride solution to represent the displacement to completion brine. The cell is then pressurised to 300 psi with carbon dioxide and statically aged for 16 hours at 80°C to reproduce the mildly acidic environment associated with well production. After the second period of ageing, the brine leak off rate through the filter cake is measured at 100psi nitrogen differential pressure at 80°C, to see how the filter cake has changed compared to control formulations.

Although this test is less complex than other techniques available today for determining the effects of the filter cake on production it does offer the advantage of reliable comparative results[12]. These include spurt and fluid loss measurements that are used to determine the bridging characteristics of the fluid. These are critical parameters that need to be monitored and controlled to minimise formation damage[13,14] when drilling in a high permeability reservoir.

4.1.1 Comparative Results: Reversible OBF vs. Conventional RDFs. The results in figure 4 show the expected trends in terms of bridging and fluid loss characteristics with the water base fluid giving the highest spurt and fluid loss values and oil base the lowest values. The reversible fluid had properties that were very similar to the conventional oil base fluid.

The results in figure 5 demonstrate what happens to the filter cake after 16 hours exposure to carbon dioxide and brine. It can clearly be seen that the conventional oil base and water base fluids give very low leak off values suggesting that the conditions have had little effect on the cake. This can be contrasted with the leak off value for the reversible fluid, which was very high, (>50 ml), within 17 minutes. This plainly shows that the cake

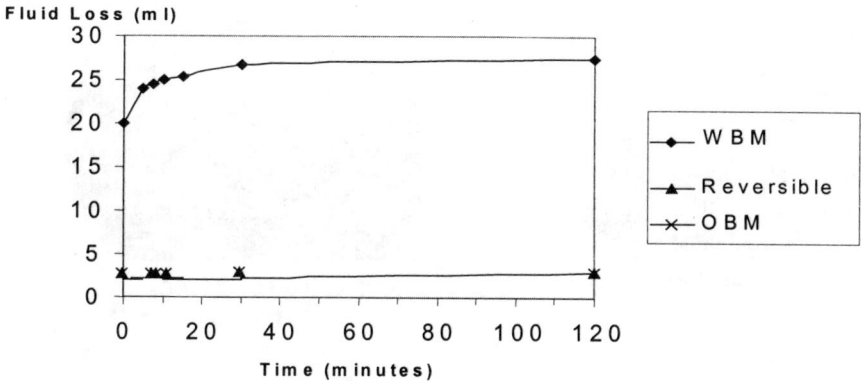

Figure 4   *HTHP fluid loss profiles of a water base, oil base, and reversible oil base reservoir drill in fluid*

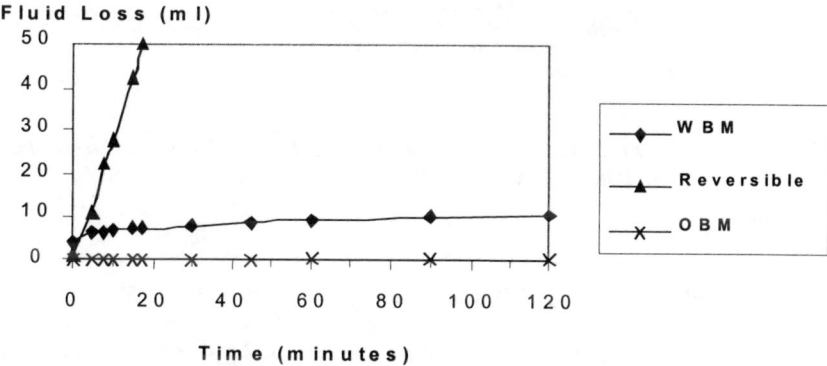

Figure 5   *Brine leak off profiles of a water base, oil base, and reversible oil base reservoir drill in fluid.*

from the water base fluid is thin and integral and that the cake from the conventional oil base fluid has become a thick, emulsified agglomeration of material. On the other hand, the cake from the reversible fluid that gave the highest leak off value was thin and broken. These results strongly suggest that cakes produced by the reversible fluid will be weakened under downhole conditions, minimising production restrictions. The use of this type of fluid may frequently allow high completion efficiency without pumping any acid treatment. Alternatively, it could improve production from areas in the well that have not been effectively treated.

WBM Cake (Post 3% KCl/CO$_2$)

OBM Cake (Post 3% KCl/CO$_2$)

Reversible OBM Cake (Post 3% KCl/CO$_2$)

Figure 6   *Images of Filter Cakes After Exposure and Leak Off Test to Carbon Dioxide Containing Potassium Chloride Brine*

4.2 Assessment of the Bridging Polysaccharide

As for the previous tests, a version of the leak off test was used to test the degradability of the Bridging Polysaccharide in the filter cake in the presence of carbon dioxide and brine. A conventional RDF was tested for comparison. The fluid formulations are shown in Table 3. Again, the test involved growing filter cakes from heat aged fluids on 10-micron pore diameter ceramic discs, at 120°C and 500psi nitrogen differential pressure, for 2 hours. After the fluid loss test, the cell was emptied and the fluid replaced with 3% potassium chloride solution to represent the displacement to completion brine. The cell was then pressurised to 300 psi with carbon dioxide and statically aged for 16 hours at 120°C to simulate the mildly acidic environment associated with well production. After ageing, the brine leak off rate through the filter cake at 100psi nitrogen differential pressure at 120°C was measured to see how the filter cake had been affected compared to that from the conventional formulation.

4.2.1 Comparative Results: Bridging Polysaccharide Fluid vs. Conventional Water Based RDF.   The results in figure 7 show that the bridging polysaccharide fluid gives a higher fluid loss than the conventional fluid. It produced a spurt loss of 4ml and a 120-minute fluid loss of 31ml as opposed to 2.5ml and 19ml for the conventional fluid. This

Table 3   *Fluid Formulations for Typical Reservoir Drill In and Bridging polysaccharide Base RDF*

Conventional Water Base RDF (9.3 ppg)		Bridging polysaccharide Base RDF (9.0 ppg)	
Product	g per lab barrel (350 ml)	Product	g per lab barrel (350 ml)
Fresh Water	321.2	Fresh Water	324.1
Potassium Chloride	27.0	Potassium Chloride	27.0
Xanthan Gum Viscosifier	1.0	Xanthan Gum Viscosifier	1.0
Starch Fluid Loss Additive	5.0	Starch Fluid Loss Additive	2.5
Sized Calcium Carbonate	34.0	Bridging polysaccharide	7.5
Defoamer	0.2	Sized Calcium Carbonate	13.5
Biocide	0.2	Defoamer	0.2
MgO pH buffer	0.4	Biocide	0.2
		MgO pH buffer	0.4

should have produced a thicker filter cake. However, the images in figure 8 show that after aging with carbon dioxide and brine the bridging polysaccharide cake has been reduced to a thin layer, whereas the conventional cake is still intact with the calcium carbonate particles and polymers forming a resilient elastic material. This is reflected in the brine loss results in figure 9, these show that the bridging polysaccharide filter cake gives a high leak off value (>50ml in 3minutes) signifying that it has been degraded compared to the conventional water base RDF, which is only 15ml after 2 hours.

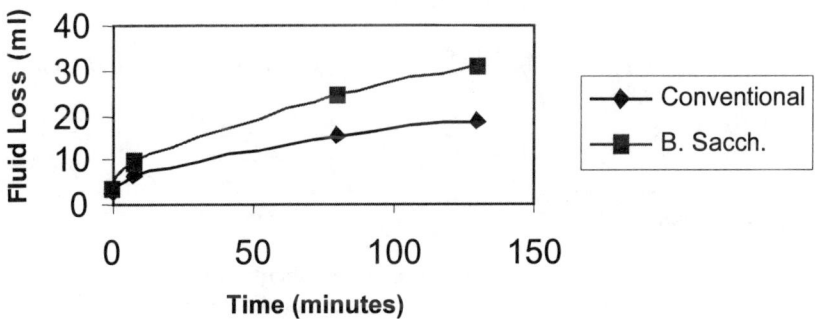

Figure 7   *HTHP Fluid Loss Profiles of a Water Base and Bridging polysaccharide Base Reservoir Drill in Fluid.*

Figure 8   *Images of Conventional (Left) and Bridging Polysaccharide Base (Right) Filter Cakes After Exposure and Leak Off Test to Carbon Dioxide Containing Potassium Chloride Brine*

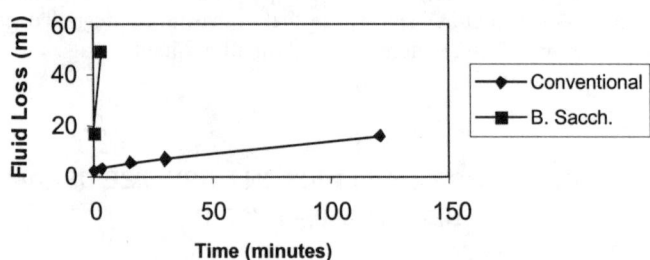

Figure 9   *Brine Loss Profiles of a Water Base and Bridging polysaccharide Base Reservoir Drill in Fluid.*

5 CONCLUSIONS

The geometry of modern well completions, and the hardware utilised downhole, has placed new demands on the performance of reservoir drilling fluids. In particular, the ability to avoid plugging by solid residues such as filter cakes has now become even more important.

This paper has established the concept of RDF additive design to cause the filter cake to breakdown spontaneously under mildly acidic well-producing conditions. With such additives the process of pumping well clean up treatments can be avoided.

Specifically, it has been shown that:

a) The filter cake from a reversible oil base fluid can be degraded and rendered permeable by the action of carbon dioxide and brine.

b) The use of a polysaccharide bridging solid is now feasible that degrades to water-soluble products in the presence of carbon dioxide.

Acknowledgements

Trevor Jappy, Alison Fraser, Mike Hodder and the Management of MI for allowing us to present this information

References

1   1 http://www.tbc-brinadd.com/products/drillin/thixsal.html.
2   Thomas C, US Patent *4620596, Well drilling, workover and completion fluids, Mondshine*, Texas United Chemical Corp.
3   http://www.tbc-brinadd.com/products/sysadd/solubrig.html.
4   http://www.tbc-brinadd.com/products/sysadd/ultrbr.html.
5   http://www.deep-south-chemical.com/wellclean.htm
6   Frick, Petroleum Production Handbook, *Calculated Reservoir Fluid Compositions 34-19*, Hydrocarbon Analysis of Gas Condensates **36-3**.
7   F.M. Sweeney and S.D. Cooper, SPE 25159, *The Development of a Novel Scale Inhibitor for Severe Water Chemistries*. SPE International Symposium on Oilfield Chemistry held in New Orleans, Louisiana, U.S.A., March 2-5, 1993.
8   F.M. Sweeney, S.D. Cooper, D.A. Ballard, C.A. Sawdon, International Patent Application WO 00/06664, *Wellbore Fluid and Acetal Crosslinked Solid*.
9   L. Bailey, G. Meeton and P. Way, SPE 39429, *Filtercake Integrity and Reservoir Damage, Symposium on Formation Damage Control* held in Lafayette, Louisiana, February 18-19, 1998.
10 M.E. Brady, A.J. Bradbury, G. Sehgal and F. Brand, S.A. Ali, C.L. Bennett, J.M. Gilchrist, J. Troncoso, C. Price-Smith, W.E. Foxenberg and M. Parlar, SPE 63232, *Filtercake Cleanup in Open-Hole Gravel-Packed Completions: A Necessity or A Myth?* SPE Annual Technical Conference and Exhibition held in Dallas, Texas, 1-4 October 2000.
11 M.E. Brady et al, SPE 63232. *Filtercake Cleanup in Open-Hole Gravel-Packed Completions: A Necessity or A Myth?*
12 D. Marshal, R.Gray, M. Byrne, SPE 54763, *Return Permeability: A detailed Study*, European Formation Damage Conference, The Hague, 31-May –1 June 1999 .
13 E. Pitoni, G. Ripa, R. Lorefice, and D. Formisani, SPE 38157, *High Productivity Open Hole Gravel Packs in Depleted Permeable Sands*, European Formation Damage Conference, The Hague, The Netherlands, 2-3 June 1997.
14 Jiao, Di and Sharma, SPE 23823, *Formation Damage Due to Static and Dynamic Filtration of Water-Based Muds*, Symposium on Formation Damage Control held in Lafayette, Louisiana, February 26-27, 1992.
15 A. D. Patel, F. B. Growcock, SPE 54764, Reversible Invert Emulsion Drilling Fluids: Controlling Wettability and Minimizing Formation Damage,1999 SPE European Formation Damage Conference held in The Hague, The Netherlands, 31 May-1 June 1999.
16 Corrosion of Oil and Gas Well Equipment. Nace and API (1958)

A CHEMICAL PACKER FOR ANNULAR ISOLATION IN HORIZONTAL WELLS

B. Lungwitz[1], Keng Seng Chan[1], Radovan Rolovic[2], Feng Wang[3] and David Ward[1]

[1]Schlumberger Oilfield Chemical Products, Sugar-Land Product Center, 110 Schlumberger Dr., MD#3, Sugar-Land, TX 77478, USA
[2] Schlumberger Conveyance and Delivery, 555 Industrial Blvd, Sugar Land, TX 77478, USA
[3] Schlumberger RD Marketing in North & East of P.R.China, P.O. Box 566, No.2 Zha Bei Road, Donggu, Tanggu, Tianjin 300452, People's Republic of China

1   INTRODUCTION

Water and gas control in horizontal wells that are completed with slotted liners or gravel packed screens is difficult due to the lack of annular isolation and uncertainties with conventional fluid placement. The development of a new technique that addresses these problems is presented in this paper. Most of the horizontal wells were completed with slotted liners or gravel-packed screens. Traditional techniques like External Casing Packers (ECP) can provide a mechanical zonal isolation but they are proven to be unreliable and cost intensive.

A solids-free gel packer has been developed not only for remedial treatment of leaking ECP, but also for local annular isolation enabling effective reservoir stimulation or water and gas shut-off treatment.

The chemical and mechanical properties of a gel packer are discussed. We also present results of mixing, pumping through a coiled tubing, and gel placement in large-scale horizontal wellbore models. In addition, the use of the annulus gel-packer as an integrated part of a dynamic water management service for improving oil production in oil fields with high water cuts is discussed.

2   BACKGROUND

Production from horizontal wells has increased significantly in recent years. Typical horizontal drains may be over 3000 ft in length and be completed with slotted liners[1]. Such fields may have overlying gas caps and underlying aquifers both of which influx into the oil zone. The consequence is a restriction in oil production due to the high volumes of produced gas or water. Commonly gas is produced both through coning at the heels of horizontal wells and through fissures/faults which are intersected by the wells. The presence of fissures may be identified from mud losses during drilling and/or production logging data[2]. These applications require a selective placement of the packer behind the liner.

This has lead to the development of the *Annular Gel Packer* (AGP) concept that allows zonal isolation with coiled tubing (CT) deployed packers or bridge plugs.

The AGP technology involves placement of a novel gel based fluid into the annular space between an uncemented slotted liner and the formation. The fluid is conveyed to and injected at the treatment zone using coiled tubing and a straddle packer assembly. Stimulation or remedial treatments in horizontal wells are most effective when the treatment zone is isolated from the remainder of the wellbore. In cased holes, and to a lesser extent in openholes, this is achieved by mechanical means, for example, by using inflatable packers. When a screen or liner has been run but not cemented, such mechanical devices are ineffective in isolating the open annular space left behind the pipe.

The goal of the AGP concept is to achieve full circumferential coverage ranging from relatively small lengths - tens of feet – to larger intervals in magnitude of several hundreds feet while leaving the liner free of material that would obstruct flow or tool passage through the section. For placement, the fluid has a sufficiently low viscosity to be pumped through CT, through a straddle assembly, and out through the small slots in the pipe (Figure 1).

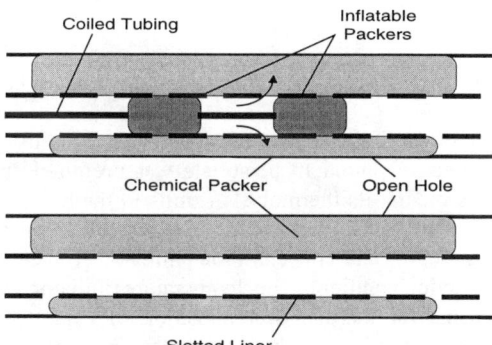

Figure 1   *Schematic view of the AGP placement in horizontal wells; the slotted liner is not centralized*

3   CHEMISTRY OF THE GEL PACKER

We designed a fluid system with properties suitable for use as a gel packer. It has two main components: 1) A thixotropic carrier fluid acting as a placement matrix; 2) A precursor system, which ultimately yields to the permanent gel.

For the gel packer Hydroxyethylcellulose is used as a preferred carrier fluid. A viscous fluid system without solids is more desirable for a carrier fluid than a system containing particles. In high-permeability reservoirs, a highly crosslinked fluid is needed to achieve good fluid-loss control. Hydroxyethylcellulose (HEC) is a linear non-ionic polysaccharide polymer based on a cellulose backbone with a molecular weight over >250,000 and is known for its low residue content[3]. In HEC, the OH groups of the cellulose unit are on adjacent carbons on the opposite sides of the ring (trans orientation). HEC can be crosslinked at pH of 10 to 12 with Zr(IV) compounds[4] or with lanthanide complexes[5].

The precursor system is based on monomer, crosslinker and a catalyst for the initiation of the radical polymerization. The monomer used for chemical packer is diallyldimethylammonium chloride (DADMAC) (Figure2). The polymerisation of nonconjugated 1,6 diene monomers might be expected to afford polymer chains with pendant unsaturation and ultimately, on further reaction of these groups, crosslinked insoluble polymer networks. However it is reported the polymerisation of (1) yields to linear polymers[6]. The presence of cyclic structures within the polymer chain was confirmed by degradation experiments[7]. Spectroscopic studies have shown that the 5-membered ring formation is the preferred pathway during cyclopolymerization of simple diallyl compounds[8,9,10,11]. Methylenebisacrylamide (MBA) is added in low amounts (0.3 – 0.5 wt/%) as crosslinking agent. As the concentration of the cross linking decreases, the gel will absorb water from this wet environment, and it will expand becoming weaker.

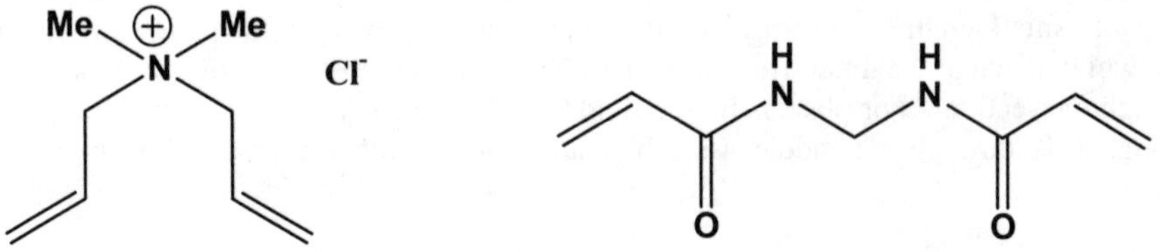

Figure 2   *Monomer DADMAC (1); Crosslinker, Methylene bis acrylamide MBA (2)*

The decomposition of the initiator into the active species is triggered by thermolysis under downhole conditions. Ammonium persulfate was preferred due to its solubility in aqueous or part aqueous media. Its thermolysis results in the homolysis of the O-O bond and formation of two sulfate radical anions[12,13]. The rate of decomposition is a complex function of pH, ionic strength, and concentration. Initiator efficiencies for persulfate are low and depend upon reaction conditions (i.e. temperature, initiator concentration)[14].

A number of mechanisms for thermal decomposition of persulfate in neutral aqueous solution have been proposed[11]. They include unimolecular decompostion (Figure 3) and various bimolecular pathways for the disappearance of persulfate involving water and concomitant formation of hydroxy radicals (Figure 4).

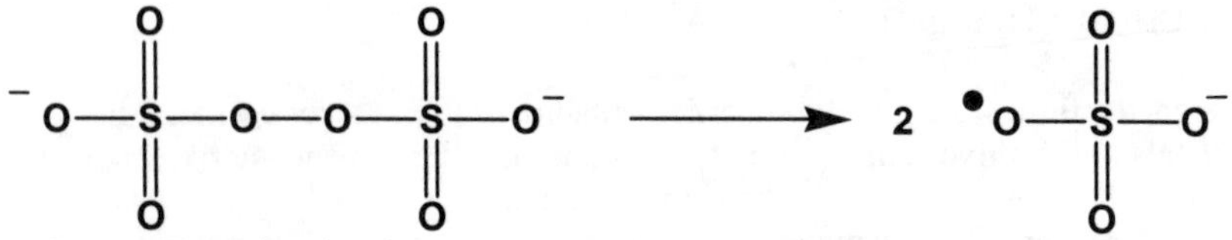

Figure 3   *Unimolecular decomposition of the persulfate anion*

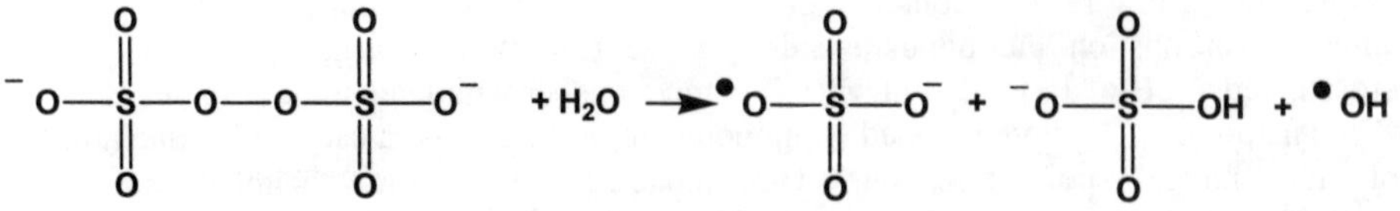

Figure 4   *Bimolecular decomposition of the persulfate anion*

The formation of polymers with negligible hydroxy end groups is evidence that the unimolecular process dominates in neutral solution. Heterolytic pathways for persulfate decomposition can be important in acidic media.

The schematic structure of the resulting polymer is shown in Figure 5. The polyacrylamide polymer has a cationic backbone based on five membered rings. The backbones are connected by incorporation of the MBA crosslinker into the framework.

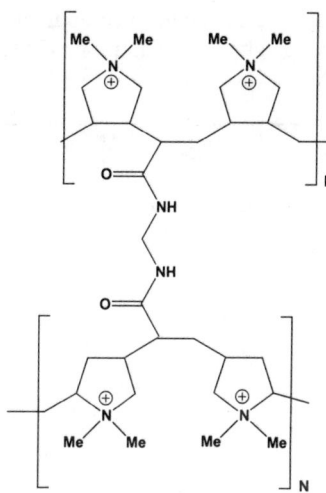

Figure 5   *Schematic structure of the crosslinked acryl amide polymer*

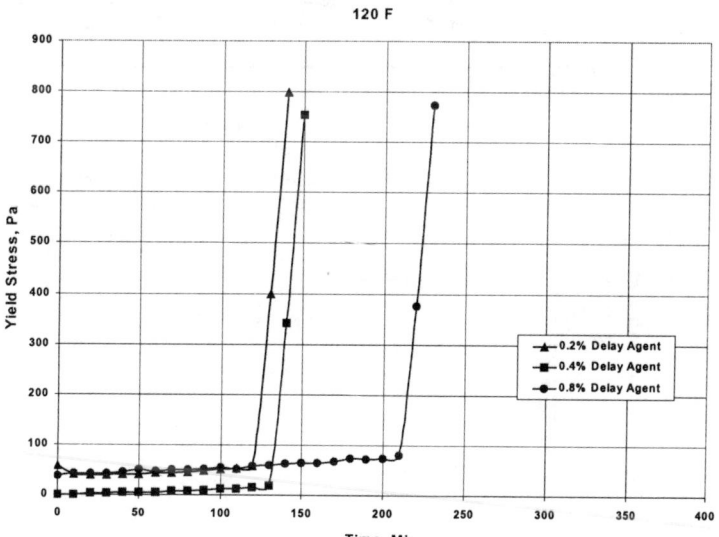

Figure 6   *Gelling time of the gel with various concentrations of delay agent at 120 °F. The gellation process was monitored with a Vane viscometer. Ammonium persulfate was used as radical initiator*

The gellation time is independent of monomer concentration but is directly dependent on the temperature and concentration of the catalyst and inhibitor. Potassium ferricyanide ($K_3Fe(CN)_6$) is used as delay agent. The influence of the delaying agent concentration on the curing time is illustrated in Figure 6. As expected the curing time of the gel can be adjusted from 120 to 240 min. After initiation the polymerisation reaction proceeds within minutes.

The effect of the delay agent concentration depends on the temperature (Figure 7). At lower temperatures the working time can be delayed up to 500 min, whereas at 186 °F, the gellation time is about 90 min. Higher concentration of the delay agent will affect the integrity of the polymer.

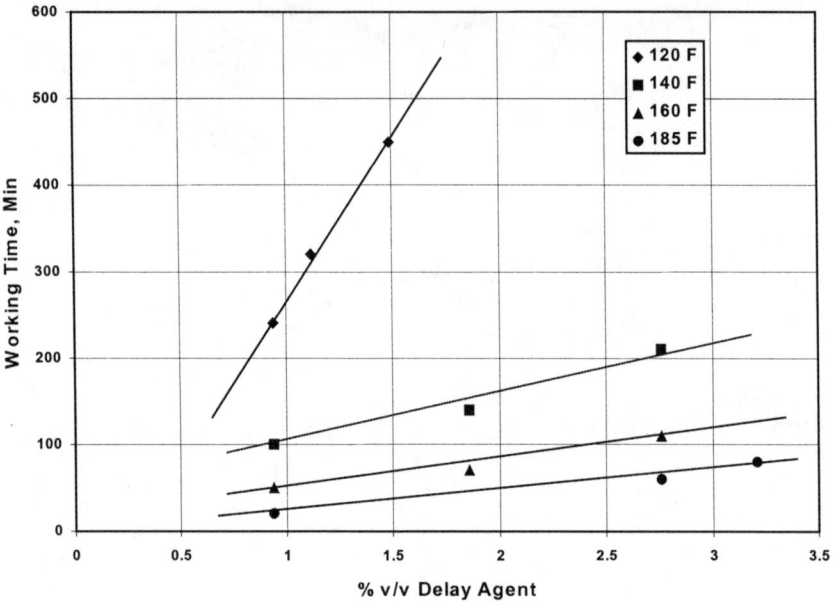

Figure 7   *Curing time of the gel at various temperatures with various concentrations of delay agent*

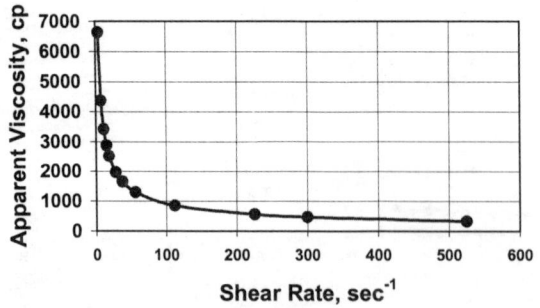

Figure 8   *Rheology of the freshly prepared gel packer at 120°F*

The rheological characteristics of the freshly prepared gel packer are outlined in Figure 8. A high viscosity at lower shear rates (< 10 s^{-1}) contributes to the thixotropic characteristic of the gel packer.

A full-scale test was performed with the AGP (Figure 9). The gel was pumped throughout a slotted liner with 3/8" perforations and squeezed into the annular space. Based on its thixotropic nature the gel filled first the inner space of the slotted liner (Figure 9a) before it was squeezed into the annulus (Figure 9b).

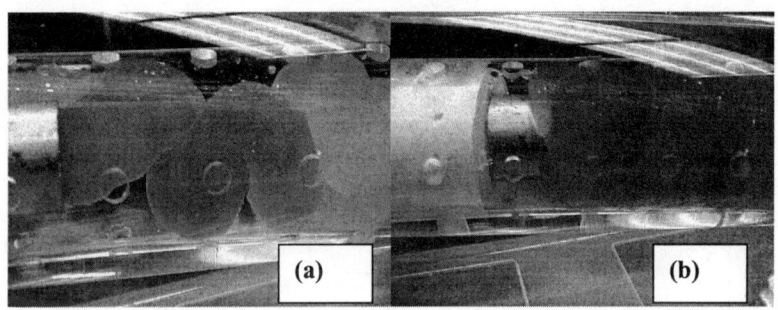

Figure 9   *The gel packer is pumped into a slotted liner of a full-scale model. Its thixotropicity is seen in picture a. After filling the inside of the slotted liner, the gel is squeezed through the slots to fill the annular space*

4   PHYSICAL PROPERTIES

Acrylamides are considered permanent[15]. The gel is unaffected by exposure to chemicals, except for very strong acids and bases, which are not naturally found in subterrean formations. Gels are, however, subject to mechanical deterioration when exposed to alternating drying and/or freezing cycles. If submersed in water or under saturated soils,

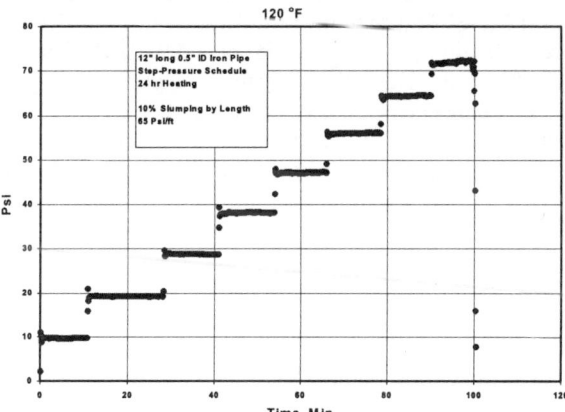

Figure 10   *Displacement of the gel through a 12" iron pipe with 0.5" ID*

the water in the gel remains in place, and the gel undergoes little or no volume change. However, gel exposed to a dry environment will lose water and shrink about 5% of the original gel volume. The extrusion pressure was measured by displacing the gel from 12" iron pipe with 0.5" ID (Figure 10).

The extrusion pressure is strongly dependent on the smoothness of the pipe wall and ranges from 35 psi/ft (PVC) to over 65 psi/ft (iron pipe). If solids, like sand are mixed into the gel the extrusion pressure can exceed over 500 psi/ft. The extrusion gradient for various pipe diameters (PVC) is shown in Figure 11.

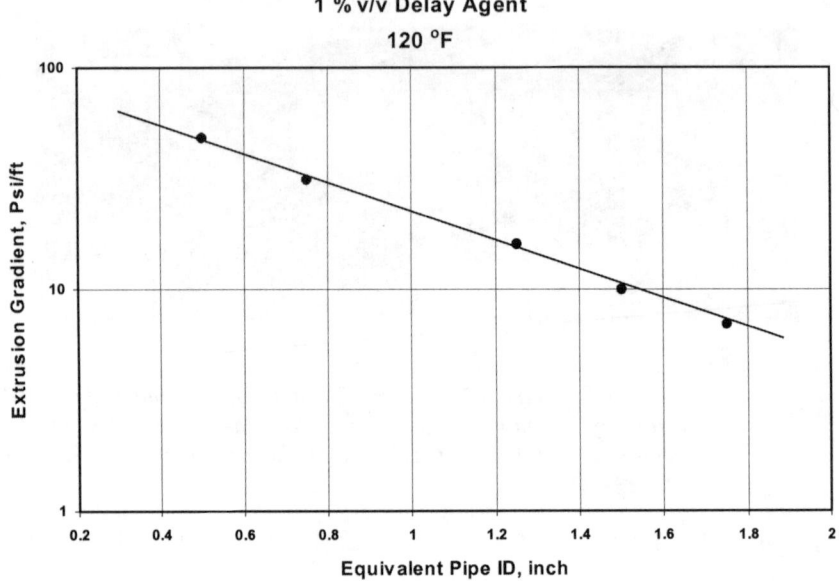

Figure 11   *Gelling time of the gel with various temperatures with various
concentrations of delay agent*

The friction pressure drop of the annular AGP was measured with full scale CT equipment at the "Lost Mule"-Test Side in Rosharon/TX. For a 1.5" CT string with an ID of 1.15", we observed a friction pressure drop of about 310psi/1000 ft with a pump rate of 0.25bbl/min. The AGP is mixable with standard field equipment, pumpable in coiled tubing and moveable in tubing even after a 1 hour static period.

The flow initiation pressure at various break times is shown in Figure 12. It indicates the additional pressure necessary to start the gel flow after various stop times.

5   APPLICATIONS

The major applications for AGP technology are isolation of horizontal wells for water and gas shutoff, remedial treatment of damaged screens or selective stimulation. The use of AGP is also attractive when a major recompletion is not economically feasible. Benefits are: AGP is a technology that can be used to effectively isolate horizontal well sections

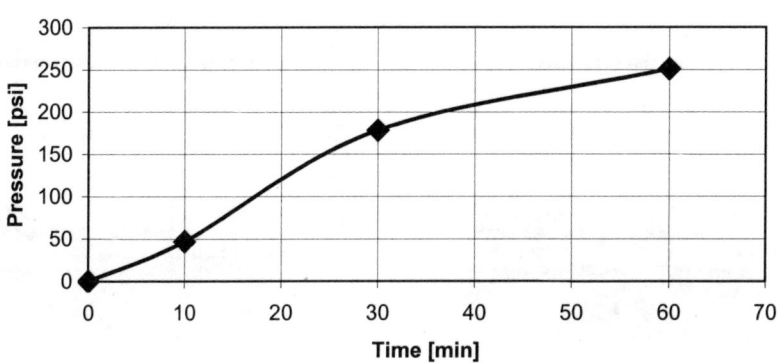

Figure 12   *Flow initiation pressure at various break times*

completed with gravel packed screens or slotted liners. Additional features are: lower cost alternative to major well recompletion, reduced pumping time and increased ability to isolate deeper well zones due to lower friction pressure.

Some applications may require straddle packers for selective placement behind a slotted liner to fill the annulus over a selected interval. It is designed to set in this position forming a permanent, impermeable high-strength plug, fully isolating that volume of the annulus.

5.1 Water Shut Off in Horizontal Wells

Figure 13 demonstrates the typical steps for a water shut off in a horizontal well. A typical procedure for AGP would include: 1) Identify water inflow; 2) Run in hole with straddle packers on CT and inflate packers; 3) Injection of the gel packer into the annular space; 4) Removing of the gel inside the liner; 4) Unset packers and circulate remaining AGP fluid from inside slotted liner.

5.2 Gas Shut Off in Horizontal Wells

Gas shut-off applications are more challenging due to the high mobility of gas, which can easily bypass isolated annular sections. The basic technologies for gas shut off described here are the Annular Chemical Packer (ACP) together with deep penetrating flowing gels or the utilization of naturally occuring barriers. Several options are presented to provide the flexibility for a full field strategy. However, appropriate candidate identification is essential in effective gas shut-off. This will require close collaboration between the service company and operator. All strategies are based around these two modes of application:

A placement of a permanent annular seal to directly shut off gas using a cement or gel in the annular gap close to a shale barrier (Figure 14). The seal will be completed with a bridge plug effectively extending the impermeable shale barrier across the wellbore. The section of well beyond the shale barrier will be lost with this application.

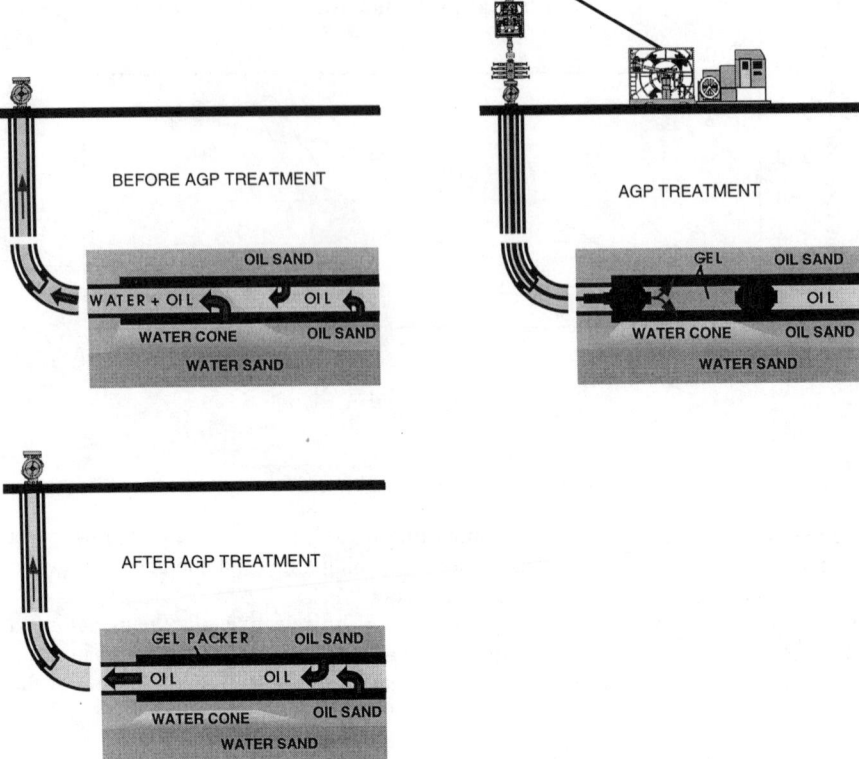

Figure 13   *Illustrated here is an AGP treatment for water shutoff. Before treatment, there is major water production from the heel of the horizontal section due to water coning. Then AGP is placed inside the completion and in the annulus behind the completion along the water-producing section. Once the AGP has solidified, it is removed from inside the completion; and oil is produced from the toe of the well.*

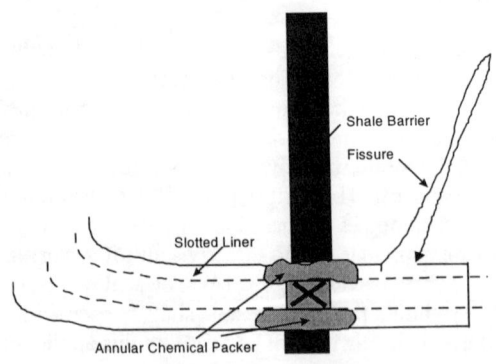

Figure 14   *Utilization of naturally occurring shale barrier*

The second alternative is to place an appropriate sealing fluid directly into fissures/fractures. Based on the nature of the fissures the sealing fluid can be the gel packer or a deep penetrating flowing gel (Figure 15).

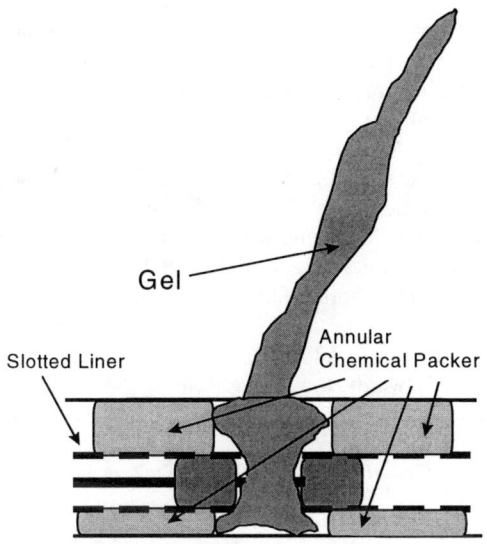

Figure 15   *Use of the Anular Gel Packer with deep penetrating gels*

6   CONCLUSIONS

The development of annular gel packer provides a new technique for isolating the annular space of horizontal wells that are completed with gravel-packed or slotted liners. This technique, which can also be used for some more traditional zonal isolations, can be a key component of remedial treatments for some horizontal wells. A solid-free polymer system is pumped through coiled tubing and placed in the annular space. Based on its unique rheological characteristics, the polymer system solidifies after a few hours and creates a perfect elastomeric type seal - an annular gel packer, which seals the selected annular space permanently.

References

1   J. Elphick, P. Fletcher and M. Crabtree: "Techniques for Zonal Isolation in Horizontal Wells", presented at the Production Engineering Association, Reading, UK, Nov. 4-5, 1998.
2   J. Elphick and Randy Seright: "A Classification of Water Problem Types" presented at the PNEC 3rd International Conference on Reservoir Conformance, Profile Control, Water and Gas Shut Off, Mexico City, Mexico, Oct. 25-27. 1999.

3   R.D. Carico and F.R. Bagshaw: "Description and Use of Polymers Used in Drilling, Workovers and Completions," *SPE 7747* presented at the SPE Production Technology Symposium, Hobbs, New Mexico, USA, Oct. 30-31, 1978.

4   D.R. Underdown, K. Das and H. Nguyen: "Gravel Packing Highly Deviated Wells with a Crosslinked Polymer System", *SPE12481*, presented at the SPE Formation Damage Symposium, Bakersfield, California, USA, Feb. 13-14, 1984.

5   H.T. Dovan and R.D. Hutchins: "Crosslinked Hydoroxyethylcellulose and Its Uses", U.S. Patent, No. 5,207,943, May 4, 1993.

6   G.B. Butler and R.J. Angelo, *J. Am. Chem. Soc.,* 1957, **79**, 3128.

7   G.B. Butler, A. Crawshaw and W.L. Miller, *J. Am. Chem. Soc.,* 1958, **80**, 3615.

8   D.H. Solomon, *J. Polym. Sci., Poly. Symp.,* 1975, **49**, 175.

9   A.L.J. Beckwith, D.G. Hawthorne and D.H. Solomon, *Aust. J. Chem.,* 1976, **29**, 995.

10   D.H. Solomon, *J. Macromo. Sci. Chem.,* 1975, **A9**, 97.

11   D.H. Solomon and D.G. Hawthorne, *J. Macromol. Sci., Rev. Macromol. Chem.,* 1976, **C15**, 143.

12   D. A. House, *Chem. Rev.,* 1962, **62**, 185.

13   E.J. Behrman and J.O. Edwards, *Rev. Inorg. Chem.,* 1980, **2**, 179.

14   Rudin, M.C. Samanta and B.M.E. Van Der Hoff, *J. Polym. Sci., Polym. Chem. Ed.,* 1979, **17**, 493.

15   R. H. Karol, *Chemical Grouting,* 2nd ed., Marcel Dekker Inc., New York, 1990.

INCREASED OIL PRODUCTION FROM WET WELLS IN SANDSTONE RESERVOIRS BY MODIFYING THE RELATIVE PERMEABILITY

R.J.R. Cairns

High Larch, Lewis Lane, Chalfont Heights, Gerrards Cross, Bucks SL9 9TS

1 INTRODUCTION

Field experiments have shown that changing the wettability of a sandstone formation, within the vicinity of a well producing both oil and water, significantly alters the relative volumes of the produced fluids. In particular, oil production has been more than doubled, on a time average basis.

Because the treatment is not permanent, the rock returns to its normal wettability after a period of time and the production returns to its starting point. The average duration of the effect was observed to be five months.

The formulation of the treatment is important and depends upon the characteristics of the well. The nature of the material with which the well is treated is, of course, crucial.

Various potential treatment materials were synthesised and laboratory tested but this paper will focus only on the material chosen for use in field trials.

2 RESERVOIR CHARACTERISTICS AND TWO-PHASE FLOW

In order to understand two-phase flow in reservoirs, the characteristics that must be considered are *the absolute permeability of the rock, the relative permeability of the system and the way in which it varies with fluid saturation,* and *the mobility ratio.* Taking these in turn:

2.1 Absolute Permeability

The absolute permeability is a property of the rock and not of the fluid flowing through it. It may be simply defined as the ease with which a homogeneous fluid can move through a porous rock. In 1856, based upon experiments on the flow of water through unconsolidated sand filter beds, Henry Darcy formulated a law which bears his name. Darcy's Law states that, in laminar flow conditions, the velocity of a homogeneous fluid in a porous medium is proportional to the pressure gradient and inversely proportional to the fluid viscosity:

$$v = - k/\mu.dp/ds \tag{1}$$

In 1, v is the *apparent* velocity and is equal to q/A where q is the volumetric flow rate and A is the *apparent* or total cross-sectional area of the rock. Thus A includes the area of the rock material and that of the pore channels. μ is the viscosity of the fluid and dp/ds is the pressure gradient measured in the same direction as v. When v is in cm per second, μ is in centipoise and dp/ds is in atmospheres per cm, then k, the permeability, is in darcies. Thus in a rock of 1 darcy permeability, a fluid of viscosity 1 centipoise would move at 1 cm per sec under a pressure gradient of 1 atmosphere per cm.

Since this is a large unit for most reservoir rocks, permeability is normally expressed in millidarcies (mD). Commercial reservoirs have permeabilities varying from a few millidarcies, at the tight end, to several thousand.

In the field trials, the permeability ranged from 47 to 1,150 mD.

It should be borne in mind that Darcy's Law only applies to laminar flow conditions; however, these conditions are met by all but the largest producing wells and are certainly met by the wells in the field trials.

2.2   Relative Permeabilty

The Relative Permeability is defined as the ratio of effective permeability to absolute permeability.

In a rock containing more than one fluid, the effective permeability of any one fluid is less than the absolute permeability. Furthermore, the sum of the effective permeabilities of all the fluids is alway less than the absolute permeability.

The effective permeability can be calculated using Darcy's Law, given knowledge of flow rate, fluid viscosity and pressure gradient. If k_w and k_o are the effective permeabilities to water and oil respectively, at a *given saturation* of each fluid, and k is the absolute permeabilty then k_{rw} and k_{ro}, the relative permeabilities of water and oil respectively, are given by:

$$k_{rw} = k_w/k \qquad\qquad (2)$$

and

$$k_{ro} = k_o/k \qquad\qquad (3)$$

The relative permeabilty varies as a function of relative fluid saturation in a non-linear way and tends to be reservoir-specific. Reducing the water saturation and hence increasing the oil saturation leads to a shift in relative permeability favouring oil.

For rock of a given pore structure, wettability and interfacial tension determine the relative permeability characteristics.

2.3   Mobility Ratio

The mobility ratio is defined as the flowing water to oil ratio. As well as depending upon the effective permeabilty ratio, it also depends inversely on the vicosity ratio. If M is the mobility ratio, q_w and q_o are the flowing water and oil volumes respectively, and μ_w and μ_o are the viscosities of water and oil respectively then:

$$M = q_w/q_o = k_w.\mu_o/k_o.\mu_w = k_{rw}.\mu_o/k_{ro}.\mu_w \qquad\qquad (4)$$

3 THEORETICAL CONSIDERATIONS

Given the foregoing theory, what steps might be taken to change the mobility ratio in favour of oil?

Inspection of equation 4, shows that to reduce the value of M, we could *either* increase μ_w, the viscosity of water, *or* change the relative permeabilities. (We could also decrease μ_o, the oil viscosity but this is only a method of choice for wells containing so-called heavy oils where the viscosity is lowered by heating). Increasing μ_w could be done, for example, by adding a water soluble polymer. Changing the relative permeabilities requires a reduction of the interfacial tension between water and oil *or* a change in the wettability of the reservoir formation in the vicinity of the well. Surface active agents have been used to change the interfacial tension between oil and water in some enhanced recovery projects. They are expensive because they must be continuously injected and their use can lead to problems due to emulsion formation. Changing the wettability of the formation so that it becomes less hydrophilic not only alters the mobility ratio in favour of oil in the vicinity of the well but also shifts the relative saturations in the treated pore space in favour of oil thus, from considerations of relative permeabilty, enhancing the effect.

It is this last choice that has been used to achieve the results found in the field trials.

4 FIELD TRIALS

4.1 Materials Used

4.1.1 Sidox. The active material used is called Sidox. It is the subject of a patent application (ref. 5312501/HJF). In essence, the material is a mixture of hydrophobic silica (A), the hydrophobicity being achieved by reaction of a siliceous material such as certain sands with a dialkyl dihalosilane, in particular dimethyl dicholorosilane, *and* a polysiloxane (B), in particular dimethylpolysiloxane. The particle size of A lies in the range 10 to 80 microns. A may be used alone.

4.1.2 Carrier liquid. The carrier liquid is a light petroleum distillate. In these trials, kerosene was used.

4.1.3 Retention material. This is a highly dispersible inactive mineral. Its purpose is to retain the active ingredients in the well formation. In these trials, bentonite was used.

4.1.4 Pre- and Post-treatment fluid. In order to reduce the local water saturation, crude oil, preferably that produced from the well, is pumped into the formation, ahead of the treatment. In order to ensure that the active material is pushed sufficiently far into the formation, crude oil is also injected post-treatment.

4.2 Location

The field trials were carried out in Russian oil fields. The wells treated were located in sandstone reservoirs at depths ranging from 1070 to 2113 metres. The characteristics are given in Table I.

4.3 Method

The well to be treated was inspected to ensure that it was free from impairment by scale, rust or other blocking materials.

Table I   *Well Characteristics*

No	Depth metres	Payzone metres	Permeability mD	Porosity %
1	1080	3.4	47	7
2	1022	8.2	530	16
3	1802	5.0	93	8
4	1817	3.7	250	9
5	1751	4.1	380	11
6	1538	5.2	490	14
7	1509	3.3	300	10
8	1917	3.1	210	8
9	2113	5.6	680	13
10	1572	4.8	230	11
11	1883	2.4	90	7
12	1525	3.8	190	12
13				
14	1602	2.1	870	15
15	1753	4.3	310	9
16	1771	3.9	430	13
17				
18	1837	4.4	470	10
19	1488	7.2	1070	17
20	1070	12.0	512	12
21	1472	10.8	1150	24

The concentration of active materials was in the range 0.5% to 1.5%, generally, and that of the retention material was 1 to 5 %, depending upon the permeability of the formation. The treatment volumes used were based on well characteristics, principally porosity and payzone thickness.

The well was left for a maturation period in order to allow the active materials to interact with the formation. In these trials, the period was generally 24 hours with two wells being given a 48 hour maturation period. The well was then brought back on stream.

The actual treatment is given in Table II.

interact with the formation. In these trials, the period was generally 24 hours with two wells being given a 48 hour maturation period. The well was then brough back on stream.

The actual treatment is given in Table II.

4.4   Monitoring

The oil and water production from the treated wells was measured on a daily basis.

5   RESULTS

The results of the field trials are given in Table III.

In this table, the post treatment values for both oil and water are time-averaged. The production increase is, therefore, a time-average value and it has been calculated with respect to the change in volume of oil produced. The water cuts are given as percentages of the total production, consistent with normal oilfield practice. The *Duration* is the time between treatment of the well and the production from the well returning to its pre-treatment levels.

Table III gives the overall results but, naturally, the actual oil and water production, post-treatment, varied with time, the well eventually returning to its pre-treated state.
Figures I to III show the results obtained.

Table II *Well Treatment*

No.	Oil Inj pre m^3	Dispersion m^3	Oil inj post m^3	Maturation hrs
1	4	6	25	24
2	8	12	40	24
3	5	12	48	48
4	4	6	32	24
5	4	6	25	24
6	5	7.5	24	24
7	4	6	18	24
8	3	5	20	24
9	6	9	38	24
10	5	7.5	35	24
11	3	4.5	22	24
12	4	6	24	24
13				
14	3	4.5	20	24
15	5	7.5	25	24
16	4	6	24	48
17				
18	5	7.5	28	24
19	7	12	35	24
20	9	20	52	48
21	11	18	50	24

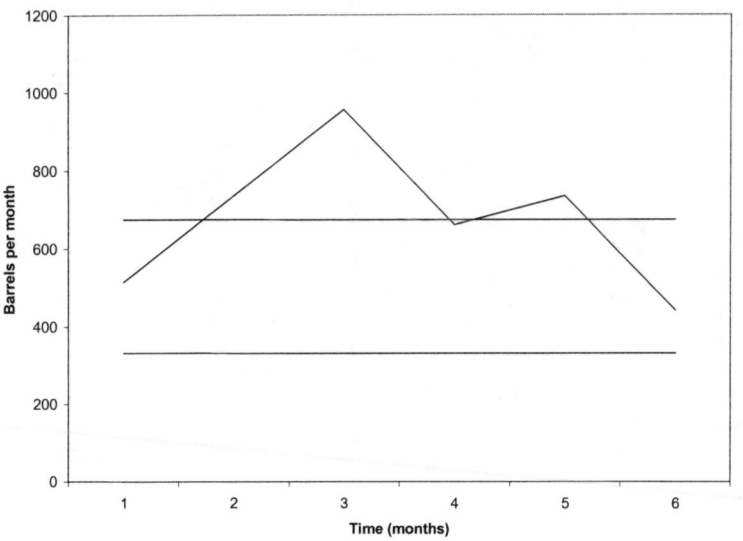

Figure I *Variation of production with time*

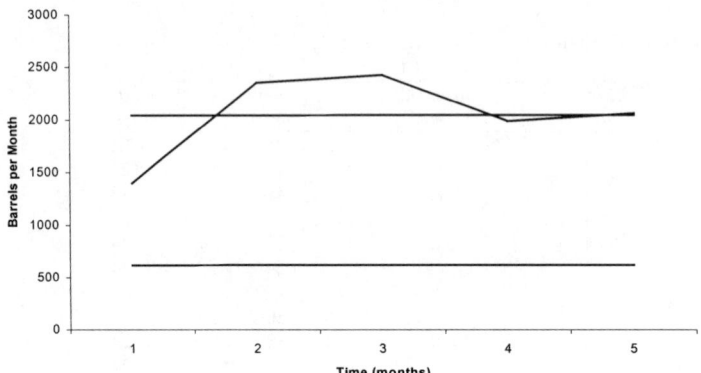

Figure II *Variation of production with time*

Table III *Overall results of field trials*

No	Oil pre bopd	Oil post bopd	Production Increase %	Water pre %	Water post %	Duration months
1	16.9	34.5	104	85	67	4
2	27.9	45.6	63	88	79	5
3	28.6	55.1	91	89	78	4
4	12.5	19.1	53	92	88	7
5	15.4	31.9	106	86	74	5
6	23.5	54.4	131	87	70	4
7	16.1	34.5	114	92	84	3
8	10.3	22.1	114	88	75	5
9	21.3	39.7	86	91	85	5
10	15.4	27.2	76	79	63	4
11	5.1	8.6	67	88	82	6
12	8.1	22.1	172	78	50	3
13	14.0	42.7	205	92	67	
14	15.4	37.7	144	87	72	5
15	13.2	27.0	104	74	53	6
16	11.0	22.6	104	63	31	6
17	16.9	55.9	230	90	59	
18	8.1	14.7	81	90	82	3
19	20.6	68.1	230	82	49	5
20	46.3	152.9	230	92	64	6
21	161.7	338.1	110	90	72	4

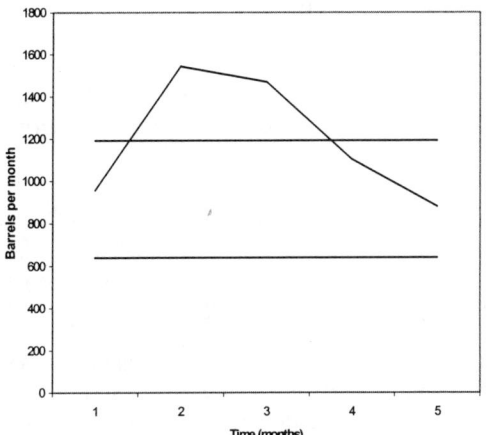

Figure III   Variation of production over time

6 DISCUSSION

The results of the field trials are generally consistent with theory, showing that a shift in wettability of the formation brings about a change in relative permeability and hence a change in the ratio of oil to water produced. It is particularly interesting to note the rapid increase in production rate, as shown in figures I to III, indicating that a shift in favour of oil is self-reinforcing because of the dependence of relative permeability on relative saturation, as set out in Section II.

The improvement in well performance appears to be independent of the depth within the range covered in the trials. It also appears to be independent of the characteristics of the reservoir.   It must be borne in mind, however, that the treatment was specifically tailored to the well by taking into account its characteristics. The inference that can be drawn is that the formulation of the treatment appears to be well-founded.

The water cut in the wells ranged from 63% (only 1 well in the 60 to 70% range) to 92% with most wells having water cuts in the high eighties and low nineties. There is no obvious dependence of the results within this range. From considerations of relative permeability, the wells most likely to benefit are those in which the water cut exceeds 50%.

The lowest increase in production was 53% and the highest was 230%. The wells with lower increases in oil production post-treatment also have the smaller reduction in water cut, as might be expected. This may mean that there was some other effect at play such as the presence of fine fractures providing a means by which the water could continue to flow.

The effect eventually dies away and the well returns to its pre-treatment level of production.   The reason for this must be a shift in the wettability of the injected hydrophobic silica due to its exposure to salt water at relatively high temperatures. It might be expected that the well could be re-treated with similar results but no such re-treatments have been carried out.

The value lies in the additional oil produced. If we consider a well producing B barrels of oil per day (bopd) pre-treatment and we take the average results of the trial, then the added value, V, is:

$$V = 2.23B.153.p \qquad (5)$$

where p is the price per barrel; the factor 2.23 represents the incremental oil production; and the factor 153 is the average number of days for which the treatment lasts. Thus, simplifying equation 5 and using \$25 as the price per barrel, we arrive at the following relationship:

$$V = 8,530B \qquad (6)$$

7   CONCLUSIONS

Theoretical considerations lead to the conclusion that changing the wettability of a reservoir to make it less hydrophilic shifts the relative permeability in favour of oil. This shift causes a reduction in the mobility ratio and hence an increased flow of oil. The increased flow of oil results in the relative saturations of water and oil to move towards oil and this further increases the relative permeability to oil thus re-inforcing the effect.

Field trials on 21 wells in sandstone reservoirs have verified these theoretical conclusions.

The material used is the subject of a patent application and is generally described in Section 4.

The formulation worked out for the treatment of a water-producing oil well in accord with its characteristics has been shown to be well-founded.

Additional oil recovery averaged 123% with all wells having at least a 50% improvement and the effect lasts for 5 months, on average.

Because the effect is brought about by changing the relative permeability, there is a possibility that the treatment may lead to additional oil recovery as well as faster recovery.

The wells returned to their pre-treated state with no evidence of reservoir impairment as a result of this treatment.

The treatment is straightforward and the monetary value added, as shown by equation 6, is considerable.

Flow Assurance

LIFE CYCLE MANAGEMENT OF SCALE CONTROL WITHIN SUBSEA FIELDS AND ITS IMPACT ON FLOW ASSURANCE, GULF OF MEXICO AND THE NORTH SEA BASIN

M.M. Jordan[1], K. Sjuraether [2], I.R. Collins[3], N. D. Feasey[4] and D. Emmons[5]

[1]ONDEO Nalco Energy Services Ltd., Tern Place, Denmore Road, Bridge of Don, Aberdeen, AB23 8JX, UK
[2]ONDEO Nalco Energy Services Norge AS, Luramyrveien 23, Postpoks 1064, 4301 Sandnes, Norway
[3]BP, Upstream Technology Group, Chertsey Road, Sunbury on Thames, Middlesex TW16 7LN, UK
[4]ONDEO Nalco Energy Services Ltd., Cadland Road, Hardley, Hythe, Southampton, SO45 3NP, UK
[5]ONDEO Nalco Energy Services LP., 7701 Highway 90-A, Sugarland, Texas, 77487, USA

1 INTRODUCTION

The economic production of crude oil is based upon effective management of flow assurance. In essence, flow assurance can be considered to be the ability to produce petroleum fluids economically from the reservoir to a production facility over the lifetime of a development. Whilst hydrate control and corrosion control are important issues, scale control is one of the key aspects of flow assurance. The challenges presented by simple vertically drilled wells have been superseded by the evolving challenges presented by the increasing number of subsea fields together with deepwater production. The complexity of new well completions in terms of horizontal and multi-lateral wells, sub sea tiebacks and commingled flow present particular challenges. Any intervention for scale inhibitor treatments in such complex wells is very costly indeed.

Scale control issues need to be addressed as part of asset life cycle management, whereby the issues are addressed prior to field development and full production (CAPEX phase) rather than being confronted in a reactive manner once water breakthrough occurs (OPEX phase). This approach allows for initial selection of appropriate economic technology. Anticipated problems may influence field development plans, for example, in terms of water injection strategies or implementing appropriate technology upon well completion.

As the wells within a field move from dry production to high water cuts, the life cycle management challenges for scale control change. This is associated with the four phases of field development – project, plateau, decline and decommission. At the project stage, scale control treatment strategies can be developed. The scale issues at subsequent stages depend upon the nature and severity of the anticipated scale problem. One class of issues is associated with natural depletion (Figure 1), where the consequent scale problems are normally restricted to calcium carbonate formation. Carbonate scale formation occurs when connate water or aquifer water passes through the bubble point and carbon dioxide is evolved. As carbon dioxide is evolved the solubility with respect to carbonate declines rapidly and a precipitate forms with divalent ions such as iron, and more commonly calcium, as outlined in the following equation (Equation 1)

$$Ca(HCO_3)_2 = CaCO_3 + CO_2 + H_2O \qquad (1)$$

Where water injection (seawater, aquifer) is used sulphate scales can form when the injection water contains sulphate ions (Equation 2 and Figure 2).

$$Ba^{2+} \text{ (or } Sr^{2+} \text{ or } Ca^{2+}) + SO_4^{2-} = BaSO_4 \text{ (or } SrSO_4 \text{ or } CaSO_4) \qquad (2)$$

Sulphate scales have a wide range of solubility depending upon which divalent cation is present (Table 1). In a production operation where barium is present within the formation water and seawater is the pressure maintenance fluid, the principal flow assurance risk is barium sulphate scale. This is owing to its very low solubility relative to calcium carbonate and the sulphate scales of strontium and calcium. Where severe sulphate scaling is anticipated, the potential for sulphate-reduction (desulphation) of injection water should be considered[1]. A review of possible scale control technologies alongside scale risk and deployment economics, along with an assessment of potential new

Table 1   *Common scale minerals, their composition, relative solubility and physical conditions that cause their formation*

Mineral type	Composition	Relative Solubility (mg/l)	Causes of Solubility Change
Calcite	$CaCO_3$	196	PCO_2, Total pressure, TDS, Temperature
Siderite	$FeCO_3$	100	PCO_2, Total pressure, TDS, Temperature
Barite	$BaSO_4$	44	Pressure, Temperature, TDS
Celesitie	$SrSO_4$	520	Pressure, Temperature, TDS
Anhydrite	$CaSO_4$	3270	Pressure, Temperature
Gypsum	$CaSO_4.2H_2O$	6300	Pressure, Temperature

Conditions 100°C, solution 1m NaCl, pH 7, Anhydrite present at over 100 °C, Gypsum present at less than 100°C

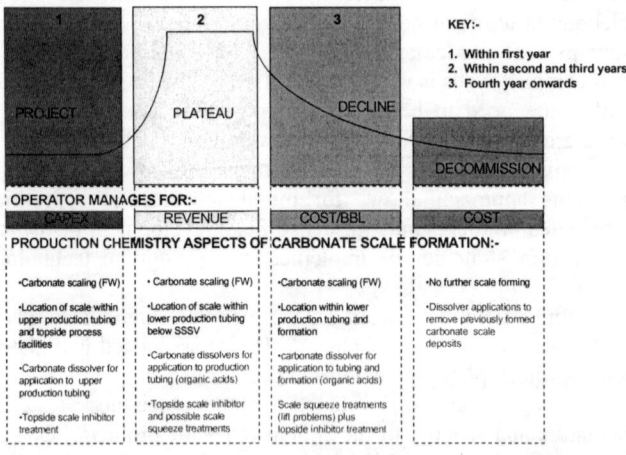

Figure 1 *Schematic of Scale challenges through a fields or wells life cycle during natural depletion either through gas cap drive or water drive*

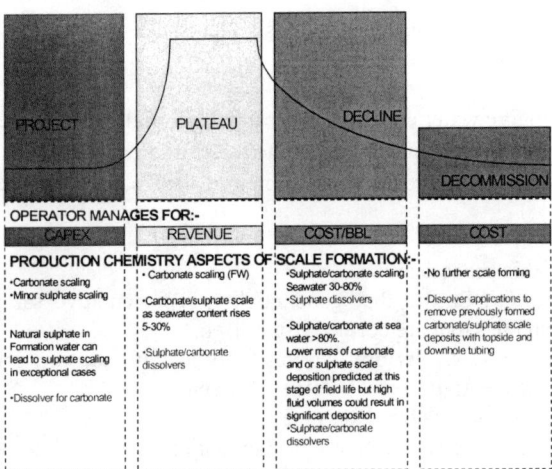

Figure 2 *Schematic of scale challenges through a fields or wells life cycle during secondary recovery via seawater injection*

technologies for scale control, can provide guidance for new developments based upon historical data. As well as providing guidance for new developments, it can also aid option selection for currently producing fields and additionally highlight areas where technology gaps exist.

An example of the development of two high temperature & high pressure fields by BP offshore Norway illustrates the options chosen at the planning stage with technology available at that time, and how new technology was subsequently introduced to improve scale control as the fields moved through the asset lifecycle.

The process of evaluating the risk of scale in a field under appraisal is outlined below and the factors to be taken into account when evaluating the risk of scale formation and control are described in detail along with presently available technologies and the gaps that exist. In addition to risk analysis, legislative requirements, for example in the North Sea, must be met for scale control chemicals. These increasingly stringent schemes encourage / require the use of more environmentally benign chemicals.

2 EVALUATING SCALE RISKS

If it is assumed that current scale inhibitor chemistries will be able to control scale for developments, then the question becomes 'can scale inhibitors be deployed and if successfully deployed will the lifetime of the treatment be economic?' The answer to this question should be assessed on both the basis of current squeeze / scale control technology and on the basis of new and near-commercialised technologies. The economics of these strategies, if found to be at all feasible should then be compared with the economics of using desulphated seawater in the waterflood (which owing to the presence of 40 ppm of sulphate in the desulphated water will still require some form of scale control[1]); the sulphate level must be reduced to < 10 ppm in order to eliminate barite scale entirely from the whole production system). Of course this option is only applicable to sulphate scale. Carbonate scale would require scale inhibitor treatments in their own right.

In order to accurately determine the risk that scale poses to a development, the following process should be followed:

- Obtain representative water chemistries for input into field development
- Make an assessment of the magnitude of the scale problem (if any)
- Review possible control processes (completion design, chemical squeeze, sulphate-reduction etc)

The components of this process are:
1. Sample acquisition. Ensure that good quality, representative samples are obtained (preferably downhole) that are correctly stabilised.
2. Key component analysis. Perform ten ion analyses with accurate measurement of the bicarbonate ion concentration and volatile fatty acid content.
3. Expert review of the water chemistry data.
4. Determine whether there has been any contamination of the samples with drilling muds / completion fluids etc.
5. Perform scale predictions using a recognised simulation package.
6. Assess the nature of scale and the mass predicted.
7. Review the development and determine which scale control technologies might be appropriate.

In order to simplify the initial assessment of the scale risk, it is useful to 'broad band' the risk into 5 categories. These categories are based upon the scaling tendency calculated for the brines assessed. The category designations are outlined in Table 2.

To successfully achieve scale management, scale inhibition technology must be successfully deployed in the field. The difficulty of deploying scale inhibition technology is a function of the type of development (e.g., water depth, position of the wellheads etc) and of the well (type of well and completion type). A recommended method of characterising the severity of the scale inhibitor deployment issues, a scale inhibitor deployment (well intervention) index has been calculated, based upon industry experience, which ranks the difficulty of performing a squeeze treatment *and* successfully placing chemical. The index is a composite of two factors encompassing, first, the difficulty of accessing the wells; and secondly, the nature of the completion. The numerical values assigned to these factors are outlined in Table 3.

The overall intervention difficulty, or index, is taken as the both factors multiplied together, giving a scale of 1 to 12.

North Sea fields on the continental shelf are predominantly low angle wells and are generally cased and perforated. Future developments in deepwater will generally be sub-sea. The wells will be mixture of moderate angle (60°) ranging to very high angle. Many of these wells will be open hole gravel packs and open hole frac packs.

A range of BP and industry data has been compiled and plotted in the form of a risk matrix of scale risk against intervention difficulty. These data are shown in Figure 3.

It is clear from these data that the upstream industry will face great challenges in the future in attempting to control downhole scale by intervention-based scale control technologies. New technologies can impact this area by providing longer life treatments. It must, however, be remembered that extending the lifetime of treatments for scale control(or indeed) eliminating them completely can have a significant value to assets currently using conventional squeeze treating for scale control. Here the value lies in reducing deferred oil costs and intervention costs by extending the time between scale-

Scaling Tendency (Supersaturation)	Scale Risk
3-29	1
30-99	2
100-199	3
200-299	4
300+	5

	Well Type	Difficulty Factor
Access	Platform	1
	Sub-sea, dry tree	1.5
	Sub-sea wellhead	3
Completion	Cased and perforated, vertical	1
	Cased and perforated, highly deviated	1.5
	Cased and perforated, gravel pack. Short interval.	1.5
	Cased and perforated, gravel packed. Long interval	2
	Short OHGP, low angle well	2.5
	Long OHGP, high angle well	4

Table 2
Supersaturation scale index and scale for associated risk of different well types as a function of access to the well and completion type

Table 3
Assessment of the difficulty of controlling scale formation/deposition

related well interventions. The value of all treatments should be quantified in terms of $/bbl of oil protected to ensure that the treatment itself is not exceeding the lifting costs of the oil. This approach allows more expensive treatments to be considered on the basis of reducing the 'per barrel protected' cost measure. The downside of this approach is that such a metric can only be fully quantified after the treatment has run its course, which can be a significant time in the future. However, historical and laboratory data can be brought together to make a prediction of how this metric will look which when taken with a quantification of the risk involved, especially if the technology is highly immature, allows the development of a risk-based solution to the scale problem.

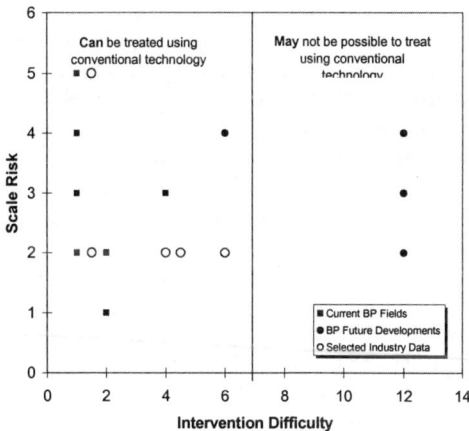

Figure 3 *Risk of scale formation impacting production Vs intervention difficulty for BP current fields, selected industry data and future BP developments*

2.1 Environmental Impact

Over the past few years concerns have been raised with regard to the environmental impact of production chemicals entering the environment during the discharge of water over-board in offshore operations. In particular the use of products such as scale inhibitors has been regulated for the North Sea. Prior to the implementation of the CHARM system the situation can be summarised as follows.

Norwegian regulations follow the OSPARCOM protocols. The operator has a frame licence in which they can produce. Any amendments must be to lower environmental hazard.

The regulations for testing of offshore chemicals were amended during 1998 and the revised requirements include the following:

- Substances with a MW>600 must be tested for bioaccumulation – otherwise they will be treated as bioaccumulating.
- Chemicals and their breakdown substances with biodegradation < 20% and Log Po/w > 5 are banned.
- Other chemical types are also banned, including nonylphenol based materials.
- Products must be phased out (in the long term) if any component is categorised as having:
    Biodegradation < 20% or
    Biodegradation < 60% and Log Po/w > 3 or
    Listed on Parcom Annex A Part II

The Norwegian legislative bodies have drawn up a set of guidelines which operators in Norway are endeavoring to follow. These guidelines differ from those of the UK sector and have been seen as the new challenge for the service sector to meet in terms of environmentally friendly or "Green" scale inhibitor for topside and downhole application.

The existing UK scheme is the OCNS (Offshore Chemical Notification Scheme) where products are categorised by CEFAS. Test data is submitted on a standard HOCNF form. Although the scheme is not mandatory, all UKOOA members have signed up to it, making it obligatory in practice. Discharge limits are cumulative and CEFAS awards letter categories of A to E for products. Category E chemicals are the most benign and use of these is permitted up to 1000 tonnes/year for production chemicals. Above this quantity of use, the operator must consult with the Government in advance. For less favourable categories, smaller amounts for use are specified; for a category A product the trigger quantity is 40 tonnes/year. The category awarded is based on the whole product toxicity to a range of organisms (algae, crustacean and fish or sediment dweller). It is then fine–tuned by assessing the likely persistence of each component. A material that has a log Po/w < 3 and is readily biodegradable can lower the category by two – from C to E. Conversely, material with log Po/w > 3 and biodegradation < 20 % will increase it to an A.

For 2001 onwards a new system is proposed for the UK sector of the North Sea. This is a risk assessment using the CHARM (Chemical Hazard and Risk Management) model. A Hazard Quotient (HQ) is derived based on the PEC/PNEC ratio where PEC = predicted environmental concentration and PNEC = predicted no effect concentration. This HQ data, issued by CEFAS, is then used by an operator to derive a risk quotient (RQ) which is platform specific. In general terms substitution of substances thereby lowering the HQ is desirable, such that operators select chemicals with the best HQ for a given application.

The United States Environmental Protection Agency (EPA) is responsible for considering guidelines for determining the potential degradation of the marine environment by oil production produced water discharges in the Gulf of Mexico as well as other deep water production areas in federal waters. The EPA provides regulations for determining the frequency of testing based on the total water production discharged at any given discharge point as well as the proximity of that discharge point to other discharge points and volumes of those discharge points. Testing generally includes oil and grease, toxicity and bioaccumulation. Test procedures are specified in the National Pollutant Discharge Elimination System (NPDES) General Permit for New and Existing Sources in the Offshore Subcategory of the Oil and Gas Extraction Category for the Western Portion of the Outer Continental Shelf of the Gulf of Mexico (GMG290000). Treatment chemicals can effect the results of required environmental testing. Therefore it is important to understand how to minimize the toxicity of treatment chemicals when formulating these products.

The Minerals Management Service (MMS) is another United States Federal Agency responsible for ensuring that all aspects of oil, gas, and sulphur leasing, exploration, development, production and abandonment activities conducted on the Outer Continental Shelf (OCS) are performed in a safe and clean manner. The MMS, Gulf of Mexico OCS Region, has recently published a study[34] which provides information on both the current and future predicted usage of chemical types and volumes in both shallow and deep water Gulf of Mexico. The study summarises the hazardous chemicals used and provides conceptual models using a range of chemical spill scenarios to predict environmental impacts. Chemical classes include products used in drilling and completion, stimulation and work over, as well as production operations. The report concludes that most chemicals do not pose a threat to the environment, if there were an accidental release at an offshore facility.

The positive ecological impact that treatment chemicals currently have is to reduce the levels of dispersed oil being discharged overboard. This is accomplished by helping clarify the discharge water and also by controlling solids using dispersants and scale inhibitors. Water clarifiers help remove dispersed oil by assisting in the agglomeration of oil droplets. Scale inhibitors can eliminate the formation of solids. These solids tend to oil wet and increase the amount of oil discharged when they are carried overboard in the discharge water.

2.2 Scale Management Economics

Simple economic models can be used to assess various scale management options during field life. The following is an example of developing such a model for a new development: In it, the various costs for all scale-related interventions are considered and ranked against development type. The model is based upon a total of 11 production wells and is for a deepwater development where the options are for sub-sea wells, dry trees and possibly sulphate reduction to help mitigate sulphate scale.

Environmental cost factors can also be considered as part of the overall economics. Operators can attribute varying costs to discharged chemicals according to their OCNS rating. An example of this is shown in Table 4. Thus category 'E' substances are assigned the lowest costs per tonne for discharge to sea. Ranges rather than exact figures are given to allow for the alternative ways of deriving such figures.

Table 4   *BP Environmental class for scale control chemicals vs. cost/value for the UK sector, North Sea*

OCNS Category	£/Tonne Discharged to Sea
E	100
D	100 – 1000
C	1000 – 10000
B	10000 – 100000
A	100000 - 1000000

2.3 Scale Inhibitor Squeeze Treatments

In wells completed over short reservoir intervals, it is possible to place the treatment effectively by bullheading the treatment (adsorption, enhanced adsorption or precipitation) from the surface.   These types of treatment have been the subject of numerous publications.[2,3,4,5,6,7,8] Where the completed interval is longer, it is often necessary to enter the well and use coiled tubing to spot the inhibitor or use another method of fluid diversion.[9,10] The design of such treatments has focused on the use of placement software which allows treatment volumes to be evaluated and economic evaluation of the treatment life versus cost to be carried out.[11,12,13]

The frequency of the inhibitor squeezes will depend on the Minimum Inhibitor Concentration (MIC) needed to inhibit scale formation. This is linked to the severity of the scale problem. Table 5 lists a summary of scale inhibitor treatment frequencies for North Sea Fields with seawater injection.

Table 5   *Examples of predicted mass of scale, squeeze and intervention frequency from the North Sea*

Field	Predicted Mass of $BaSO_4$ (mg/L) at Bottomhole Conditions*	Bullhead Squeeze Frequency (int. / well-year)	Minor (coiled tubing) Frequency (int. / well-year)	Major(sidetrack) Frequency (int. / well-year)
Statfjord	85	0.461	0.011	0.019
Gullfaks	30			
Forties	200-350	ca. 0.4	No data available	ca. 0.011
Magnus	128 - 318	ca. 0.4	~15 jobs out of 17 wells over field life (12-17 yr)	no data available

The life of the scale inhibitor squeeze treatment will also depend on the scaling tendency of the fluid mixture, which changes depending on the ratio of seawater and produced water, Figures 4 and 5. The Magnus data is for a well at its maximum scaling tendency. Because of this, even with the use of squeeze extenders and state of the art phase separation chemistry, the squeeze lifetime was only 250,000 bbls of water (this was equivalent to less than one month on production). The next squeeze treated 1 MMbbls of water because the scaling environment was less severe and the minimum inhibitor concentration considerably lower. Current squeezes are protecting *ca.* 4.5 MMbbls of water to a MIC of 2 ppm. In order to achieve squeeze lifetimes of this magnitude, wax divertor must be employed to ensure that the scale inhibitor is uniformly placed in the productive interval and a squeeze life extender is used in conjunction with a phase separation (precipitation) scale inhibitor.[5,6,7,10]

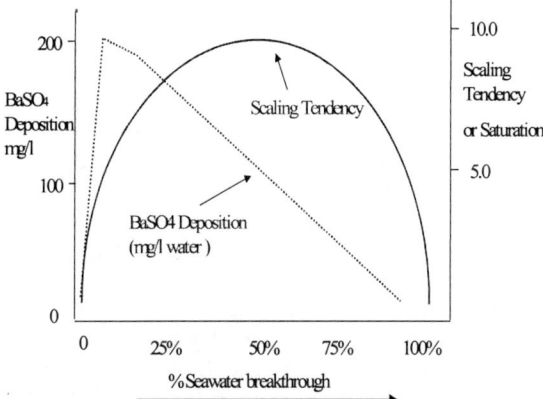

Figure 4 *Scaling tendency (mass and supersaturation) for barium sulphate scale during a well's life cycle (Seawater injection)*

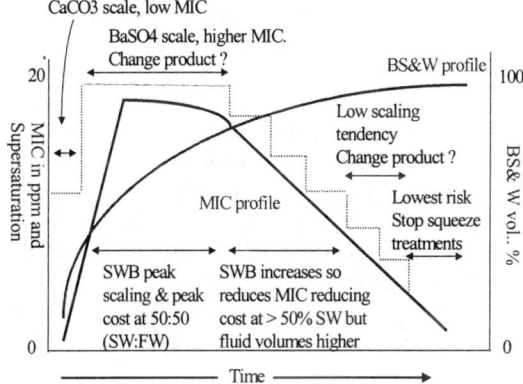

Figure 5 *Changing factors that affect scale control during well life cycle*

The use of divertors in gravel packed wells is a relatively immature technology. Statoil and Gullfaks have used oil-soluble resins a number of times on Stafjord but concede that the treatments have damaged the gravel packs. The wax beads discussed above have proven to be totally non-damaging in cased and perforated wells[14] owing to their total solubility in hot oil and thus would seem to be a good choice for gravel packs if it was not for the lack of zone isolation that solid diversion displays in such completions. In such cases where gravel packs or sand exclusion screens are used foam or gel divertors agents have proved more effective.[15] For worse scaling environments, the worse case scaling conditions can be very difficult to manage owing to high supersaturation resulting in high minimum inhibitor concentration (MIC) values and difficulty in chemical placement.

The reduction in scaling potential resulting from the use of low sulphate seawater would have the consequence of extending the lifetime of squeeze treatments. This is because the MIC required to prevent scale formation is considerably lower. For example, for a MIC of 30 ppm, the squeeze lifetime may be 1,000,000 bbls, whereas for 3 ppm, the lifetime could achieve 6,000,000 bbls of protected water. Furthermore, low MIC requirement's allow scale inhibitors to be selected that have the properties of giving long lifetimes at low MIC (for example pentaphosphonates).[16,17] This will reduce the well intervention frequency required to control scale. If all of the Magnus wells had low MICs of the order of 2-3 ppm, the average squeeze intervention frequency would be, on average, 1 per well per year.

Figure 6 gives an indication of the range of squeeze lives that has been achieved in a North Sea field with the volumes of water protected ranging from 300,000 bbls to 6.7 MMbbls.

The scale squeeze intervention frequency is obtained by multiplying the time water is produced (e.g., 2/3) by the time the scaling index is expected to be a problem (e.g., 0.85) by the number of scale squeezes per well per year.

Because the composition of the seawater/formation brine mixture is constantly changing, the scaling potential is also changing, Figures 4 and 5. At some mixing ratios, the scaling indices will be low enough that scale is not likely to form, Figure 5. Furthermore, if sulphate-reduction is used to help manage sulphate scale, the scaling potential will be affected by the concentration of sulphate in the injected seawater, which will vary with time.

In addition to the above, the well architecture will also dictate how the treatments are applied. For example, for wells that have dry tree access, it is likely that a larger percentage of the wells would be high angle wells with longer completion intervals. These are more difficult to bullhead squeeze effectively and it is likely that coiled tubing would need to be used to place the scale inhibitor treatment. Therefore in the following economics, it has been assumed that 50% of the scale inhibitor treatments would be bullheaded and 50% would be placed using coiled tubing.

For subsea wells, more of the wells are likely to be subvertical with shorter step outs. Therefore a higher percentage (75%) could be effectively bullheaded from surface and only 25% would require an intervention with coiled tubing to effectively place the scale inhibitor. Given this assumption, the subsea facilities would need to be designed to allow for bull heading scale inhibitor treatments.

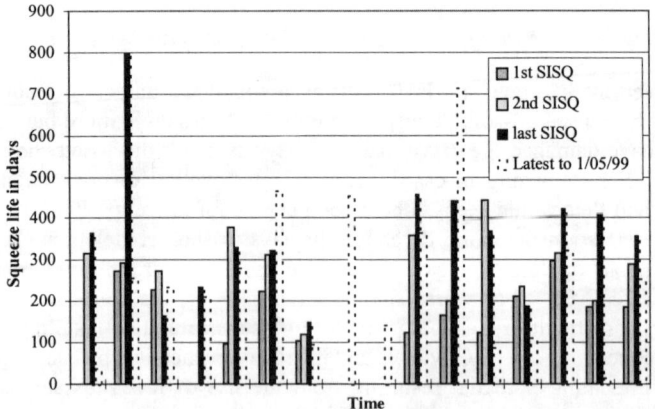

Figure 6   *Range of squeeze life times (days) for typical production North Sea production wells using convention scale squeeze technology*

Typical treatment times can be calculated for these two scenarios. The duration of the bullhead scale squeeze treatments is around four (4) days for a dry tree and six (6) for subsea wells. This takes into account the time to treat the well and bring it back onto full production. The time associated with the coiled tubing squeeze treatments for dry tree and subsea wells are outlined in Table 6.

Table 6 *Estimated intervention duration for CT scale inhibitor placement and scale milling*

Operation	Dry XT Well	Subsea Well
Mobilize & Position MODU	NA*	72
Run Drilling / Intervention Riser	NA	48
Rig up CT	24	24
RIH & mill out scale or place inhibitor	96	96
Rig Down CT	14	14
Pull Drilling / Intervention Riser	NA	36
Demobilize MODU	NA	48
Total (No NPT / WOW) hrs	134	338
NPT / WOW Multiplier (1/1-0.25)	179	451
Total Time (days)	*7 days*	*19 days*

* *Assumes SIMOPS can be conducted on the DDCV with drilling and CT*

* *Assumes SIMOPS can be conducted on the dry XT facility with drilling and CT*

2.4 Scale Milling

Scale milling operations are intended to allow access to the wells with downhole tools and not necessarily intended to recover lost productivity. The coiled tubing milling intervention frequency, in this example, has been estimated to be 0.0367 int. / well-year. (The intervention frequency is based upon 11 wells / 300 well-years = 0.367 int / well-year using historical BP data which suggest that, on average, one coiled tubing milling operation is performed per well over field life).

The duration for scale removal operations (coiled tubing scale milling) is listed in Table 6.

2.5 Scale Related Sidetracks

The scale-related sidetrack frequency is estimated to be 0.0062 int. / well-year. This number was obtained as follows:

- North sea operator data indicates that approximately 7.5% of the wells were plugged by a combination of scale and/or fines. Owing to the well-sorted nature of the North Sea sands, it is likely that the majority of the plugging is due to scale formation.
- Based on the number of plugging (scale and/or fines) failures and the number of well-years, the plugging frequency of the Statoil data is 0.02 int / well-year. It was assumed that the majority of this (75%) was the result of scale. This gives an intervention frequency of 0.015 int. / well-year.
- The number of major sidetrack operations expected owing to scale plugging is determined by applying the Statoil plugging frequency, associated with scale, to the number of well-years for those wells where scale will be a problem. This calculation (see below) results in three (3) sidetrack interventions for scale management, corresponding to an intervention frequency of 0.0085 int / well-years. ((0.015 int / well-year) x (2/3 well life producing water) x (85% time in scaling region) = 0.0085 int / well-year).

Table 7 summarises these data that highlights the difference in intervention costs between a development based upon dry trees and sub-sea trees.

Clearly, these values are only part of the overall economic assessment that must be performed. However, they do show the sensitivity of scale management costs to development type and to treatment type and lifetime. Such an analysis can be performed at any point during an asset's lifecycle to evaluate scale management options and the impact that new technology can have upon scale management costs and the environment. Analyses performed later in life obviously benefit from an abundance of good historical data.

3 SCALE TECHNOLOGY OPTIONS

Over the past 35 years, different technologies have been developed to reduce the risk of scale formation, to control scale formation and to remove it if formed within the downhole and topside oil/gas facilities both onland and offshore. Most presently available technologies are focused on use in the OPEX phase of field life. However evaluation of the scale risk and the combination of engineering and chemical solutions in the CAPEX phase can greatly improve the long-term field economics. Tables 8, 9 and 10 outline the different technologies and give a brief description of what is presently available to the industry, what is being developed, and what technologies are still required (the gaps). In Figures 7, 8 and 9 the location of these technologies within a range of production scenarios is used to illustrate the presently available scale control/monitoring technologies and the gaps that still require a solution. In all three scenarios, the scaling environment is the same, with early scale formation owing to production of formation water (carbonate scale) followed by a more severe scale observed on seawater breakthrough (sulphate/carbonate). Though a single production well is presented in the illustration, it is envisaged that this well forms part of a cluster of wells tied into a common manifold that feeds a single flow line to the platform.

Figure 7 outlines the location where current technology can be placed within a vertical drilled, cemented, perforated completion designed to recover oil from a sandstone reservoir. The interval is envisaged to be less than 300ft of moderate permeability (150 to 250mD), porosity (15 to 20%) sandstone. At present, a wide range of technologies can control the formation of scale within the reservoir. Some systems can be deployed prior to water breakthrough, such as solid inhibitor deployed within completions, solid scale inhibitor placed in the rat hole, (or gravel pack if completed in that way). All can be deployed in the CAPEX phase of the project.

The commercial manufacture of solid scale inhibitors presents a number of technical challenges. Whilst it is possible to manufacture solid phosphate and phosphonate compounds for the detergent industry which dissolve within a few minutes within a domestic appliance, the requirement to make solid scale inhibitor for oil field application, which would have a limited solubility such that they would dissolve slowly over many months, is more of challenging. The manufacturing process normally involves the formation of calcium/magnesium salts of phosphate, phosphonate, or polymer-based scale inhibitors. The manufacturing route is either through bulk precipitation or via a spray drying process. The resulting solid is then tableted or extruded when mixed with a suitable binder. The solubility of the compound can be varied with differing ratios of calcium/magnesium within the complex. Solubility of the formulation is also affected by the brine salinity in which it is placed. Low salinity brines result in higher product

Table 7   *Actual cost of scale control/management via scale squeeze, scale milling and sidetracking production wells*

	Wells	Water part of life	Scale Problem, %	Field Life, yrs	Scaling years	Scale Squeeze				Scale Milling			Sidetracks		
						Interventions, per res/yr	Wet Cost (CT) $m/y @ $4.9 MM/sq.	Dry Cost (CT) $m/y @ $0.86MM/sq	Bullheading Cost (Dry or Wet) $m/y @ $0.51MM/sq	Interventions, per res/yr (once /well life)	Wet Cost (CT) $m/y @ $4.9 MM/sq.	Dry Cost (CT) $m/y @ $0.86MM/sq	Interventions, per res/yr (0.02/well/yr)	Wet Cost, $m/y @ $12.5 M/well	Dry Cost, $m/y @ $8.0 M/well
Field 1	4	0.667	85%	15	8.5	2.27	11.11	1.95	1.16	0.07	0.33	0.06	0.08	1.00	0.64
Field 2	3	0.667	85%	15	8.5	1.70	8.33	1.46	0.87	0.07	0.33	0.06	0.06	0.75	0.48
Field 3	4	0.667	85%	15	8.5	2.27	11.11	1.95	1.16	0.07	0.33	0.06	0.08	1.00	0.64
Field 4	4	0.667	85%	15	8.5	2.27	11.11	1.95	1.16	0.07	0.33	0.06	0.08	1.00	0.64
Total	15					8.50	$41.65	$7.31	$4.34	0.27	$1.31	$0.23	0.30	$3.75	$2.40

Total intervention cost= $18.72   m/yr   **Wet Trees**
                        $8.45   m/yr   **Dry Trees**

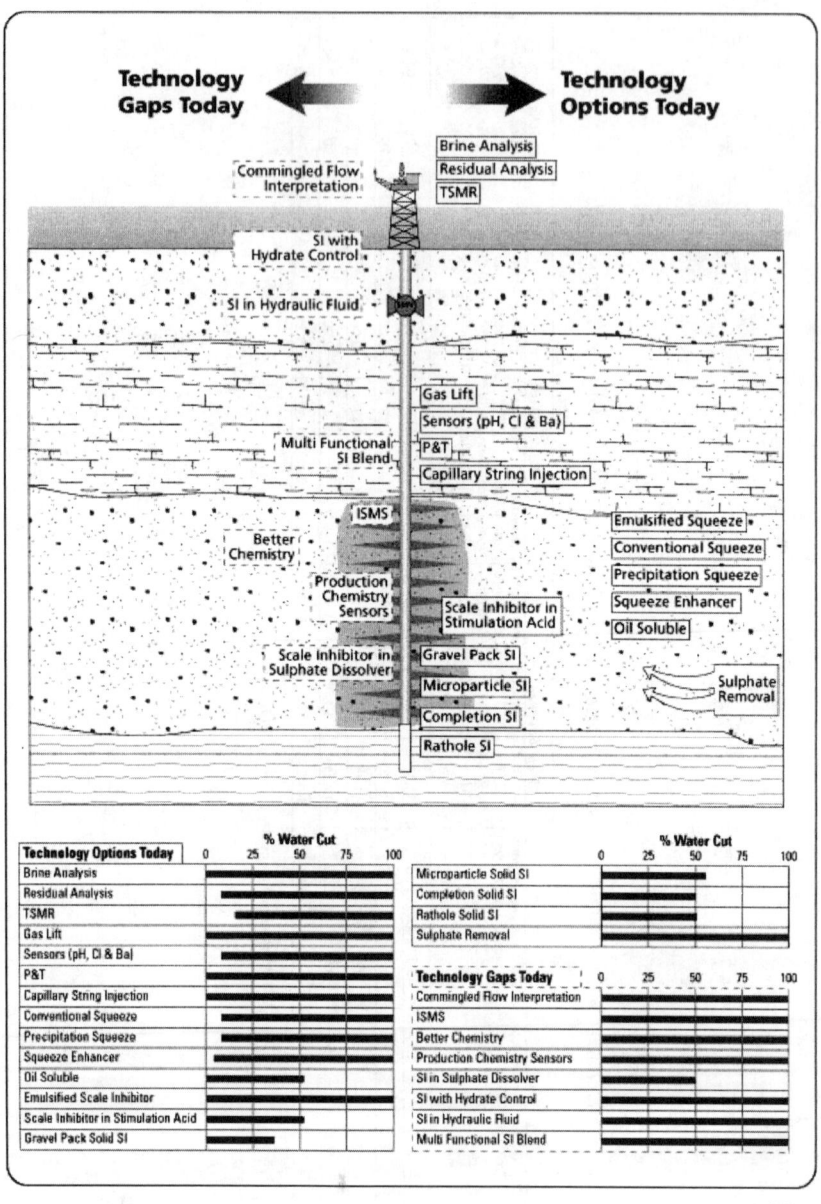

Figure 7 *Technology options in use today plus technology gaps to control/manage downhole and topside scale within a vertically drilled, cemented, perforated production well*

solubility. The brine temperature and the flow rate of the fluids over the solid particles and the size of the solids particles are also factors that have a very large impact on the compound solubility. Higher brine temperatures normally suppress solid scale inhibitor solubility; high flow rates can result in lower solubility than the desired range whereas too slow a flow rate might result in very high solubility approaching equilibrium (rapid dissolution). The size of the particles also has an impact with smaller surface areas resulting in a lower solubility than particles with a larger surface area. Given these variations and the fact that the MIC for any application will also vary as the seawater content increases within the produced fluid, a real challenge is posed to produce a single formulation that can cover all such variations and yet still give a release rate that is close to the required MIC. Environmental concerns have also been raised by such technology in term of the requirement to register the calcium/magnesium salt as new compounds on the TOSCA and IONEX register. At present while the sodium salts of phosphonate and polymer compounds are registered, the calcium and magnesium salts are not. The cost of this registration >$100,000 has reduced the economic viability for the generation of a wide range of such formulations.

Table 8   *Currently available scale control/management technology*

Technology	Comments
Conventional Squeeze Treating	Adsorption of polymer or phosphonate scale inhibitors onto mineral surface from which they desorb into the produced brine
Precipitation Squeeze	Precipitation of a calcium or calcium-magnesium complex with polymer or phosphonate based scale inhibitors which dissolves into the produced brine
Squeeze Extenders	Mutual solvent chemicals used to enhance chemical adsorption and retention; also aid in well clean up rates
Oil-Soluble Scale Inhibitor (OSI)	Scale inhibitor formulated with mutual solvent which gives the aqueous formulation oil dispersible properties, thus eliminating any increase in near wellbore saturation (and also aids well clean up rates)
Gas Lift Deployed Scale Inhibitor	Aqueous or oil soluble scale inhibitor formulated such that it remains in the liquid phase when applied with gas lift gas
Sulphate Removal	Nano filtration of injection quality seawater results in reduced sulphate levels of between 40 and 120 mg/l.
Solid Inhibitor: • Rathole Deployed Scale Inhibitor	Solid scale inhibitor deployed into the rat hole of a production well
• Scale Inhibitor Impregnated Proppant	Fracture proppant impregnated with the calcium-magnesium salt of polymer or phosphonate-based scale inhibitor which is mixed into a fracture proppant or gravel pack
• Reservoir Deployed Controlled Release Microparticles	Microparticle of solid scale inhibitor that can be injected into the matrix of the reservoir
• Completion Installed Controlled Release Beads	High strength particles containing high loadings of scale inhibitors that can be used in gravel packs, frac packs and stored in completion-installed reservoirs.
• Solid Scale Inhibitor in Completions	Solid scale inhibitor, which is located behind sliding sleeve within completion. The sleeve is opened when water production is detected to release the chemical.
Chemical Injection	Aqueous or oil based scale inhibitor can be injected into the produced fluid via a capillary string run to the lowest packer
Emulsified Scale Inhibitor	Emulsified scale inhibitor within an oil external phase allow placement of aqueous droplet which are retained via entrapment
Scale Inhibitor in Frac Fluid	Aqueous based scale inhibitor is formulated into the fracture fluid such that any fluid leak off will carry scale inhibitor into the rock matrix where it is adsorbed
Scale Inhibitor Blended with acid	Aqueous based scale inhibitor formulated with a stimulation acid so that when dissolver permeates the matrix, scale inhibitor adsorption occurs

Liquid chemical systems such as oil soluble scale inhibitors and emulsified scale inhibitor could be applied as squeeze treatments prior to water breakthrough. Again these technologies fit within the CAPEX phase of development. The application of liquid scale inhibitor can also be applied via the gas lift and capillary string run to above the packer. These completion-related scale control mechanisms rely on identification of the scale risk early enough in field evaluation to allow inclusion of this treatment option prior to installation of the completion itself. The onset of water breakthrough, and the concomitant increase in the mass of scale that can form, together with injection water breakthrough, would result in a more severe scale risk (requiring a higher MIC, or amount of inhibitor, to control the scale). For this scaling environment, conventional scale squeeze treatments using enhanced adsorption,[5,6,7] precipitation[18,19,20] or emulsion droplet entrapment[21,22] could give effective wellbore protection. A novel method of applying scale inhibitor as

Table 9   *Currently available scale monitoring technology*

Technology	Comments
Brine Analysis	10 ion analysis of principle ion such as Na, Mg, K, Ca, Sr, Ba, Fe, SO_4, HCO_3, Cl and organic acid plus gas phase CO_2, H_2S
Residual Analysis	Scale inhibitor residual via ICP, wet chemical and tagging methods
TSMR	Thickness Shear Mode Resonator work by changes in vibration frequency as a result of deposition of scale on the crystal surface
Sensors: • pH, Cl • Ions	Sensor suitable for downhole application which can measure the change in pH, chloride ions and specific scaling ions all allow the changes in water chemistry associated with injection water breakthrough to be monitored as well as the success of scale control programs
Pressure and Temperature	Sensor which can measure the change in downhole pressure and temperature to give indication of water breakthrough and also scale deposition causing flow restrictions

Table 10   *Identified gaps in scale control/management and monitoring technology*

Technology	Comments
Chemical Injection Across The Sandface Completion	Chemical delivery via capillary lines or X-pipe to modified sandscreens allowing chemical injection across the productive interval
Well Tractor Chemical Delivery	Chemical is placed in the form of a solid or a liquid via a well tractor that can reach locations in a well not possible with conventional bullhead or coiled tubing applications
Shunt Tube Conveyed Inhibitor	Chemical injection facilitated via a shunt tube used to improve the effectiveness of gravel pack operations
Better Chemistry	More efficiency, better adsorption, thermally stable, better compatibility, green chemistry
Scale Inhibitor Chemical Sensors	May be specific for given inhibitor
Commingled Flow Interpretation	Ability to detect the flow from a number of production wells mixing in a single manifold and flowing up a single flow line
Scale Inhibitor Within Sulphate Scale Dissolver	Aqueous scale inhibitor that could be applied within sulphate dissolver near wellbore stimulation
Scale Inhibitor With Hydrate Control	Scale inhibitor formulations are required that can cope with application in deep water subsea application where ingress of produced gas could result in hydrate formation within chemical injection lines
Scale Inhibitor Within Hydraulic Fluid	Aqueous scale inhibitors to be included within hydraulic fluid such that controlled release of hydraulic fluid would introduced inhibitor into produced fluid at the wellhead and also downhole.
Multifunctional Inhibitors	Scale inhibitor formulated with wax, hydrate, asphaltene, corrosion inhibitors to allow effective control of more then a single flow assurance problem (continuous injection)

solid microparticles has also recently been developed. This provides a means of applying scale inhibitor into production intervals.

Monitoring of downhole conditions such as pressure/temperature/pH, and specific ion concentrations[22] can be performed if such instrumentation was included in the CAPEX phase of the project. The monitoring of scale inhibitor residuals and brine analysis on the topside can be carried out as long as sample points are present within the process prior to injection of any "top up" scale inhibitor for topside scale control. The topside processes degree of protection, and stability of the fluid, reaching the wellhead can also be assessed using scale deposition monitoring technology, such as that based on thickness shear mode resonators (TSMR).[23,24] If scale is observed via a monitoring program, then dissolver treatments for carbonate scale removal formulated (blended) with scale inhibitor can give a dual benefit of stimulation and prevention of further scale deposition downhole.[25] The points in the well life cycle (water cut) where these technologies can be utilized are presented in Figure 7.

The gaps between technical needs and available technological solutions, revolve round the need for increased efficiency scale inhibitors that can: (i) control scale at lower MIC; (ii) have excellent thermal stability; (iii) good adsorption characteristics; (iv) excellent compatibility; and (v) ease of detection at the MIC level. The development of a scale inhibitor that could be deployed within a sulphate scale dissolver, would improve well stimulation and help to maintain production post stimulation.

The scale squeeze treatments are typically associated with a period of well clean up during which the fluids pumped into the reservoir during the treatment are displaced back to surface by the produced fluids; this period can be anything from 24 hours to 3 months before the pre-squeeze oil rate is re-established. Whilst the oil production rate decline owing to scale formation is arrested there is no production enhancement due to the squeeze treatment (Figure 10, B-A). In combining a scale inhibitor with a stimulation package (acid or alkaline) the economic benefit of the stimulation (production enhancement) given by the dissolver is maintained by the presence of the scale inhibitor (Figure 10); this gives a significant improvement in the economics of later life fields, (C+B-A).

Acid stimulation formulations can be based upon hydrochloric acid where 15% to 7.5% are typical concentration for removal of carbonate minerals and scales, blended acids such as formic and acetic acid which are less corrosive to high chromium (Cr^{27}) metallurgy and have similar rates of dissolution and capacity to dissolve carbonate minerals, and finally acid stimulation formulations which are based upon mixtures of hydrochloric acid and hydrofluoric acid for the removal of silicate minerals such as clays and feldspar. During an acid stimulation, scale inhibitor must initially overcome any risk of acid hydrolysis, which might break down the scale inhibitor molecule and reduce its efficiency to retard scale precipitation. Acid stimulation packages are not simply hydrochloric acid dissolved in potassium chloride solution or produced water but are in fact a complex mixtures of sequestering, anti sludging agents, surfactant and possible drag reducers which might cause incompatibility problems and precipitation of the scale inhibitor or the additive itself. During an acid stimulation reaction, high concentrations of calcium, magnesium and iron are normally released owing to the reaction of the acid with carbonate minerals such as siderite ($FeCO_3$), calcite ($CaCO_3$) and dolomite ($Ca_{0.5}Mg_{0.5}CO_3$) present in the formation. The presence of high concentrations of divalent ions (>10,000 ppm) in the stimulation fluid as the pH of the fluid itself is changing from pH <1 to pH >5 can result in the formation of divalent ion salts of phosphonate and polymer scale inhibitors. Scale inhibitors must be screened prior to application to check that they will be totally compatible with the reaction products of the acid stimulation treatment as well as the other additives.

The application of scale inhibitor within alkaline-based stimulation treatment such as sulphate scale dissolvers again presents problems with scale inhibitor compatibility at the high pH of the fully formulated products (pH 10 to 12), which are normally based on EDTA, the high temperature, and the presence of high concentration of divalent ions such as barium, strontium and calcium as a result of scale dissolution. Precipitation of the inhibitor and the formation of pseudoscale may result in more damage being introduced than was present within the near-wellbore prior to the stimulation treatment. Chemical compatibility testing must be carried out on the proposed final formation of combined alkali dissolver/scale inhibitor package.

Production chemistry sensors would allow detection of the chemical concentration along the wellbore, whilst ISMS (Integrated Scale Management Systems would allow effective chemical injection along the length of the production tubing. The development of deep water subsea fields and the requirement for chemical injection to achieve scale protection from wellhead to platform requires the development of a range of scale inhibitors that will not form hydrates if gas bypasses the injection values. In many of the proposed subsea developments for the Gulf of Mexico and West Africa, the fluid from a large number of wells will be flowing along a single flow line presenting problems of commingled flow interpretation i.e. determining the concentration of scale inhibitor from each of the commingled wells. Advances in each of these areas would greatly improve risk management for future field developments in deep water fields.

Figure 8 outlines the locations where current technology can be placed within a horizontally drilled, cemented, perforated or gravel packed completion designed to recover oil from a sandstone reservoir. The interval is envisaged to be less than 1000 ft of moderate to high permeability (250 to 1000 mD), porosity (15 to 20 %) sandstone. At present, a range of technologies exist that can control the formation of scale within the reservoir, some of which can be deployed prior to water breakthrough, such as: (i) solid inhibitor placed within completion; and (ii), solid scale inhibitor deployed within the gravel packed completion. Furthermore, liquid chemical systems such as oil soluble scale inhibitors and emulsified scale inhibitor could be applied as squeeze treatments prior to water breakthrough. The principal issue with the use of any liquid system is the placement of these chemicals owing to the long completion interval and high rates of fluid leak off. These technologies would need to be placed using coiled tubing prior to the onset of production. Liquid scale inhibitor can also be applied via the gas lift system and via capillary string run to above the packer. The onset of water breakthrough, and the concomitant increased mass of scale that can form with the onset injection water breakthrough, would result in a more severe scale risk owing to the formation of both sulphate and carbonate scales. The technical difficulty in treating horizontal wells once water production has commenced is covered in a number of publications over the past few years.[9,15,27,28] However, principally, the problem revolves around the location of water production and the subsequent deployment of the chemical into that region of the wellbore. For this situation, conventional scale inhibitor squeeze treatment using enhanced adsorption, precipitation or emulsion droplet entrapment could give effective wellbore protection if suitable diversion technology were applied to allow effective chemical placement.[15,28] If scale deposition is expected, or observed via monitoring of downhole pressure/temperature/pH, specific ion sensors[22] or gauges backed up by topside residual and brine analysis for specific wells, then scale dissolver treatments can be applied. As with scale squeeze treatments, chemical placement is critical and options for chemical diversion would need to be evaluated.[28] The flowlines from subsea developments and the topside process would require protection in a similar fashion as was outlined for the vertical wells (Figure 7).

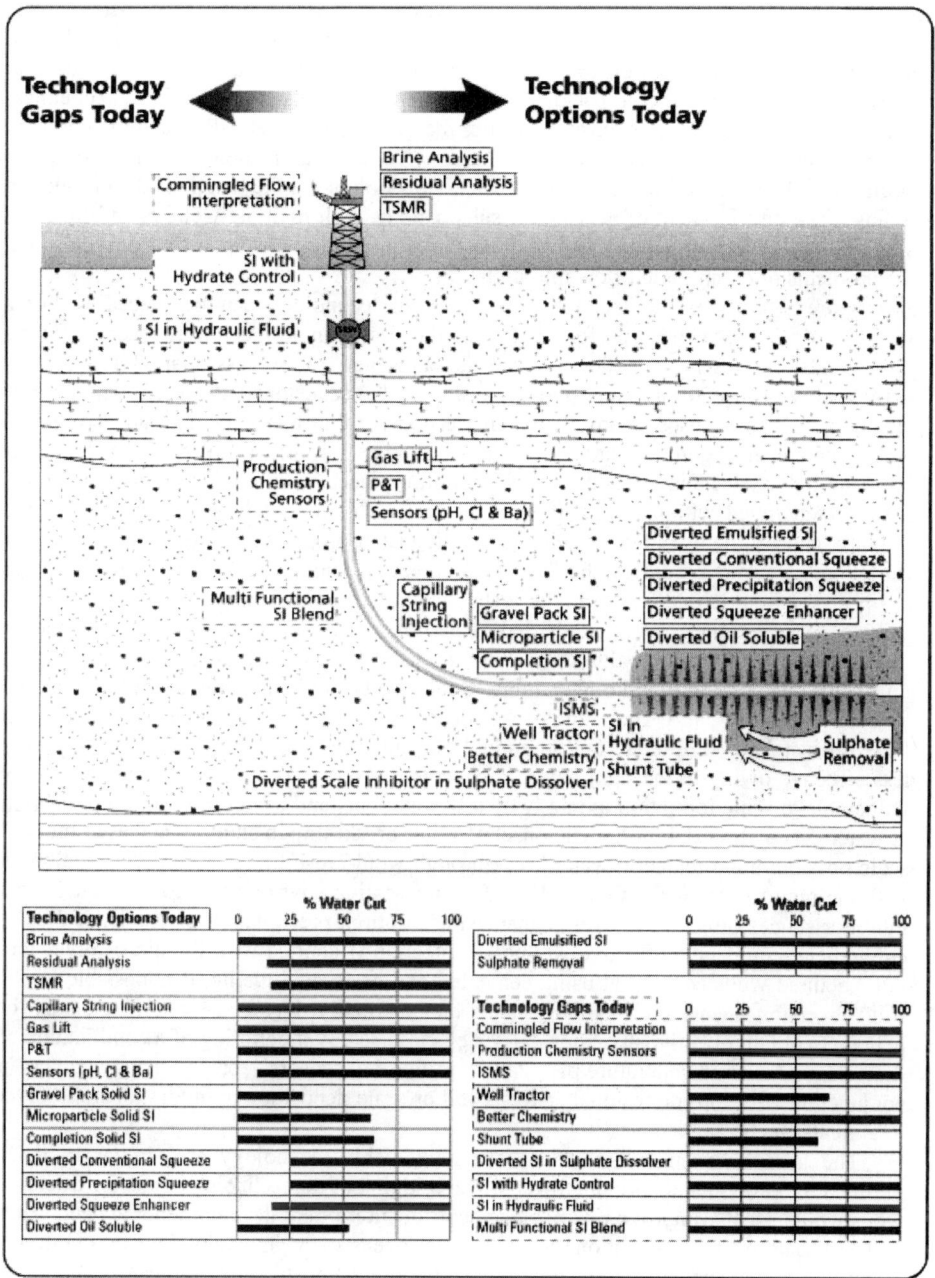

Figure 8 *Technology options in use today plus technology gaps to control/manage downhole and topside scale within a horizontally drilled, cemented, perforated or gravel packed production well*

The gaps in the available technology revolve around the need for 'better' scale inhibitors and improved chemical placement technologies. The latter could be achieved using technology such as well tractors. Chemical application via a "shunt tube" (used at present to help with gravel packing operations) would also offer effective chemical placement to prevent scale deposition within the gravel pack itself. Production chemistry sensors would allow detection of the chemical concentration along the wellbore. In addition ISMS would allow effective chemical injection along the length of the production tubing. The development of deep water subsea fields with multiple wells manifold into a single flow line presents the risk of hydrate formation within chemical injection lines and the problems of commingled flow interpretation

Figure 9 outlines the location where current technology can be placed within a vertically drilled, cemented, perforated fractured completion designed to recover oil from a tight sandstone reservoir. The interval is envisaged to be less than 300ft of low permeability (15 to 50 mD), and porosity (10 to 15 %) sandstone. At present a number of technologies can be used, many of which were applicable to the vertical and horizontal wells shown in Figures 7 and 8. Scale inhibitor can be deployed prior to water breakthrough by techniques such as solid inhibitor placed within completions, solid scale inhibitor deployed within the rat hole, solid scale inhibitor retained within the propped fracture, and by the use of scale inhibitor impregnated proppant.[30] Liquid scale inhibitor can either be deployed during the fracturing operation within the fracture fluid itself[31] or up-front as a liquid system prior to the 'data' fracture. Liquid chemical systems such as oil soluble scale inhibitors and emulsified scale inhibitor[23] could be applied as squeeze treatments prior to water breakthrough, during post fracture operation, or during well clean up. The application of liquid scale inhibitors can also be applied via the gas lift system and by capillary string run to above the packer or to the inlet of an electric submersible pump (ESP) if installed. ESP's are quit commonly installed in wells where pressure support is less effective owing to the tight nature of fractured reservoirs. These pumps can be very vulnerable to scale formation at relatively low water cuts and as a result require protection at water cuts as low at 1 %[32]. The onset of water breakthrough and the increased mass of scale possible upon injection water breakthrough would result in a more severe scale risk with the requirement for a higher MIC. For this scaling environment, conventional scale inhibitor squeeze treatments using enhanced adsorption, precipitation or emulsion droplet entrapment, could give effective wellbore protection if applied with the aid of diversion. For fractured wells completed using cemented and perforated tubing, it is possible to use solid diversion agents such as ball sealers and thermally degraded sized wax beads (or pellets) to obtain effective diversion. If scale is inferred by monitoring of a combination of downhole pressure/temperature/pH, and specific ion sensors[22] or by using gauges in conjunction with topside residual analysis and or scale deposition monitors, then dissolver treatments can be applied.

The gaps between technical needs and available technology are similar to those outlined for vertical and horizontal wells. The bridging of the technology gaps for fractured wells would greatly improve risk management for future field developments in deep water or large-scale onshore operations where many dozens of wells could flow into common manifolds, and where access may be limited owing to the hostile environment.

It is clear that a large number of technologies already exists to reduce the risk of scale formation on water breakthrough and to manage it post water breakthrough; and, moreover, to remove it if uncontrolled precipitation takes place. However, while much of this technology has a track record in the North Sea basin where many of the presently available technologies have been applied, the future deployment in the Gulf of Mexico and West Africa will require the 'smart' application of current technology to ensure effective

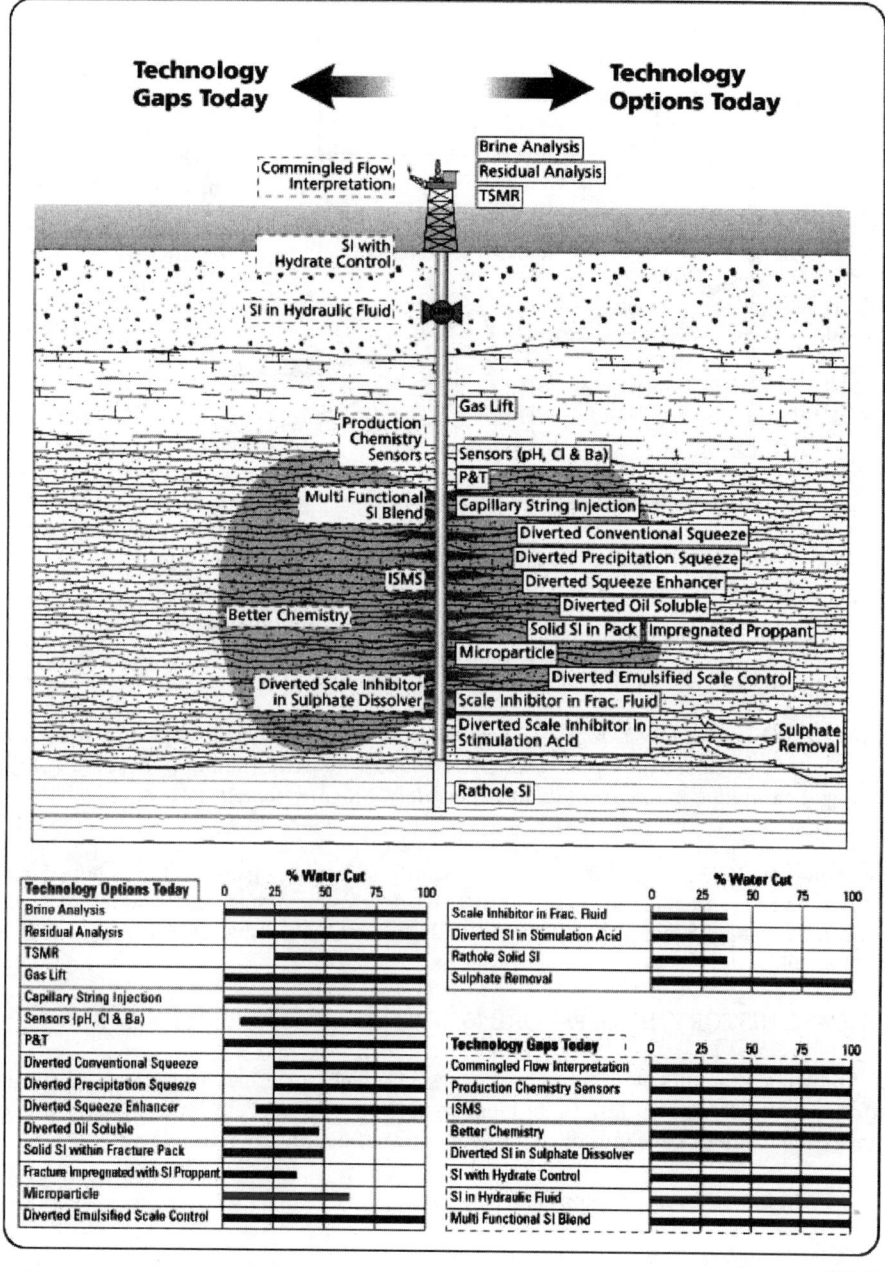

Figure 9 *Technology options in use today plus technology gaps to control/manage downhole and topside scale within a vertically drilled, cemented, perforated, fractured production well*

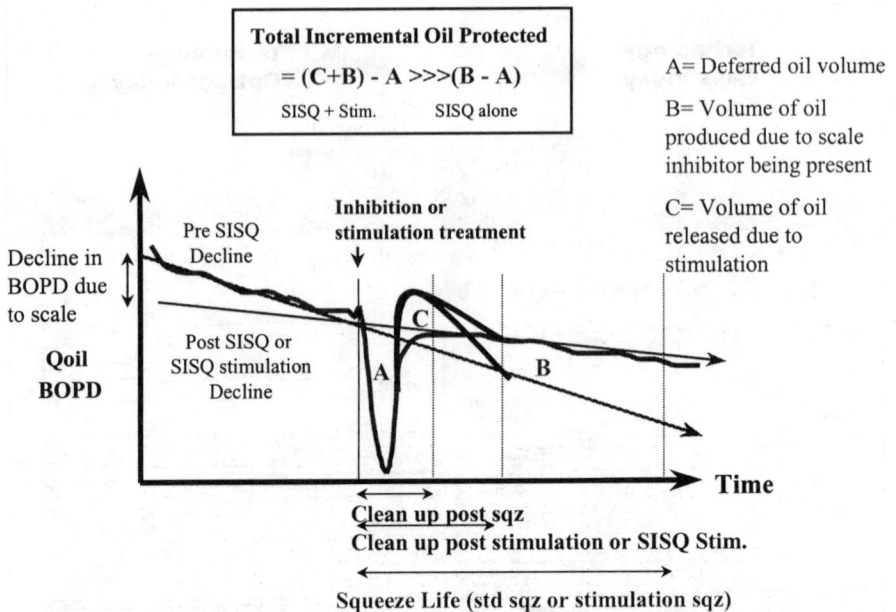

Figure 10   *Oil Production rate decline due to natural reservoir pressure and scale induced damage plus impact on oil rate of scale inhibitor squeeze (SISQ), stimulation and a stimulation combined SISQ treatment*

flow assurance. The filling of the presently identify technology gaps, will, it is hoped, also improve the range of tools available during the development stage of an asset such that scale can be managed more effectively within both the CAPEX and OPEX phases of field development.

4 CASE HISTORY: BP OFFSHORE NORWAY

The following examples will outline the ways in which scale has been managed on the platform and will discuss how a full understanding of the scale type and the use of optimization methods have improved the economics of hydrocarbon recovery.

4.1   Field history and Development

Field A is situated south west of Stavanger, Norway. The reservoir trap is a dip-closed salt dome structure in Upper Jurassic Sandstone - 160 million years old. Pressure support for field A's production is enhanced through both water and gas re-injection and some of the production wells require gas lift for continuous production. In total their are 18 well slots of which 7 are producers, 4 are water injectors and one is a gas injector. Production commenced in October 1986 and average oil production in 2000 was 21,100 BOPD at a water cut of approximately 78 %.

Field B is located south west of Stavanger and 28 km south east of field A. The reservoir trap is again in Upper Jurassic Sandstone. The field is now supported by seawater injection. In total there are 32 well slots of which 12 are for producers with a further 9 wells dedicated to water injection. Production commenced in June 1990 and average oil production in 2000 was 20,800 BOPD at water cut of approximately 42 %. As for field A, field B has a lifetime of approximately 20 years.

The production chemistry issues associated with production of hydrocarbon from the two assets have been dominated by the need to control the scaling potential of the aquifer water when it is co produced with injection seawater. The severity of the scale potential will be described below but suffice to say the control of scale has been the largest single operating cost for the assets.

4.2 Water Chemistry and Scale Types

Typical formation water chemistry plus reservoir pressure/temperature values are shown in Table 11 as are the operating conditions at the wellheads for both fields. The compositions of these brines, when the field was discovered and went into production were some of the most severe in the North Sea basin. Brine composition, combined with the high reservoir temperatures, created a range of scale types, which challenged existing production chemicals due principally to the very high calcium level within the formation water.

Table 11 *Formation waters and injection water for fields A and B*

Ion Type & Property	Field A (ppm)	Field B (ppm)	Injection Water (ppm)
Na	52555	65340	10890
K	3507	5640	460
Mg	2249	2325	1368
Ca	34675	30185	428
Sr	1157	1085	8
Ba	91	485	0
Cl	153025	167400	19700
SO$_4$	44	0	2960
CO$_3$	0	0	0
HCO$_3$	134	76	124
pH	5.4	5.46	8
Res T	143	150	-
Res P	410 bar	450 bar	-
WHFT	110	68	-
WHFP	20 bar	25-60	-

Seawater injection has been utilized to maintain reservoir pressure and improve oil recovery. The introduction of sulphate from the seawater, with the high calcium, strontium and barium within the brines (especially field B), presented a range of different scales that were predicted to form during the life cycle of each production well on both platforms. The types of scale possible from thermodynamic modeling of typical brines from both fields, under a range of production conditions, are presented in Figures 11 to 14. Carbonate scale is not considered to be a problem downhole for field A and is only considered to be an issue for field B during late seawater breakthrough, (Figures 11, 12, 13

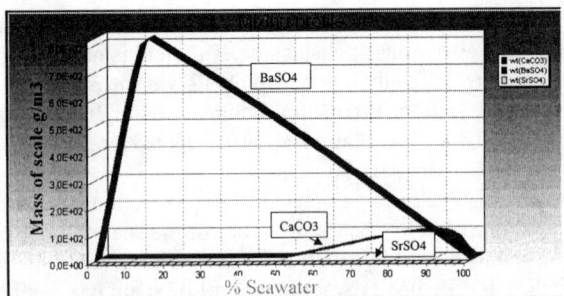

Figure 11   *Mass of scale (gram/m³) predicted to form under reservoir B conditions with increasing proportion (v/v) of injection water (seawater) breakthrough*

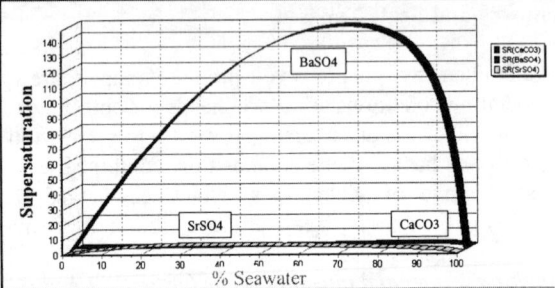

Figure 12   *Supersaturation index under reservoir B conditions with increasing proportion (v/v) of injection water (seawater) breakthrough*

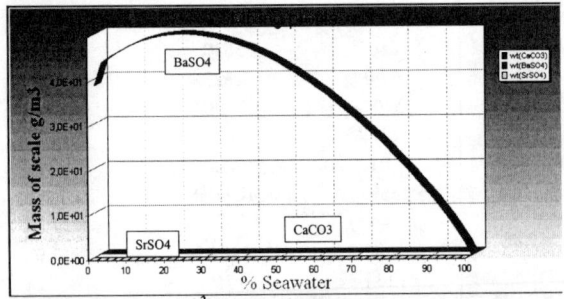

Figure 13   *Mass of scale (gram/m³) predicted to form under reservoir A conditions with increasing proportion (v/v) of injection water (seawater) breakthrough*

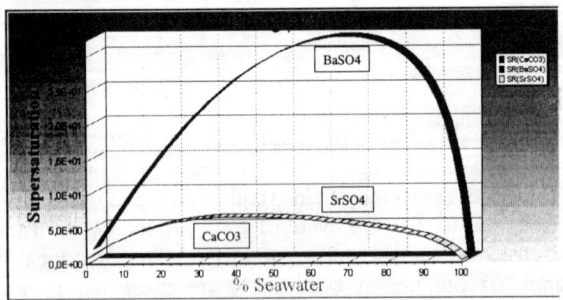

Figure 14   *Supersaturation index under reservoir A conditions with increasing proportion (v/v) of injection water (seawater) breakthrough*

and 14). The maximum mass of barium sulphate scale is predicted to be deposited at 12% seawater breakthrough for field B and 20 % for field A at reservoir conditions (Figures 11 and 13). However the maximum mixed brine supersaturation is predicted to occur at about 65 % seawater for both brines (Figures 12 and 14). The supersaturation values for field A brine are much lower than for field B with the result that inhibition of scale formation will be more difficult for field B brines.

If left uninhibited, production of formation water, and the subsequent mixture of seawater/formation water, would result in the deposition of sulphate scale followed by carbonate scale. The deposition of scale could occur in perforation tunnels and production tubing, causing flow restrictions and possibly compromise the effectiveness of subsurface safety values. As the fluid falls in temperature, it is predicted that the supersaturation, with respect to sulphate scale, will rise under topside process conditions. The observation of the scale types within the two reservoirs follows the scale prediction calculations very closely. Carbonate scale was not observed downhole in either reservoir until a higher seawater cut occurs and is regarded as a problem in the later stages of a well's life cycle.

4.3   Life Cycle Scale Control

Scale squeeze treatments have been deployed to control downhole scale within both fields as water cut has increased with time. The selection procedure for the scale inhibitor to control scale as the water chemistry evolves with time is outlined in a previous publication.[33] Figure 15 shows the development of the downhole scale management strategy for the fields, as new scale inhibitor molecules were developed to meet the challenges of changing water chemistry for both fields. It is clear that that the technology development and deployment within these fields was principally focused on bullhead or coiled tubing application of aqueous based scale inhibitors. Optimization of the scale squeeze program was driven by the development of a total cost of operation (TCO) reduction model within an alliance agreement between BP and the service company.[33] The development of TCO reduction was one of the principal drivers for the introduction of new technology to extend squeeze lifetime. In addition, TCO drives a proactive approach to scale management involving continuous review of water chemistry and the appropriateness of the scale management options to the *current* production situation.

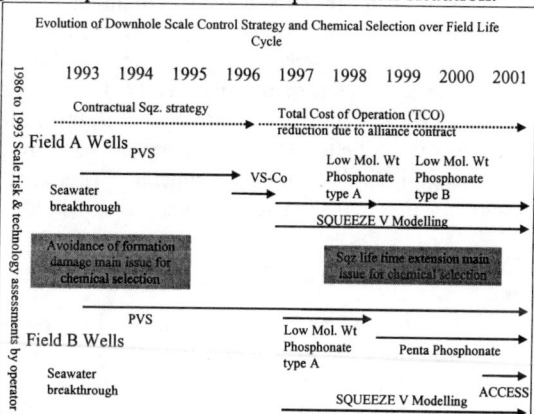

Figure 15   *Evolution of downhole scale control strategy and chemical selection over field life cycle*

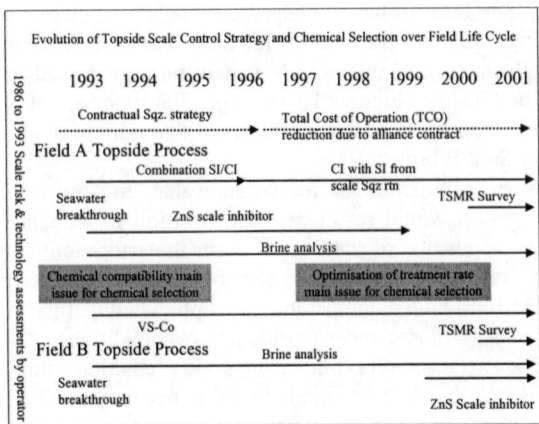

Figure 16   *Evolution of topside scale control strategy and chemical selection over field life cycle*

The development of scale control technology within the topside of both platforms is outlined in Figure 16. Topside scale control was implemented across the process facility on both platforms shortly before seawater breakthrough occurred. A combined scale/corrosion inhibitor was utilized to provide additional inhibition to the produced fluid which already contained scale inhibitor from the squeeze treatments for field A whereas a scale inhibitor alone was applied in field B as corrosion was eliminated due to metallurgy selection. Additional topside chemical was needed in the early life of the field as not all the wells had been squeezed. The untreated wells effectively diluted the inhibited water and reduced the concentration of the downhole scale inhibitor to such an extent that protection was no longer adequate. As more downhole production wells were squeezed, the need for additional topside scale inhibition became less and the treat rate was reduced. Facility integrity was maintained by monitoring known scale deposition hot spots. The treat rate of the combined product was reduced until further reduction would have compromised corrosion control. At present, a corrosion inhibitor is injected on the topside and scale control is managed with downhole squeeze chemical residuals. The effectiveness of this type of scale management is greatly improved by frequent monitoring of inhibitor residuals and determination of the stability of the produced brine across the process system using the TSMR system. Based on this process, the discussion can be made to re-introduce a scale inhibitor to the topside when new wells are brought on line or when production conditions change both downhole and topside. This monitoring method has achieved a considerable reduction in the amount of chemical deployed.

4.4   What Would We Do Differently Today?

If fields A and B were to be developed today, the technology options available today would allow more flexibility in scale management. For example, completion installed scale inhibitors is an option that can be used to defer squeeze treating the wells. Chemical application via capillary string injection lines to below the subsea safety valve would have reduced the incidences of valve failure owing to zinc sulfide deposition. Likewise, the wells could have been protected against scale on first water breakthrough by pre-emptive squeezing (i.e., employing a proactive rather than reactive approach to scale management). The application of oil soluble scale inhibitors or emulsified scale inhibitor would have allowed deployment of chemical prior to water breakthrough without impacting well productivity. Production chemicals that have only been available during the later stage of field life could have, if available during the late 80's to early 90's allowed longer squeeze life times to be achieved. Indeed, sulphate reduction of the injection water is a management option available today that was not available when fields A and B were originally developed. The requirement to lift wells back into production with coiled tubing could be eliminated if gas lift had been installed in both fields during the CAPEX stage of development. The potential values of these principle technologies are outlined in Table 12. Furthermore, currently available monitoring options could be deployed in the wells, and on the topside, to gain a better understanding of the in-situ scaling environment over field life, which can have a dramatic impact on TCO. Better monitoring could help by identifying the point at which a squeeze treatment needs to be re-applied and topside treatment rates optimized to take changing downhole control of scale inhibitor into account.

In essence, today, there are a far greater number of scale management options available than was the case in when fields A and B were at the development (CAPEX) stage. Thus if such fields were to be developed today, a full risk assessment and economic analysis of the various options available for managing scale, as outlined above, would need to be performed to allow a risked development and production scheme to be implemented.

Table 12 *Estimated value of current technology to fields A & B if available from 1993 onwards*

Technology	Estimated Value	Field Code
Gas Lift Possibility (no need for CT)	3000 kNOK (2 CT Operations)	A
Gas Lift Possibility (no need for CT)	12000 kNOK (8 CT operations)	B
Emulsified SI (no need for CT)	300 kNOK (2 CT Operations)	A
Emulsified SI (no need for CT)	12000 kNOK (8 CT operations)	B
Emulsified SI (reduced deferred oil)	43,000 bbls oil	B
OSI (pre-emptive squeezing)	1000 k NOK	B
Capillary Injection ability	7000 k NOK	A
TSMR (prolong life)	14000 k NOK	A and B

Assumptions:
- Covers period from 1993 – 2000 for both field A and B
- Coil Tubing (CT) Operation costs 2000 kNOK
- BullHead (BH) Operation costs   500 kNOK
- Assume that Emulsified SI could be used instead of CT
- Emulsified SI improved clean-up time reduces deferred oil by 1,000 bbl oil compared to water-based
- Use of pre-emptive squeezing assumes ability to protect against all scales in early field life, the value of this technology is due to the elimination of the need for DHSV / Production tubing scale removal, est. operation cost 250 kNOK, when WC is low < 5.0%
- Capillary injection ability assumes that we are able to prevent scale formation across production tubing and DHSV – value is no need for DHSV change-outs / re-installation (operation cost 250 kNOK – assume 75% of failures to be scale related, for field A 28 operations)
- In early sqz days residual SI and required MIC conservative- resulted in shortening of sqz lifetime and hence higher frequency of scale squeeze then required – assume with proper scale monitoring able to reduce number of scale squeezes by 25% for both fields (112 operations x 0.75 = 84 , 28 sqz operations saved – only BH)

5 CONCLUSIONS

Scale control technology has greatly improved in recent years. Significantly the costs of implementation of new technologies at the CAPEX stage can be offset against OPEX costs during production. However even for mature fields such new technologies can provide economic ways of controlling scale. Identified technology gaps still exist for better delivery of scale inhibitors downhole, improved sensors, flow interpretation and incorporation of scale inhibitors within multi purpose application packages.

1   Effective scale management should encompass two essential processes: risk assessment and economic evaluation. Scale management risk assessment allows the severity of any predicted mineral scales to be reviewed against current and future field development options. This allows current and near-market scale management technologies to be objectively assessed and ranked allowing scale management options to be identified. Employing simple economic models of the full life of field cost of scale management for the various technology options identified in the risk assessment phase, allows decisions to be made on the choice of both hardware development options and scale management options during the CAPEX phase of a development, and, moreover, allows more cost effective scale management solutions to be identified, and justified, during the OPEX phase a field, i.e., the risk and economic aspects of scale management should be employed throughout a fields lifecycle.

2   From the case study it is clear that new and innovative technology can significantly reduce the total cost of scale control and deferred oil associated with scale squeeze treatments and well/process cleaning operations. While presently available technology can reduce current operating costs gaps in technology still exist which, if closed, have the potential to deliver even greater savings.

3   If the current technology in scale control and monitoring had been available during the CAPEX phase of both fields then significant savings in operating costs could have been achieved. Of most significance is the use of gas lift to aid in well clean up (reduce the use of coiled tubing), scale squeeze treatment prior to water breakthrough (reduce production decline due to low water cut scale) and longer squeeze life time based on better chemistry/carrier systems (reduce deferred oil costs).

4   Economic as well as ecological impact of discharging scale inhibitors into the North Sea basin (UK and Norway) and the Gulf of Mexico has been reviewed. The development of new technologies, which limit such discharge and ecological impact, will have a high value to both operating and service companies.

Acknowledgement

The authors would like to thank BP Exploration and ONDEO Nalco Energy services for permission to publish this work. We also acknowledge the help and co-operation of members of the asset teams for their assistance in carrying out the evaluations and treatments described in this paper.

References

1   J. A. Hardy and I. Simm, *Low Sulphate Seawater Mitigates Barite Scale*, Oil and Gas Journal, December 2, 1996.

2   M.M. Jordan, K.S. Sorbie, P. Jiang, M.D. Yuan, A.C. Todd, and K.E. Hourston, *Scale Inhibitor Adsorption/ Desorption and the Potential for Formation Damage in*

Reconditioned Field Core, presented at the SPE International Symposium on Formation Damage Control, Lafayette, Louisiana, 7-10 February 1994.

3   R. Børeng, K.S. Sorbie and M.D. Yuan, *The Underlying Theory and Modelling of Scale Inhibitor Squeezes in Three Offshore Wells on the Norwegian Continental Shelf*, presented at the 1994 NIF International Oil Field Chemicals Symposium, Geilo, Norway, Mar. 20-23.

4   K.S. Sorbie, M.D. Yuan, M.M. Jordan and K.E. Hourston, *Application of a Scale Inhibitor Squeeze Model to Improve Field Squeeze Treatment Design*, paper SPE 28885 presented at the 1994 SPE European Petroleum Conference (Europec94), London, UK, Oct. 25-27.

5   I.R. Collins, N.J. Stewart, S.R. Wade, S.G. Goodwin, J.A. Hewartson and S.D. Deignan, *Extending Scale Squeeze Lifetimes Using a Chemical Additive: From the Laboratory to the Field*, presented at the 1997 Solving Oilfield Scaling Conference, 23-24 January.

6   I.R. Collins, L.G. Cowie, M. Nicol and N.J. Stewart, *Field Application of a Scale Inhibitor Squeeze Enhancing Additive*, paper SPE 54525 presented at Annual Technical Conference, San Antonio, Texas, 5-8 Oct. 1997.

7   H.M. Bourne, S.L. Booth and A. Brunger, *A.: Combining Innovative Technologies To Maximize Scale Squeeze cost Reduction*, SPE 50718, presented at SPE International Symposium on Oilfield Chemistry, Houston, 16-19 February 1999.

8   G.E. King and S.L. Warden, *Introductory Work in Scale Inhibitor Squeeze Performance: Core Test and Field Results*, SPE18485, presented at the SPE International Symposium on Oilfield Chemistry, Ananheim, California, 20-22 February 1991.

9   J. E. Pardue, *A New Inhibitor for Scale Squeeze Applications*, SPE21023, presented at the SPE International Symposium on Oilfield Chemistry, Anaheim, California, 20-22 February 1991.

10 P. D. Ravenscroft, L.G. Cowie and P.S. Smith, *Magnus Scale Inhibitor Squeeze Treatments – A Case History*, SPE 36612, presented at SPE annual Technical Conference and Exhibition, Denver, Colorado, 6 – 9 October 1996.

11 K.S. Sorbie, A.C. Todd, R.M.S. Wat and T. McClosky, *Derivation of Scale Inhibitor Isotherms for Sandstone Reservoirs*, Royal Society of Chemistry Publications: Chemicals in the Oil Industry – Developments and Applications, Edited by P.H. Ogden, 1991.

12 H. Zhang and K.S Sorbie, *SQUEEZE V: A Program to Model Inhibitor Squeeze Treatments in Radial and Linear Systems – User's Manual*, Department of Petroleum Engineering, Heriot-Watt University, Edinburgh, December 1997.

13 H. Zhang and K.S. Sorbie, *ASSIST II User's Manual*, (1997) Department of Petroleum Engineering, Heriot-Watt University, Edinburgh.

14 M.M. Jordan, M.C. Edgerton, J. Cole-Hamilton and K. Mackin, *The Application of Wax Divertor to Allow Successful Scale Inhibitor Squeeze Treatment to Sub Sea Horizontal Wells, North Sea Basin*, paper SPE 49196 prepared for the SPE Annual Technical Conference and Exhibition, New Orleans, Mississippi, 28-30 September 1998.

15 E.J. Mackay, A. Matharu, K.S. Sorbie, M.M. Jordan and R. Tomlins, *Modelling of Scale Inhibitor Treatments in Horizontal Wells: Application to the Alba Field*, SPE39452, presented at the 1998 SPE International Symposium on Formation Damage.

16 M.M. Jordan, K.S. Sorbie, M.F. Marulier, K. Taylor, K. Hourston and P. Griffin, *Implication of Produced Water Chemistry and Formation Mineralogy on Inhibitor Adsorption/ Desorption in Reservoir Sandstone and their Importance in the Correct*

Selection of Scale Squeeze Chemicals, presented at the RSC 6th International Symposium on Chemistry in the Oil Industry, Charlotte Mason College, Ambleside, Cumbria, 14-17 April 1997.

17 M.M. Jordan, K.S. Sorbie, P. Griffin, S. Hennessey, K.E Houston and P. Waterhouse, *Scale Inhibitor Adsorption/Desorption vs. Precipitation: The Potential for Extending Squeeze Life While Minimising Formation Damage*, SPE 30106, presented at the European Formation Damage Conference, Hague, Netherlands, 15-16 May 1995.

18 B.L. Carlberg, *Scale Inhibitor Precipitation Squeeze for Non-Carbonate Reservoirs*, SPE 17008, presented at the SPE Production Technology Symposium, Lubbock, TX, 16-17 November 1987.

19 A, Malandrino, M.D. Yuan, K.S. Sorbie and M.M. Jordan, *Mechanistic Study and Modelling of Precipitation Scale Inhibitor Squeeze Processes*, SPE29001, presented at the SPE International Symposium on Oilfield Chemistry, San Antonio, TX, 14-17 February 1995.

20 H.M. Bourne, I.R. Collins, L.G. Cowie, C. Strachan and M. Nicol, *The Role of Additives on Inhibitor Precipitate Solubility and its Importance in Extending Squeeze Lifetimes*, presented at the 1997 IBC Technical Conference on Advances in Solving Oilfield Scaling, Aberdeen, January.

21 I.R. Collins, M.M. Jordan and S.E. Taylor, *The Development and Application of a Novel Scale Inhibitor System for Deployment in Low water cut, water Sensitive or Low Pressure Oil Reservoirs,* Presented at the SPE Second International Symposium on Oilfield Scale, January 2000. Aberdeen, UK.

22 S. Brabon, P.A. Fennell and R. Peat., *Permanent Monitoring of pH/Chloride using Electrochemical Sensors*, Presented at SPE Third International Symposium on Oilfield Scale, January 2001. Aberdeen, UK.

23 I.R. Collins, M.M. Jordan, N.D. Feasey and G.D. Williams, *The development of Emulsion Based Production Chemical Deployment systems*, SPE 65026. SPE International Symposium on Oilfield Chemistry, Houston Texas, 13-16 February 2001.

24 D.H. Emmons, G.C. Graham, S.P. Holt, M.M. Jordan and B. Locardel, *On-Site, Near-Real-Time Monitoring of Scale Deposition*, SPE 56776, presented at SPE Annual Technical Conference and Exhibition, Houston, Texas, 03-06 October 1999.

25 N.D. Feasey, R.J. Wintle, E.R. Frieter and M.M. Jordan, *Field Experience with a Novel Near Real time Monitor For Scale Deposition in Oilfield System*, NACE 2000, Orlando Florida, March 2000.

26 P.S. Smith, L.G. Cowie, H.M. Bourne, M. Grainger and S.M. Heath, *Field Experiences with a Combined Acid Stimulation and Scale Inhibitor Treatment,* SPE 68312 presented at the SPE Third International Symposium on Oilfield Scale, Aberdeen, 30-31 January 2001.

27 E.J. Mackay and K.S. Sorbie, *Modelling Scale Inhibitor Squeeze Treatments in High Crossflow Horizontal Wells*, paper SPE 50418 presented at the 3rd SPE International Conference on Horizontal Well Technology, Calgary, Alberta, 1-4 November 1998.

28 M.M. Jordan, M. Egderton and E.J. Mackay, *Application of computer Simulation Techniques and solid Divertor to improve Inhibitor Squeeze treatments in horizontal Wells*, SPE 50713, presented at SPE International Symposium on Oilfield Chemistry, Houston, 16-19 February 1999.

29 M.M. Jordan, I. Hiscox, K. Mackin and I.R. Collins, *The Design and Deployment of Enhanced Scale Dissolver/Squeeze treatments in Platform and Subsea Horizontal production wells, North Sea Basin,* presented at the 12th NIF International Symposium on Oilfield Chemicals, Geilo, Norway, April 1-3 March 2001.

30 P.J.C. Webb, T.A. Nistad, B. Knapstad, P.D. Ravenscroft and I.R. Collins, *Advantages of a New Chemical Delivery System for Fractured and Gravel-Packed Wells*, SPE Production and Facilities, **14**, 210-219 (1999)..

31 J.P. Martins, R. Kelly, R.H. Lane, J.D. Olson and H.L. Bannon, *Scale Inhibition of Hydraulic Fractures in Prudhoe Bay,* presented at the 1992 SPE International Symposium on Formation Damage, Lafayette, Louisiana, February 26-27.

32 M.M. Jordan, C. Graff and K. Cooper, *The Development and Deployment of a Scale Squeeze Enhancer and Oil Soluble Scale Inhibitor to Reduce Deffered Oil Production Losses During Squeezing of Low Water Cut Wells, North Slope, Alaska.*, resented at the SPE International symposium on Formation Damage Control, Lafayette, Louisiana, February 2000.

33 M.M. Jordan, K. Sjuraether, G. Seland and H. Gilje, *The Use of Scale Inhibitors Squeeze Placement Software to Extend Squeeze Life and Reduce Operating Costs in Mature High Temperature Oilfields*, presented at NACE 2000, Orlando Florida, March 2000.

34 Deepwater Program, *Literature Review, Environmental Risk of Chemicals Used in Gulf of Mexico Deepwater Oil and Gas Operations*, Volume I: Technical Report and Volume II: Appendices OCS Study MMS 2001-011 and MMS 2001-012.

NEW METHODS FOR THE SELECTION OF ASPHALTENE INHIBITORS IN THE FIELD

Hans-Jörg Oschmann

TR Oil Services Limited, Howe Moss Avenue, Kirkhill Industrial Estate, Dyce, Aberdeen, Scotland AB21 0GS. E-mail: Hans.Oschmann@trosnet.com

1 INTRODUCTION

Formation damage, production decline or the loss of transport capacity due to the precipitation and subsequent deposition of asphaltenes is costing the oil industry millions of dollars each year. Even amounts of only 50ppm of precipitated asphaltenes can cause complete blockages in producing wells[1]. The correct selection and application of a suitable chemical can reduce or eliminate asphaltene deposition and associated problems. However, before any test work is started, it is essential to confirm that the deposition is really caused by asphaltenes and has not been confused with other deposits of similar appearance like paraffin or iron sulphide. Asphaltene depositions as observed in the field vary in appearance. Sometimes they form black, brittle and shiny deposits which look similar to graphite, in other cases they may form deposits, which are brown, soft and sticky. The fact that asphaltenes can co-precipitate with paraffin or even scale, makes it even more important to check the composition of an assumed asphaltene deposit.

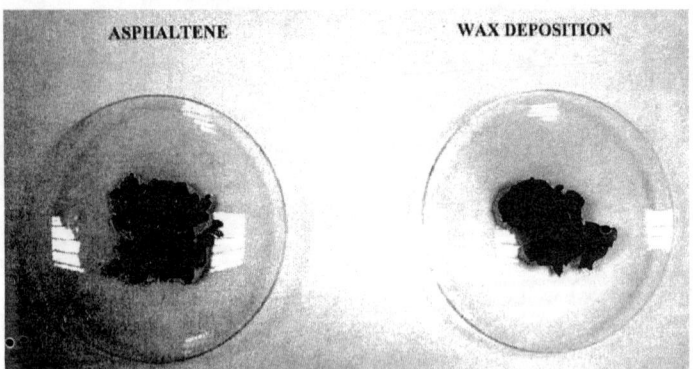

Figure 1   *Asphaltene and wax depositions are very similar in appearance, though their chemical composition is entirely different*

Paraffin is a chemically well defined substance consisting of the saturated components of crude oil with a carbon number higher than 18. Asphaltenes however are more generally defined as crude oil components soluble in toluene and insoluble in lower molecular weight n-alkane[2,3], so their exact chemical composition varies widely in different crude oils. Therefore there is no such thing as "the asphaltene" with a defined chemical structure, although through the years detailed research[4,6] has been able to establish common properties, with subsequent identification of typical molecular groups and structures, which are common in all asphaltenes regardless of origin.

It is known that asphaltenes are dispersed as colloids in the crude[5] which are stabilised by other natural crude oil components. This led to the development of models for the colloidal state of asphaltenes in crude and how other crude components are arranged to them. The most common model postulates an asphaltene core consisting of poly-condensed aromatic structures, polar components and salts[6,7]. This core is surrounded by resins, who consist of a polar/aromatic part attached to the asphaltene core, and a non-polar alkane part, pointing away from the core into the surrounding crude. These resins work as stabilising agents for the asphaltenes, as they keep the asphaltenes dispersed in the crude and separated from each other. Therefore the higher the ratio of resins to asphaltenes is in the crude, the less likely the asphaltenes will agglomerate, flocculate and deposit.

Although the asphaltenes are in equilibrium with the resins under reservoir conditions, changes in pressure and temperature during production or the mixing with lower molecular n-alkanes (aliphatic condensate) results in stripping of the resins from the outer asphaltene core followed by colloid flocculation. Asphaltene flocculation may also be initiated by a number of other operations which include acid or steam treatments (pH-change, temperature change, oxidation of resins etc.) and other EOR processes[8].

2 CHEMICAL SOLUTIONS TO ASPHALTENE PROBLEMS

The best option to avoid asphaltene related problems would be to design the system to operate under conditions where asphaltenes do not form. Although this might be theoretically possible for some problematic fields[1] it means compromising production and is rarely economic. Other options include cyclic dissolver treatments or mechanical removal of asphaltenes, but if the deposition is more severe, a range of asphaltene inhibitors and dispersants provide the easiest and most common solution to deposition problems. Chemical applications range from continuous dosage to batch/squeeze treatments. It is known that based on the differences between the "asphaltenes" of different origins the chemicals performance is crude oil specific and that for even the best chemicals exists a number of crude oils, where they have little to no effect, so testing and evaluation is essential for the selection of the appropriate inhibitor or dispersant.

2.1 Asphaltene Inhibitors versus Dispersants

There are generally two possible ways for chemicals to prevent deposition in oilfield fluids: Inhibition or Dispersion

The first, inhibition, increases the stability of components which would otherwise precipitate and allows operating under more severe conditions. Typical examples for other commonly used inhibitors would be paraffin inhibitors and thermodynamic or kinetic hydrate inhibitors.

The second, dispersion, influences particle size and agglomeration behaviour, to insure that any precipitated solids are dispersed over the volume of the liquid and unable to deposit. In this case, examples of other commonly used chemicals would include paraffin dispersants, PPDs, or hydrate anti-agglomerants.

In case of asphaltenes, the use of an asphaltene inhibitor would stabilise the asphaltene micelles in the crude by supporting the action of natural crude oil resins, who protect the outer layer of the asphaltene colloid.

The use of an asphaltene dispersant however, would not necessarily increase the stability of the asphaltenes, but disperse any asphaltenic agglomerates and keep them from forming depositions.

3 EVALUATION METHODS FOR ASPHALTENE INHIBITORS AND DISPERSANTS

A number of tests methods have been suggested, including tests based on viscometry[9], conductivity[10], fluorecsence spectroscopy[11] and microscopy[12] just to name a few, but none of them is suitable for tests in the field, as these tests are designed for a modern lab environment. This is even more true for common flocculation point testing or solid detection systems, which is not only bulky, and heavy, but fragile which makes it unsuitable for field use. So all tests in the field are generally still based on simple precipitation "bottle" tests, where crude is precipitated with a fixed amount of precipititant and the volume of precipitated asphaltenes obtained on an untreated sample is compared with the volume obtained on treated crude. Less volume of precipitated asphaltenes suggests a more efficient chemical. Although this method seems simple and practical, it includes a number of in-built errors and can at best be used as an indication for potential efficiency. The method is for instance unsuitable for the ranking of asphaltene dispersants, which would try to keep asphaltenic particles dispersed other the sample volume, and, in direct contradiction to the design of the test, keep them from sedimentation. Therefore, in the worst case, the least effective dispersant would appear the most suitable chemical according to this test.

So if the simple test is sufficient and the accurate equipment difficult to transport, why not send the crude to central labs for evaluation?

Even today, the sending of samples from oil-field location may, depending on the country of export, take a long time. In some cases, crude samples take months to travel from a third world country to a central lab in Europe, which results in an unacceptable response time. Also the transport of crude to distant locations may result ageing of the crude[13], as the asphaltenes aim to reach the state of equilibrium with other crude components again. So the best practise would be the selection of the appropriate chemical directly in the field.

The demand for the use of lightweight, portable field testing equipment, has led us to the development of two pieces of field equipment. One of them is an automated fibre-optic system for the determination of crude stability and the selection of asphaltene inhibitors, which we call Flocculation Point Tester. The other is a modified small particle-sizer, with fast response times, which only needs 30 min per chemical tested, which we call Laser Dispersant Tester.

3.1 Determination of the Flocculation Point

The Asphaltene Flocculation Point is defined as the onset of asphaltene flocculation, caused either by a PVT change, or by any other change that affects the crude oil composition. Non PVT induced changes of crude oil composition in the field usually occur as a result of mixing crude oils of different origin, which may sometimes even mean different horizons in the same reservoir.

In laboratory test work, either a pressure-drop experiment, which requires a live crude sample, or titration with a standard precipitant (usually n-pentane or n-heptane)[14] is used to determine the flocculation point of crude oil.

The lower the pressure can be reduced to or the more precipitant can be added to the crude oil before flocculation occurs, the more stable are the asphaltenes in the crude. So an increase in the flocculation point correlates with an increase in asphaltene stability.

Therefore the efficiency of Asphaltene Inhibitors can be assessed by the comparison of the flocculation point of an uninhibited crude oil with the result of a test performed with inhibited crude.

The first set-up we developed was designed for pressure and titration experiments and allowed us to observe the flocculation of asphaltenes under conditions of 200bar and 200°C max. Although not designed for use in the field, we already started to miniaturise all essential electronic parts for the fibre-optic sensor system, to allow us the development of an even smaller version for use in the field

Figure 2 *Flocculation Point Test Equipment (Autoclave version). The two grey boxes under the Metrohm dosimat contain the electronic of the fibre-optic sensor*

The PVT range of this initial set-up was only limited by the max. temperature and pressure required for safe operation of the autoclave initially used, later experiments on a higher rated autoclave allowed the extension to pressures up to 600bar.

The set-up consists basically of a double-mantled autoclave where the oil sample is continually stirred with a torque-controlled stirrer. An attached thermostatic bath provides the thermostatic fluid for the double mantle. The computer is attached to the electronic control unit which determines the light transmission through the crude oil sample with a fibre-optic sensor, and is also used to control temperature and pressure. For titration experiments at ambient pressure, a Metrohm dosimat is used, whereas for titration at elevated pressures and temperatures, a Knaur HPLC pump is employed.

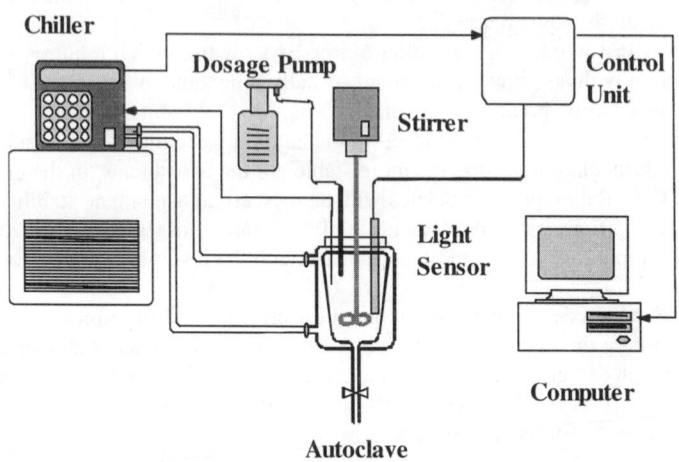

Figure 3   *Shows a schematic diagram of the asphaltene flocculation apparatus*

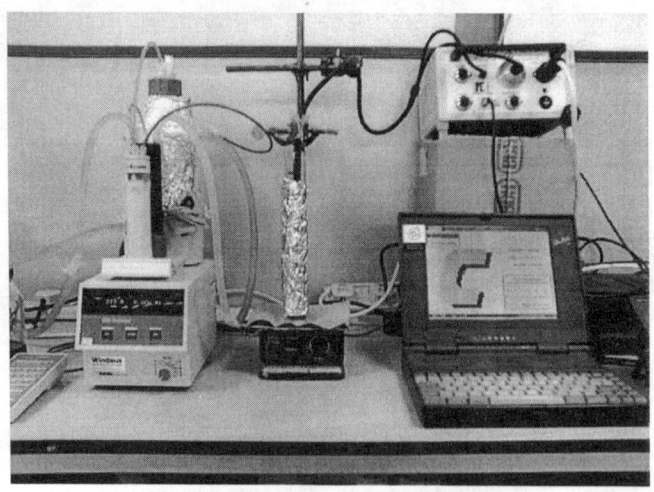

Figure 4   *Shows the transportable Flocculation Point Tester during a field evaluation*

The equipment developed for use in the field had to be considerably smaller and is lacking the massive autoclave of the stationary unit, as field labs generally lack the facilities to deal safely with high pressure / high temperature samples. Instead it is fitted with a 12V transformer and adapter for a car cigarette lighter, which makes it possible to operate it even if a proper lab is missing. Also the software was designed to use a minimum of resources, so even an old 486 kB laptop computer can be used for data retrieval. The whole equipment is fitted into an aluminium case and the total weight of equipment, box and accessories is less than 12 kg.

During a titration test, the precipitant is continually added. This has two effects on the transparency of the crude oil.

1. The crude is diluted which leads to a continuous increase of transmittance
2. The colloidal stability of the asphaltenes in the crude decreases

After a certain amount of added precipitant, the stability is decreased so far that small flocs of asphaltenes are precipitating and the light transmission decreases. A typical result is shown in Figure 5. The maximum at 45.95 ml represents the flocculation point.

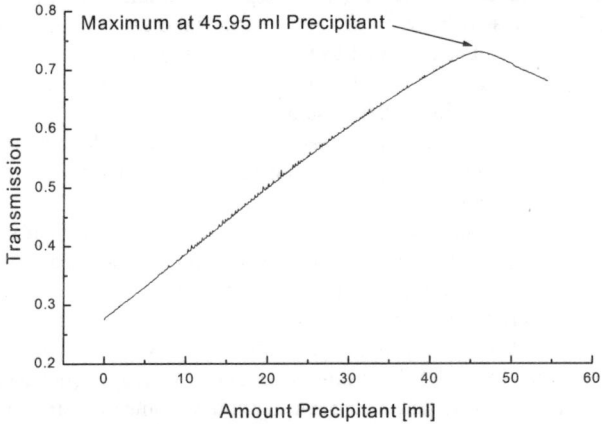

Figure 5   *Determination of the flocculation point by titration with n-heptane (North African Crude, asphaltene content <1%)*

3.2 Limitations of the Flocculation Test Method

Although we have not yet found an asphaltenic crude oil sample which was unsuitable for the test method, results obtained on relatively stable crudes with less than 2% asphaltenes are in some cases difficult to analyse, as they tend to develop a plateu instead of a well defined maximum, as the amount of asphaltenes available for precipitation is to small to overcome the diluting effect of the precipitant.

Tests on nearly a hundred samples with an asphaltene content varying between 2-38% (the latter a number of vacuum residues) gave a maximum error of 2.5%, with tests on crudes having an asphaltene content between 5-15% showing a maximum error of 1.5% We were positively surprised, that even tests on vacuum residues to predict bitumen stability were giving good results.

Note: After initial tests, we found it disadvantageous to dilute the crude with toluene to reduce viscsity, as has been suggested by other authors. Diluted crude needs to equilibrate for at least two weeks before the solution can be used[13]. Since then we have only tested on undiluted samples, in cases of highly viscous fluids like vacuum residues at elevated temperature up to 85 °C.

3.3 Determination of asphaltene dispersant efficiency:

Asphaltene dispersants are substances, which do not affect the flocculation point, but reduce the particle size of flocculated asphaltenes, thereby reducing operational problems. It has been frequently observed, that a number of commercial products, including most surfactant type chemicals, do show little or no efficiency in flocculation point tests, although they are effective in the field[15]. So a different techniques had to be developed to allow accurate comparison of asphaltene dispersants.

Past efforts to develop techniques, which are more reliable than the bottle test mentioned earlier, included the use of UV/VIS spectrophotometers. In this case, a sample is mixed with a fixed dose of precipitant, and after a fixed time during which a part of the asphaltenes flocculated and sedimented to the bottom, the transmission of the supernatant liquid was determined. Low transmission indicated a larger amount of asphaltenes dispersed in the solution than high transmission rates.

Still this test did not give any information about the particle size of the asphaltenes, which had precipitated, although this is of importance as smaller particles are less likely to result in operational problems.

Other test methods, which were purely aimed on the determination of particle size distribution, give no information on the amount of asphaltenes remaining in the solution. As these methods generally employ large commercial particle-sizers, which also require gas connections, they are not suitable for field application.

The new method tries to combine the best of both worlds in one portable piece of equipment, which only requires a minimum amount of sample (normally 1ml of crude sample per test). As the test is automated and computer controlled, it is less susceptible to human error.

During the test, a near infrared laser scanner analyses transmission and back scattering of light. The apparatus operates in both static and kinetic mode and is able to detect and quantify particle size variation or particle migration in colloidal systems. The sample under analysis is placed into a flat-bottomed cylindrical cell. The reading head comprises a pulsed near infrared light source and two synchronous detectors; the transmission detector receives the light which goes through the sample, whilst the back scattering detector receives the light back scattered by the sample. The reading head scans the entire length of the sample, acquiring transmission and back scattering data every 40μm. The integrated microprocessor's software handles data acquisition, analogue to digital conversion, data storage, motor control and computer dialogue.

A typical result for tests performed is given in Figure 7, in which the effect of various inhibitors on the stability of the asphaltenes can be seen. The blank shows precipitation after a few minutes with most of the asphaltenes sedimenting to the bottom. The reference standard, a chemical developed through a JIP project for the development of asphaltene inhibitors, shows a good performance, keeping the asphaltenes partially dispersed in the solution. Chemicals A and B however show excellent performance on the crude and keep all the asphaltenes dispersed with no sedimentation.

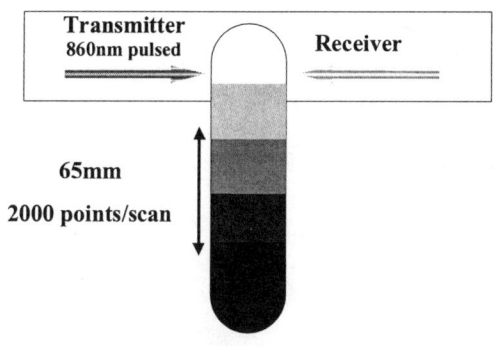

Figure 6   *Laser Dispersant Test*

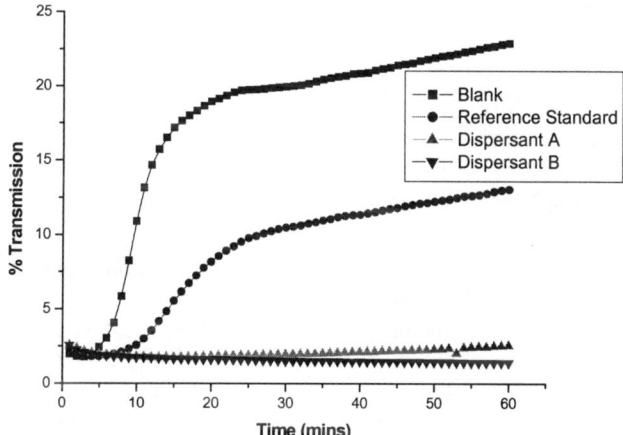

Figure 7   *Performance Evaluation using LDT tests*

Both the Flocculation Point Tester and the Laser Dispersant Tester are also suitable for the monitoring of chemical performance in the field, as the shift in the flocculation point or the decrease in sedimentation and transmission during an LDT test can be correlated to the efficiency of the chemical in the crude. The easiest way to correlate is the development of a calibration graph prior to the treatment, as presented in the next figure.

Increasing amount of asphaltene dispersant lead to a decrease of transmission and particle sedimentation, which in turn correlates with particle size.

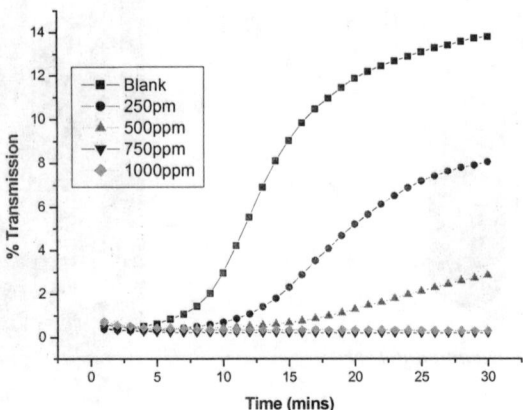

Figure 8   *Monitoring the effectiveness of asphaltene chemicals on crude oil*

The results of the use of both pieces of equipment to monitor the efficiency of an asphaltene squeeze treatment will be presented in another paper in near future.

4 CONCLUSIONS

Selection of the appropriate chemical to mitigate asphaltene problems heavily depends on the correct test procedure for the specific type of chemical used on the crude oil. Whereas the determination of the flocculation point is the best option for asphaltene inhibitors, asphaltene dispersant require a special test which produces information on particle size and the grade of dispersion of the asphaltenic particles in the crude.

For both types of tests, lightweight, automated test equipment has been developed and tested under field conditions.

In addition for the use as tools for the selection of chemicals, the equipment also offers the potential to be used as monitoring tools for chemical efficiency during trials.

5 ACKNOWLEDGEMENTS

The author would like to thank Dr. Alfred Hase from the German Petroleum Institute for his valuable contribution to the development of these test methods and his enthusiasm. I would also like to thank Dr. Iradj Rahimian and Prof. D. Kessel for their guidance, and continuous support. The cooperation with the German Petroleum Institute has been very fruitful for both parties and is a good example for successful knowledge transfer between industry and research.

References

1   Keinitz, W. and Andersen, S.I., Paper presented at the 8th Symposium on *Oil Field Chemistry*, Geilo, Norway, March 1997.

2   Although n-heptane is recommended in all present versions of ASTM, IP and DIN/ISO n-pentane is still commonly used.

3   ASTM D6560-00, DIN 51595 or IP 143.

4   Speight, J.G., Moschopedis, Sp. E., *Some observations on the molecular 'nature' of petroleum asphaltenes*, Symposium Chemistry of asphaltenes, Washington, 1979.

5   Neumann, H.-J., *Bitumen – neue Erkenntnisse über Aufbau und Eigenschaften*, Erdöl Kohle Erdgas Petrochemie; 34; 1981; **336-342**.

6   Yen, T.F, *Structure of petroleum asphaltenes and its significance*, Energy Sources 1; 1974; 4; **362-365**.

7   Jovanovic, J.A., *Model of a micelle of the petroleum colloid system*, J. Serb. Chem. Soc.; 59(9);1994; **619-625**.

8   Novozad, Z. and Costain, T.G., *Experimental and Modeling studies of Asphaltene equilibria for a reservoir under CO2 Injection*, SPE 20530 Paper presented at the 65th Annual Technical Conference and Exhibition of the Society of Petroleum Engineers, New Orleans, *LA, Sept. 23-26, 1990.

9   Escobedo, J., Mansoori, G.A., *SPE Production and Facilities*, 1995, **115.**

10   Fotland, P., Arnfridsen, H., Fadness, F.H., *Proceedings 6th FPECPD* 92 – Cortina, July 12-24, 1992.

11   MacMillan, D.J., Tackett Jr, J.E., Jessee, M.A., Monger-McCluer, T.G., *J.Pet. Tech.*, 1995, **788.**

12   Buckley, J.S., *Fuel Science Technology International* 14, 1996, **55.**

13   Hase, A., *Asphaltenausfällung in verschiedenen Löser-/Fällersystemen bei unterschiedlichen Druck- und Temperaturbedingungen und Möglichkeiten zu ihrer Verhinderung*, dissertation, TU-Clausthal, 1995.

14   Bouts, M.N., Wiersma, R.J., Muijs, H.M. and Samuel, A.J., *An Evaluation of new Asphaltene Inhibitors*: Laboratory Study and Field Testing, *Journal of Petroleum Technology*, Sept. 1995, P.782-786.

15   de Boer, R.B., Leerloyer, K., Eigner, M.R.P., Mij, N.V., van Bergen, A.R.D., *Screening of Crude Oils for Asphalt Precipitation: Theory, Practice and the Selection of Inhibitors*, SPE 24987, Paper presented at the EPC in Cannes, France, 16th – 18th November 1992.

THE DEVELOPMENT OF ADVANCED KINETIC HYDRATE INHIBITORS

Bob Fu

ONDEO Nalco Energy Services, 7705 Highway 90-A, Sugar Land, Texas 77478, USA

1 ABSTRACT

New advanced kinetic hydrate inhibitors (KHI) have been developed for applications in oil and gas fields. These new-generation inhibitors demonstrated the enhanced performance resulting from the improvement of polymer chemistry and reformulation. The polymer chemistry was improved by a number of concerted approaches that included the synthesis of new polymers, copolymerization, control of molecular weight and distribution, and subtle changes in the polymerization process. The resulting polymers were further formulated with performance-enhancing additives, using synergistic compounds and solvent packages. The improved performance of advanced KHI was verified in the autoclave and rocking cell tests. All advanced KHI passed the benchmark of extending induction time for hydrate formation to more than 48 hours at 13°C subcooling.

2 INTRODUCTION

Control of hydrates is a major challenge for the production and transportation of oil and natural gas. Natural gas hydrates are crystalline, ice-like solids that form when small hydrocarbon molecules are trapped in hydrogen-bonded water cages under high pressure and low temperature conditions.[1,2] Hydrate formation can be inhibited by a number of established methods, such as dehydration, heat management and chemical inhibiton.[3] The preferable method is often selected based upon the operating environment and the cost of the control method. These methods can also be used conjunctionally for hydrate prevention or remediation.

Chemical inhibition is typically implemented with thermodynamic inhibitors, such as methanol and glycol. These inhibitors have the ability to shift hydrate equilibrium conditions toward higher pressures and lower temperatures. As a consequence, the operating condition is no longer in the hydrate stable region. However, a large quantity of inhibitors is usually required, especially for severe conditions. The demand for a large volume of thermodynamic inhibitors creates logistic problems and safety concerns for the handling, transportation and storage of these chemicals.

Recently, low dosage hydrate inhibitors (LDHI) have been developed as an alternative to thermodynamic inhibitors and as a relief for high dosage problems.[4-18] These chemicals are very different from the traditional thermodynamic inhibitors because they do not shift the thermodynamic equilibrium of hydrate formation. Instead, these inhibitors interfere with the process of hydrate formation via a number of proposed mechanisms that further divide them into different groups. Two of the most popular LDHI are kinetic hydrate inhibitors (KHI) and anti-agglomerants (AA). Anti-agglomerants are thought to exhibit the ability to change the size of hydrate crystals and the morphology of their agglomerates by incorporating the inhibitor molecule into the hydrate crystal lattice.[19] As a consequence, the resulting hydrates are slush-like and transportable. Kinetic hydrate inhibitors are believed to interfere with the hydrate nucleation step and/or the crystal growth process. The interference slows down the kinetics and consequently extends the induction time for hydrate formation.[20] Kinetic hydrate inhibitors are usually polymer-based chemicals.

Following a few, small-scale field trials, kinetic hydrate inhibitors were first commercially deployed in the Hyde/West Sole by BP in 1996.[21] The number of field trials and applications continued to increase over the last few years.[22-29] However, the percentage of KHI used in the field remains far lower than that for the thermodynamic inhibitors. One reason for the slow market growth is the performance limit of the older generation KHI. These previous generations are very effective in controlling hydrates up to 8°C subcooling and with induction time extension of 24 hours; however, they were less effective for applications that operate at higher degrees of subcooling.

In an attempt to improve the performance of KHI, extensive research has been conducted for the past five years. The intensive effort has led to the development of advanced kinetic hydrate inhibitors.[30-36] These new-generation inhibitors have demonstrated enhanced performance in the autoclave and rocking cell tests. Hydrate formation was successfully inhibited at 13°C subcooling for at least 48-hour shut-in protection. These new advanced KHI were developed by improvement in the polymer chemistry and reformulation with synergistic additives. This paper discusses various approaches that were taken to develop these advanced inhibitors.

3 EXPERIMENTAL

The performance of advanced KHI was evaluated in high-pressure autoclaves and rocking cells. The induction time, defined as the elapsed time from the start of the experiment to the onset of hydrate formation, was measured and used as the major performance criteria. The test was given a "pass" grade if the induction time exceeded 48 hours at 13°C subcooling. Following the onset of hydrate formation, the rate of hydrate formation and the morphology of hydrates were also measured and used as the secondary criterion. The lower crystal growth rate and the smaller hydrate crystals are preferred.

Since hydrate nucleation is a stochastic process, meaning that the induction time is a variable, replicates were conducted for each test condition to reduce the variability. The dosage of inhibitors ranged from 0.5 wt% to 5 wt%, with a typical concentration of 2.5 wt% used most frequently. The test fluid consisted of de-ionized water and/or a field condensate. Two synthetic gases based on the composition of the produced gas in Gulf of Mexico and North Sea (shown in Appendix A) were used to pressurize the test vessels. Temperature and pressure inside the test vessels were recorded throughout the experiment. Onset of hydrate formation was identified by the sudden drop in pressure due to gas uptake and the corresponding increase in temperature due to heat generated by the exothermic crystallization process of hydrate formation. Photos and video recording were taken whenever visible changes were observed through the viewing window.

The autoclaves are comprised of 500-mL stainless steel, high-pressure vessels, equipped with thermostatted cooling jacketed, sapphire window, gas inlet and outlet, magnetic stirring bar, and process monitoring devices. Exactly 200 mL of test fluids were prepared and transferred to the autoclaves that were kept at high pressure and low temperature, typically at 1000 psi and 4°C. While the temperature and pressure data were being logged, the autoclave content was visually monitored using a boroscope and a video camera connected to a monitor and a time-lapsed video recorder. The fluids were either stirred to simulate flowing conditions or kept at the static mode to simulate shut-in conditions.

The rocking cells were fabricated from a tempered glass that was enclosed in a stainless steel cylinder. Two slits, opposite to each other, were cut open on the side of the cylinder to allow viewing through the glass. The cells were mounted on a rocking plate, which was rocked gently about 10 times per minute during the experiment, simulating steady-state operation. Shut-in conditions were also simulated without the rocking motion. The cell volume was about 37 mL with one third of the volume occupied by the test fluids. The whole assembly was immersed in a thermally controlled water/glycol bath to maintain the test temperature. Clear plastic was used as the construction material for the water/glycol bath to allow direct viewing of the cell content. Photos and videos were taken whenever major changes were noted. Pictures of autoclaves and rocking cells are shown in Appendix B.

The molecular weight of polymers was determined by the GPC (Gel Permeation Chromatography) method. A polyethylene glycol solution was used as the standard. Occasionally, the measured molecular weight was verified by other techniques that included MALLS (multi-Angle Laser Light Scattering), relatively viscosity measurement and a proprietary GPC method.

4 RESULTS AND DISCUSSION

4.1 Development of Advanced Kinetic Hydrate Inhibitors

Kinetic hydrate inhibitors are typically polymer-based products formulated in various carrier solvents. The chemistry of the active polymers predominantly controls the performance of these inhibitors. In addition to the basic chemistry, the characteristics of the polymers, such as molecular weight and its distribution, can vary the KHI performance greatly. Special additives in the products, including carrier solvents, can also make a direct impact on the performance. Extensive research was conducted in an effort to determine the optimum combination of these variables. As a result of this research effort, a new group of inhibitors, referred to as advanced KHI, have been developed with outstanding performance. All advanced KHI are able to extend the induction time for hydrate formation to more than 48 hours at 13°C subcooling in both autoclave and rocking cell tests. Examples of typical autoclave test results are shown in Appendix C. The techniques used to develop advanced KHI are discussed in the following sections.

4.2 Polymer chemistry

The development of KHI has been in a steady progress for the past ten years. Approximately 80 patents on hydrate inhibitors have been issued during this period.[37-45] The split among them is quite even, with half of the patents covering KHI and the other half focusing on AA technology. Most of the KHI technology is based on the polymeric molecules. There are some non-polymer based KHI, such as phosphonates and phosphate esters.[37] However, none of these small molecules have been used in the oil industry. The chemistry of the polymers is the most important factor to determine if the molecules are effective.

A large variety of chemistry has been claimed to exhibit kinetic hydrate inhibition capability. These molecules can be ionic (cationic or anionic), neutral, amphiphilic, amphoteric or zwitterionic compounds. Some examples of building blocks for these polymers are alkylene oxides,[38] sulfonates,[39] oxyalkylenediamines,[40] imides,[41] acrylamides,[42,43] acrylacrylates,[43] carboxylic acids or esters,[44] and lactams (*e.g.*, pyrrolidone and its derivatives).[45] The effectiveness of these molecules varies with the specific type of chemistry on which they are based.

Among this large number of chemistry, only a small portion found its way to the oilfield application. Lactam-based homopolymers and its copolymers and terpolymers are the most widely used chemistry. Table 1 lists this type of polymers that have been used in the field.

Table 1 *List of lactam-based polymers used in the oilfield applications*

Polymer Type	*Polymer Chemistry*
Homopolymers	Polypyrrolidone (PVP), polyvinylcaprolactam (PVCAP)
Copolymers	Poly(pyrrolidone/vinylcaprolactam) (P [VP/VCAP]), Poly(vinylmethylacetamide/ vinylcaprolactam) (P[VIMA/VCAP])
Terpolymers	Poly(vinylmethylacetamide/ vinylcaprolactam/dimethylaminoethylmethacrylate) (P[VP/VCAP/DMAEMA])

These polymers inhibit hydrate formation by the interaction of their pendant groups with the hydrate surface. The hydrate surface can be imagined as being made up of various open cavities in which small hydrocarbon molecules can fill to form gas hydrates, The pendant groups interact in two ways with the hydrate surface. While the small alkyl group penetrates an open cavity, the amide group is hydrogen bonded to the hydrate surface via the carbonyl group. Hydrogen bonding between nitrogen of the amide group and water molecules on the hydrate surface is less pronounced according to the molecular modeling studies. Several of the alkylamide groups must interact with the hydrate surface for any significant inhibitor effect to be observed. This is why single monomer or oligomer is found to be ineffective.

The interfacial properties of the polymers are equally important. A common approach is to measure the hydrophobic and lipophobic balance, or HLB, of the polymers and coordinate it to the hydrate inhibition performance. This interfacial property is strongly coupled to the driving force for the transport of polymers from aqueous phase to the hydrate-water interface. Once at the interface, the ability of polymers to be strongly adsorbed at the interface will likely determine the induction time that each type of polymers is capable of extending.

Some advanced KHI are based on the molecules listed in Table 1, others are new polymers that have not been introduced to the market. Some examples of these new polymers are polyvinylpyridine,[34] acryloylpyrrolidine copolymers,[47] polyacrylamides,[42] and polyacrylacrylate.[43] For those advanced KHI that use the same polymers listed in Table 1, the molecules are made quite differently and consequently the performance is significantly improved. The advanced KHI molecules were developed by a few systematic approaches described below.

First, the molecular weight of most advanced KHI polymers was fine tuned to their optimum ranges. For example, the optimum molecular weight range for PVCAP and P[VP/VCAP] is found to range between 500 and 4,000,[35] preferably in the range of 500 and 2,500.[31] To produce such a low molecular weight polymers can be quite a challenge. There are several different ways to cut the molecular weight of these polymers. Each method leads to a different molecular weight distribution and to different amounts of reaction byproducts and residual (or decomposed) monomers. Although no evidence was found to indicate any interference between the polymers and the reaction byproducts or residual monomers, an excessive amount of these inactive species might take a toll on the stability of the finished product. All of these variations have shown a varied degree of impact on the polymer performance. Therefore, care must be taken to select the right combination of molecular weight and its distribution.

Secondly, the polymerization process should be adjusted to produce the polymers with the optimum molecular weight and the distribution. Several factors in the polymerization process were found to affect these properties, which affect the ultimate product performance. Examples of these process variables include reaction temperature, polymerization mechanism, the type and the amount of initiator, level of mixing and materials addition rate. Once the optimum process is established, precise control of the developed process is the key to consistently producing polymers with the desired molecular weight and distribution. For example, inactive oligomers may be produced instead of the desired low molecular weight polymers if the process runs out of control. Like non-reacted monomers, oligomers are ineffective for hydrate inhibition.

Thirdly, the solvent used for the polymerization process was found to be another important factor. Some compounds have the ability to participate in the polymerization reaction, in addition to serving as the inert solvent. Whether this ability is directly related to the improved polymer performance is unknown. For example, it was found that ethylene glycol monobutyl ether (EGMBE) is a preferable solvent for polymers listed in Table 1.[31] Polymers produced in EGMBE were found to outperform those generated in other aqueous based solvents. The unique EGMBE capability was demonstrated in a special experiment comparing two PVCAP polymers produced in different solvents. One PVCAP was made in EGMBE solvent and the other was made in isopropanol. Isopropanol in the second sample was then removed and replaced with an equal amount of EGMBE as in the first sample. The performance of the first sample, which was synthesized in EGMBE, was clearly superior to the second sample made in isopropanol.[31] It is known that EGMBE has the ability to participate in the polymerization reaction and to incorporate itself into the polymer chain. No correlation has yet been identified between the activity of the solvent and the polymer performance.

Copolymerization is another effective method to alter the chemistry of the polymers. Even a small incorporation of a second monomer into a homopolymer chain can lead to a dramatic change in the performance. Indeed, some advanced KHI gain a major boost in performance by using new copolymers in their formulations.[47,48]

The effect of copolymerization cannot be fully realized by simply incorporating a second or a third monomer. Other factors, such as the ratio of monomers and order of addition, are important to the product performance and its physical properties. One such property important to the hydrate inhibition application is the cloud point of the polymers. It is not uncommon that the polymer performance can have an opposite trend to its desired physical properties. In some cases the best performers cannot be made with the most desirable properties. As an example, P[VIMA/VCAP] with a monomer ratio of

VIMA/VCAP = 25/75 gives the best performance over any other monomer ratios. However, its cloud point is mostly lower than the same copolymers with different VIMA/VCAP ratios.[48] Therefore, it is necessary to pick the optimum monomer ratio in order to achieve the right balance between the performance and the desired properties.

4.3 Product Formulation

Formulation of polymers into finished products is as important in boosting the performance as choosing the right chemistry. Formulation can drastically enhance the performance of most production chemicals, including kinetic hydrate inhibitors. The challenge is to find the right combination of solvents, additives and polymers that are physically compatible to one another and chemically work synergistically.

It was found that glycol ethers are good solvents for the polymers listed in Table 1.[32] Low molecular weight glycol ethers containing an alkoxy group having at least 3 carbon atoms were very effective performance boosters. Table 2 lists some of the glycol ethers typically used as performance enhancement solvents for advanced KHI.

Table 2 *Glycol ethers as performance enhancement solvents for advanced KHI*

2-isopropoxyethanol	ethylene glycol monopropyl ether
propylene glycol propyl ether	ethylene glycol monobutyl ether
propylene glycol butyl ether	diethylene glycol monobutyl ether

Some glycol ethers were found to show the similar synergistic effect for the newly developed polymers. Other synergistic solvents have also been identified. For example, alcohols that contain one hydroxy group and 3 to 5 carbon atoms, such as propyl alcohol, butyl alcohol and pentanol, are considered synergistic solvents for advanced KHI.[36] It is not clear how these solvents enhance the performance of polymers. In the case of glycol ethers, it is thought that these molecules may help stabilizing the polymers in the hydrate-water interface. There is an optimum amount of glycol ethers to be used in the formulation. The optimum concentration is strongly dependent upon the ethers used and other components present in the formulation.

In addition to using a specific solvent package to enhance the performance, other proprietary additives can be used in the advanced KHI formulations to boost the performance of polymers. Using additives, such as oil-in-alcohol emulsion, has been successfully applied in the water-base drilling fluids to inhibit hydrate formation.[49-51] The performance enhancing additives for advanced KHI can be either small molecules or macromolecules like polymers. In the small molecule group, for example, tetrabutylammonium bromide (TBAB) and tetrapentylammonium bromide (TPAB) have been reported that they can enhance the performance of AA/AMPS (acrylamide and acrylamidomethylpropane sulfonate) copolymers.[46] In the polymer group, polyoxyalkylenediamine and polyoxyarylenediamine were found to be very good performance-enhancing additives for the polymers in Table 1.[33] Examples of compounds used in this sub-category include polyoxypropylene diamine and polyethoxylated tallow propylenediamine. It is interesting to point out that polyoxyalkylenediamine and polyoxyarylenediamine by themselves are good kinetic inhibitors.[40] By formulating these compounds with lactam-based polymers in Table 1 leads to much better performance than the individual components. Furthermore, the blend of these two molecules is believed to exhibit a side benefit of corrosion inhibition capability.[33]

The aforementioned example indicates that a physical blend of two different KHI can result in a better performer than its individual components. However, this is by no means a universal trend for all KHI polymers. For example, a physical blend of PVCAP and PVIMA does not lead to increased performance over its individual components, PVCAP or PVIMA.[48] Interestingly, copolymers of VIMA and VCAP, P[VIMA/VCAP], does boost the performance over their respective homopolymers.[48] Apparently, there is no correlation of the performance-boosting effect between physical mixing and copolymerization. Neither approach is the preferred way to enhance the polymer performance.

It is very important that KHI's must be formulated to meet the physical specifications for the field deployment, especially in deepwater applications. Physical properties, such as flash point, pour point and viscosity, are important factors that can determine the applicability of potential KHI. Viscosity directly affects the pressure drop across the chemical injection line and the overall pressure rating for the line. High viscosity can be a major concern for offshore applications because of the pressure limit of the long umbilical and the capacity limit of the delivery system. For this reason, most KHI packages are formulated to keep the viscosity low enough for practical use. Typical advanced KHI viscosity values range from 15-20 cP at 20°C to 40-60 cP at 4°C.

Material and chemical compatibility of advanced KHI with the system being treated is also a key factor affecting whether the specific product can be deployed. Material compatibility includes both metallic and non-metallic materials. The active species in the advanced KHI formulations often include both polymers and performance enhancing additives. Polymers in advanced KHI are inert to most materials used in the oil and gas fields. Therefore, the performance-enhancing additives and the solvent packages used in the formulations are the controlling factors for the compatibility of the finished products with the existing materials. Most of the advanced KHI are formulated in aqueous base solvents, such as water and alcohols. These are the same fluids as those encountered in the conventional systems treated with thermodynamic inhibitors. This makes the selection of the performance-enhancing additives the critical step for achieving good material compatibility. The material compatibility of most additives has been well established. Since mixing these additives with advanced KHI polymers does not alter the compatibility significantly, the material compatibility of the resulting advanced KHI is often predictable prior to the physical testing.

In contrast to the predictable material compatibility, chemical compatibility of advanced KHI with other production chemicals can vary greatly. Therefore, it is important to carry out compatibility studies to investigate the effect of advanced KHI on the performance of existing chemicals and vice versa. The most frequently used production chemicals, especially in subsea pipelines, are corrosion inhibitors, scale inhibitors and paraffin inhibitors. No chemical interference is expected to decrease the performance of either advanced KHI or production chemicals if the chemistries of these products are properly selected. However, physical compatibility between these chemicals can be a major challenge if they are to be co-injected in the umbilical.

Some advanced KHI are formulated with high boiling solvents to prevent these solvents from being stripped off by the gas. Some produced gases can be relatively "dry" because the operating temperature is above their water dew point or hydrocarbon dew point. In such cases, the dry gas has the tendency to strip all or part of the solvents away from the inhibitor, leaving behind the active species as solid residues. There are many choices of high boiling organic solvents for this type of application. The choice often depends heavily on the cost and the stability of the finished products.

5 SUMMARY

New advanced kinetic hydrate inhibitors have been developed for controlling hydrates in the production and transportation of petroleum fluids. These new-generation inhibitors demonstrated significantly improved performance in the autoclave and rocking cell tests compared to previous generations. Hydrate formation was effectively inhibited for a minimum period of 48 hours under the condition of 13°C subcooling. These new inhibitors were developed using systematic approaches that included the use of new polymers, new processes, and new formulations. Specific solvent packages and synergistic additives have been identified as performance enhancing compounds for these advanced KHI formulations.

Acknowledgement

The author would like to thank Dr. Arvind Mathur, Dr. Kirill Bakeev, Dr. Jenn Shi, Dr. Vu Thieu and Dr. David Graham of ISP for their vital contributions to the development of advanced KHI.

References

1. Sloan, E. D.: *Clathrate Hydrates of Natural Gases*, 2nd edition, Marcel Dekker, Inc., New York (1998).
2. Makogon, Y.F.: *Hydrates of Hydrocarbons*, PennWell Pulishing Co., Tulsa, Oklahoma (1997).
3. E. D. Sloan Jr. and J. B. Bloys, "Hydrate Engineering", SPE Monograph volume 21, published by SPE, Richardson, Texas (2000).
4. Sloan, E.D. et al.: "Additives and Methods for Controlling Clathrate Hydrates in Fluid Systems", WO 95/32356.
5. Kelland, M.A., Svartaas, T.M. and Dybvik, L.A.: "Control of Hydrate Formation by Surfactants and Polymers", SPE 28506 presented at the 1994 SPE 67th Annual Technical Conference and Exhibition, New Orleans, Sept. 25-28.
6. Kelland, M.A., Svartaas, T.M. and Dybvik, L.A.: "Studies on New Gas Hydrate Inhibitors", SPE 30420 presented at the 1995 SPE Offshore Europe Conference, Aberdeen, Sept. 5-8.
7. Pakulski, M.: "High Efficiency Non-Polymeric Gas Hydrate Inhibitors", SPE 37285 presented at the 1997 SPE International Symposium on Oilfield Chemistry, Feb. 18-21.
8. Delion, A.S., Durand, J.P., Gateau, P. and Velly M.: "Process for Inhibiting or Retarding the Formation, Growth and/or Agglomeration of Hydrates in Production Effluents", US Patent 5,817,898.
9. Klomp, U.C., Kruka, V.R., Reijnhart, R. and Weisenborn, A.J.: "Method for Inhibiting the Plugging of Conduits by Gas Hydrates", US Patent 5,648,575.
10. Klomp, U.C., Reijhart, R.: "Method for Inhibiting the Plugging of Conduits by Gas Hydrates", WO 96/34177.
11. Klug, P. and Kelland, M.: "Additives for Inhibiting Formation of Gas Hydrates", WO 98/23843.
12. Palermo, T. *et al.*: "Pilot Loop Tests of New Additives Preventing Hydrate Plugs Formation", paper presented at Multiphase '97 Conference, Cannes.
13. Colle, K.S., Costello, C.A., Oelfke, R.H., Talley, L.D., Longo, J.M. and Berluche, E.: "Method for Inhibiting Hydrate Formation", US Patent 5,600,044.

14. Colle, K.S., Oelfke, R.H., Costello, C.A., Talley, L.D. and Berluche, E.: "Method for Inhibiting Hydrate Formation", WO 96/41784.
15. Urdahl, O., Lund, A, Gjertsen, H.L. and Austvik, T.: "Experimental Testing and Evaluation of a Kinetic Gas Hydrate Inhibitor in Different Fluid Systems", page 498., Div of Fuel Chemistry, ACS Meeting, San Francisco, CA, April 13-17, 1997.
16. Lederhos, J.P., Long, J.P., Sum, A., Christiansen, R.L. and Sloan E.D.: "Effective Kinetic Inhibitors for Natural Gas Hydrates", *Chemical Engineering Science* (1996) **51**, No. 8, 1221.
17. Kelland, M.A., Svartaas, T.M. and Dybvik, L.A.: "A New Generation of Gas Hydrate Inhibitors", SPE 30695 presented at the 1995 SPE Annual Technical Conference and Exhibition, Dallas, Oct. 22-25.
18. Long, J., Lederhos, J., Sum, A., Christiansen, R. and Sloan, E. D.: "Kinetic Inhibitors of Natural Gas Hydrates", page 85, *Proceeding of the 73rd GPA Annual Convention*, New Orleans, March 7-9, 1993.
19. Knott, T.: "Holding Hydrates at Bay", *Offshore Engineer,* **p.** 29, Feb. 2001.
20. Sloan, E. D.: Hydrate Engineering, pp 26-30, SPE Monograph Series Vol. 21, SPE, 2000.
21. Argo, C.B., Blain, R.A., Osborne, C.G. and Priestley, I.D.: "Commercial Deployment of Low Dosage Hydrate Inhibitors in a Southern North Sea 69 Kilometer Wet-Gas Subsea Pipeline", SPE 37255 presented at the 1997 SPE International Symposium on oilfield Chemistry, Feb. 18-21.
22. Lederhos, J. and Sloan, E.D: "Transferability of Hydrate Kinetic Inhibitor Results Between Bench Scale Apparatus and a Pilot Scale Flow Loop", page 373, *Proceedings of the 2nd International Conference on Natural Gas Hydrates,* Toulouse, FRANCE, June 2-6, 1996.
23. Corrigan, A., Duncum, S.N., Edwards, A.R. and Osborne, C.G.: "Trials of Threshold Hydrate Inhibitors in the Ravenspurn to Cleeton Line", SPE 30696 presented at the 1995 SPE Annual Technical Conference and Exhibition, Dallas, Oct. 22-25.
24. Notz, P.K., Bumgardner, S.B., Schaneman, B.D. and Todd, J.L.: "Application of Kinetic Inhibitors to Gas Hydrate Problems": (a) paper 7777 presented at the 27th Annual OTC, Houston, May 1-4, 1995 and (b) *SPE Production & Facilities* (November 1996) 256.
25. Bloys, B. Lacey, C. and Lynch, P.: "Laboratory Testing and Field Trial of a New Kinetic Hydrate Inhibitor", OTC paper 7772 presented at the 27th Annual OTC, Houston, May 1-4, 1995.
26. Mitchell, G.F. and Talley, L.D.: "Application of Kinetic Hydrate Inhibitor in Black-Oil Flowlines", paper SPE 56770 presented at the 1999 SPE Annual Technical Conference and Exhibition, Houston, Oct. 3-6.
27. Talley, L.D. and Mitchell, G.F.: "Application of Proprietary Kinetic Hydrate Inhibitors in Gas Flowlines", paper OTC 11036 presented at the 1999 Offshore Technology Conference, Houston, May 3-6.
28. Fu, B., Cenegy, L. M. and Neff, C. S.: "A Summary of Successful Field Applications of A Kinetic Hydrate Inhibitor", paper SPE 65022, presented at the 2001 SPE Symposium of Oilfield Chemistry, Houston, Feb. 13-15.
29. Fu, B.: "A Novel Kinetic Hydrate Inhibitor for Hydrate Control", paper presented at the IBC *"Controlling Hydrates, Paraffins and Asphaltenes"* conference, Oslo, Dec. 7-8, 1998.
30. Rasch, A., Mikalsen, A., Gjertsen, L. H. and Fu, B.: "Evaluation of Low Dosage Hydrate Inhibitor in Hydrocarbon Fluid Systems at High Subcooling", page 239 in the

proceeding, presented at10th International Multiphase 2001 Conference, Cannes, France, June 13-15, 2001, BHR Group Limited, UK.

31. Bakeev, K. N., Chuang, J., Drzewinski, M. A. and Graham, D. E.: "Methods for Preventing or Retarding the Formation of Gas Hydrates", US Patent 6,117,929 (2000).

32. Cohen, J. M. and Young, W. D.: "Methods for Preventing or retarding the Formation of Gas Hydrates", US Patent 5,723,524 (1998).

33. Bakeev, K. N., Myers, R. T. and Graham, D. E.: "Blend for Preventing or Retarding the Formation of Gas Hydrates", US Patent 6,180,699 (2001).

34. Bakeev, K. N., Chuang, J., Winkler, T. Drzewinski, M. A. and Graham, D. E.: "Methods for Preventing or Retarding the Formation of Gas Hydrates", US Patent 6,281,274 (2001).

35. Bakeev, K. N., Myers, R. T., Chuang, J., Winkler, T. and Krauss, A.: "Methods for Preventing or Retarding the Formation of Gas Hydrates", US Patent 6,242,518 (2001).

36. Cohen, J. M. and Young, W. D.: "Methods for Preventing or retarding the Formation of Gas Hydrates",

37. Matthews, R.R. and Clark, C. R.: "Inhibition of Hydrate Formation", US Patent 4,856,593 (1989).

38. Thomas, M., Baley, A. and Durand, J.: "Process for reducing the agglomeration tendency of hydrates in the production effluent", US Patent 5,426,258 (1995).

39. Durand, J, Sinquin, A. and Valley, M.: "Method for inhibiting or retarding hydrate formation, growth and/or agglomeration", US Patent 5,789,635 (1998).

40. Pakulski, M.A.: "Method for controlling gas hydrates in fluid mixtures", US Patent 5,741,758 (1998).

41. Costello, C. A., Berluche, E., Olefke, R. H. and Talley, L. D.: "Maleimide copolymers and method for inhibiting hydrate formation ", US Patent 5,744,665 (1998).

42. Colle, K.S., Costello, C. A., Olefke, R. H. Berluche, E. and Talley, L. D.: "Method for Inhibiting Hydrate Formation", US Patent 5,600,044 (1997).

43. Colle, K.S., Costello, C. A., Talley, L. D., Olefke, R. H. and Berluche, E.: "Method for Inhibiting Hydrate Formation", US Patent 6,107,531 (2000).

44. Rojey, A., Thomas, M., Delion, A. and Durand, J.: "Process for transporting a fluid such as a dry gas likely to form hydrates", US Patent 5,816,280 (1998).

45. Sloan, D.: "Method for controlling clathrate hydrates in fluid systems", US Patent 5,420,370 (1995).

46. Cingotti, B.: "Study of Methane Hydrate Inhibition Using AA/AMPS copolymers", These genie des procedes. 2 Dec 1999. 183 p. OSTI; Available to ETDE participating countries only(see www.etde.org); commercial reproduction prohibited; OSTI as DE20052901; PURL: https://www.osti.gov/servlets/purl/ 20052901-A3Jo11/webviewable/.

47. Colle, K.S., Costello, C. A., Olefke., Talley, L. D., Longo, J. M. and Berluche, E.: "Method for Inhibiting Hydrate Formation",, US Patent 6,028,233 (2000).

48. Colle, K.S., Olefke, R. H. and Kelland, M.: "Method for Inhibiting Hydrate Formation", US Patent 5,874,660 (1999).

49. Blytas, G.C. and Hale, A.H.: "Water base drilling fluids comprising oil-in-alcohol emulsion", US Patent 5,248,664 (1993).

50. Blytas, G.C. and Hale, A.H.: "Method for drilling a well with emulsion drilling fluids", US Patent 5,085,282 (1992).

51. Blytas, G.C. and Hale, A.H.: "Alcohol-in-oil drilling fluid system", US Patent 5,072,794 (1991).

APPENDIX   A

Compositions of Synthetic Gases Used for the Performance Tests

Component	North Sea	Gulf of Mexico
Nitrogen	0.00%	0.39%
Carbon Dioxide	1.00%	0.00%
Methane	95.31%	87.26%
Ethane	2.96%	7.57%
Propane	0.53%	3.10%
Isobutane	0.10%	0.49%
Normal butane	0.10%	0.79%
Isopentane	0.00%	0.20%
Normal pentane	0.00%	0.20%
Total	100%	100%

APPENDIX   B

Figure 1 *Picture of autoclaves for evaluation of advanced kinetic hydrate inhibitors*

Figure 2 *Picture of rocking cells for evaluation of advanced kinetic hydrate inhibitors*

APPENDIX C

Autoclave Test Results @ 10°C subcooling
(35 bars and 4.5°C)

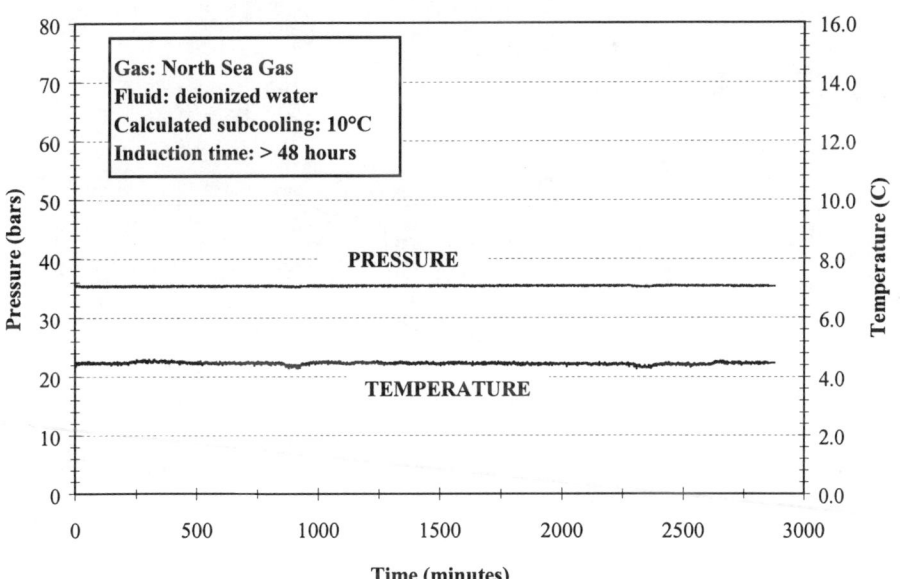

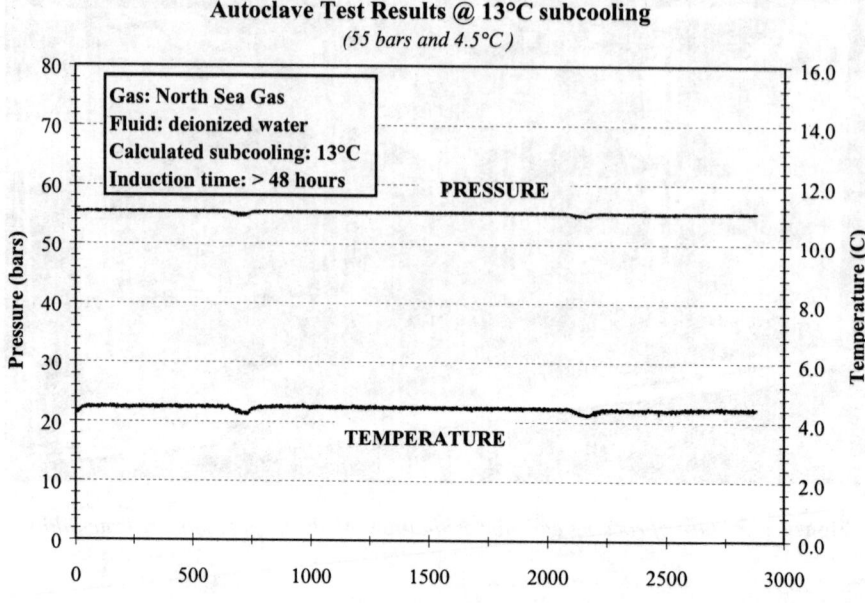

Autoclave Test Results @ 13°C subcooling
(55 bars and 4.5°C)

PREDICTION AND SOLUTION OF ASPHALTENE RELATED PROBLEMS IN THE FIELD

S. Asomaning and A.Yen

Baker Petrolite, Sugar Land, TX, USA 77478

1 INTRODUCTION

Problems associated with asphaltene precipitation and deposition have been widely reported in the petroleum production and processing industry.[1,2,3] Asphaltene precipitation and subsequent deposition in production tubing and topside facilities present significant cost penalties to crude oil production. The solubility class known as asphaltenes is the heptane-insoluble, toluene-soluble fraction of crude oils. They are made up of high molecular weight condensed aromatics with cross-linked heteroatoms and aliphatic side chains. Their exact molecular structure is a matter of speculation. Asphaltenes have a tendency to aggregate and flocculate. In crude oils, they are kept in solution by resins, which peptize them.

Asphaltene deposits in oil production are considered troublesome because they are difficult to remediate. In the past, when asphaltene deposition had been encountered, remediation measures such as mechanical cutting, scrapping, and chemical dispersants had been used. These measures were sufficient because such deposition had usually been in land and shallow water wells, or in topside production facilities.

Subsea wells have become widespread in deepwater developments. The remoteness of such wells from host platforms precludes the use of the above conventional asphaltene-deposition remediation techniques. Continuous injection of asphaltene inhibitors becomes necessary to inhibit asphaltene precipitation and deposition. Effective management of asphaltene deposition in deepwater subsea-systems requires a well-thought out strategy. At Baker Petrolite, a three prong strategy is used viz. the correct prediction of the stability of asphaltenes in a crude oil; testing and selection of asphaltene inhibitors for continuous injection; and monitoring the performance of the inhibitors once they are deployed.

This paper was written to show the three-prong strategy in action. It describes the results of work that resulted in recommending an asphaltene inhibitor for an offshore well in a Latin American field. Wells in this field had experienced chronic asphaltene deposition problems. As part of planning for new wells, continuous injection of asphaltene inhibitor was proposed. The oil from the well under consideration had an API gravity of 32 and an asphaltene content of 9%. The flowing bottomhole pressure and temperature were 6539 psia and 138°C respectively. The well had a 3.5'' casing to a depth of 1580', then a 3.5" perforated liner from 15,850'-17,958'.

2   METHODS AND PROCEDURES

2.1   Testing for Asphaltene Stability

Potential asphaltene problems in oil production systems can be predicted through an analysis of the stability of the asphaltenes in the crude. Amongst the more popular methods for determining asphaltene stability are Saturates, Aromatics, Resins, and Asphaltenes (SARA) based methods such as the asphaltene-resin ratio and the Colloidal Instability Index (CII);[4] the Oliensis spot test;[5] and microscopy methods. In the present work the stability of asphaltenes was predicted using four stability tests; two SARA based ratios – the Asphaltene-Resin ratio and the CII, the Oliensis Spot Test and the Asphaltene Precipitation Detection Test (APDT).[6]

2.1.1   Testing of Dead Oil. The CII and the asphaltene-resin ratio are SARA based stability test methods. The CII considers an oil as a colloidal system made up of Saturates, Aromatics, Resins and Asphaltenes. The index is calculated using weight percentages of the fractions from a SARA analysis as:

$$CII = \{(Saturates + Asphaltenes)/(Resins + Aromatics)\} \qquad (1)$$

A CII value below 0.7 usually indicates an oil is stable, while values above 0.9 are normally obtained for unstable oils. The SARA data for calculating both the asphaltene-resin ratio and the colloidal instability index were obtained using a modified form of IP143[7]. The asphaltenes were determined as the heptane insoluble fraction of the crude oil. The saturate, aromatic and resin components were separated using liquid chromatography.

The Oliensis Spot Test Number, is a measure of the amount of hexadecane needed to destabilize 5 g of crude oil. It is obtained by titrating hexadecane, in incremental amounts of 1 ml, into 5 g of crude oil. After each addition of hexadecane, a portion of the resulting crude-hexadecane solution is sampled onto a filter paper with numbered squares on it. Titration and sampling continues until a black ring of asphaltenes forms on one of the squares on the filter paper. The filter paper is then dried in an oven. The Spot Test Number corresponds to the amount of precipitant that gives a faint ring of asphaltenes that does not disappear upon drying. An oil with a Spot Test Number of less than 9 will be considered unstable, 9-12 will be considered moderately unstable and more than 12 will be considered stable.

The Asphaltene Precipitation Detection Test (APDT) which measures the onset of asphaltene flocculation was used to confirm the stability of asphaltenes in the dead crude oil. The APDT is done using a bench-top solids detection system that uses a near infra-red (NIR) laser to determine the onset of asphaltene flocculation during titration of an aliphatic solvent such as heptane into 16.5 grams of oil. The oil is heated to 38°C and stirred for 45 minutes. Heptane is then titrated into the oil at a rate of one ml/min and the transmittance of the NIR laser monitored. In the initial stages of titration, the transmittance of the laser increases, due to the decrease in the density of the oil resulting from the addition of heptane. When the asphaltenes begin to aggregate, the laser transmittance decreases, resulting in an inflexion in a plot of transmittance vs. volume of added heptane (Figure 1). The point of inflexion, expressed as an APDT number, corresponds to the point of asphaltene precipitation and provides a relative measure of how prone the asphaltenes are to precipitation.

2.1.2   Testing of Live Oil. The stability of the asphaltenes and hence the deposition potential of the crude oil was established beyond doubt using the live oil deposition experiments. These were done on a high-pressure PVT cell equipped with a NIR laser-based solids detection system (SDS) described previously.[8] The equipment measures the onset pressure of asphaltene precipitation, the saturation pressure of the crude oil, and both the quantity of asphaltenes precipitating out of solution and depositing on the wall.

The oil was equilibrated on a rocker to a single-phase fluid at the bottomhole temperature and an initial pressure of 12000 psia for one week. To avoid destabilizing the asphaltenes, all runs were started at this initial pressure. Forty-five milliliters of the previously conditioned oil was charged into the PVT cell, using a positive displacement pump. The temperature of the cell and its contents were kept at 138°C and 12000 psia for 8 hours to further condition the oil inside the cell. A baseline transmittance of the laser was initially established. The depressurization experiment followed with the stepping down of the pressure in 100 psia decrements. There was a relaxation time of 7 min. between pressure steps to allow the transmittance of the laser to equilibrate. As the pressure of the live oil was decreased it became less dense thus allowing more light to be transmitted through it. The power of transmitted light increases until asphaltenes begin to precipitate when the transmittance starts to decrease. The point at which the transmittance starts to decrease is the onset pressure of asphaltene precipitation and corresponds to the pressure at which asphaltene precipitation will begin in the wellbore. Figure 2 shows a typical trace for a live oil depressurization experiment. For an unstable oil, after the onset point the asphaltenes aggregate further into bigger flocs This is shown in the figure as a gradual decrease in transmittance to the saturation pressure. At the saturation pressure, the evolution of gases scatters the laser light resulting in a transmittance of zero.

After depressurising the oil, the asphaltene deposits on the cell walls, on the mixer, and endcap were washed off with toluene. The toluene was evaporated and the solids analysed gravimetrically. The fraction of asphaltenes that precipitated but that did not stick to a surface was filtered out of the oil using an attached bulk filtration set-up.

2.2   Asphaltene Inhibitor Selection

The determination of the effectiveness of asphaltene inhibitors has occupied the attention of vendors of chemicals and their clients for quite some time. It has been the desire of both parties to develop test methods that are easy to use and that mimic real oil producing systems. Traditionally, the performance of asphaltene inhibitors has been determined in the laboratory using the heptane precipitation test.[9] In this test, 4 µl of chemical is added to 10 ml of heptane and shaken. Hundred microliters of crude oil is then added to the solution of heptane and inhibitor. The solution is shaken, allowed to stand for 1 hour and centrifuged. An aliquot of liquid is taken from the top layer, diluted in 4 ml of toluene and the absorbance of the diluted solution measured on a spectrophotometer. Good inhibitors keep more asphaltenes in solution resulting in lower transmittances.

Recent innovations in asphaltene inhibitor testing include the use of the APDT and the live oil deposition test. In the APDT, described above, asphaltene inhibitors are injected into the dead oils to determine the inhibitors' effectiveness in delaying the onset of asphaltene flocculation. In the live oil test, described in section 2.1.2, the inhibitors are injected into the live oil sample prior to depressurization. A good inhibitor is supposed to reduce the amount of asphaltene deposits and precipitates formed from the depressurization of the oil. The high-pressure deposition test mimics events in the wellbore and closely resembles a real production system.

In the present work the heptane precipitation test and the APDT were used to determine the effectiveness of the asphaltene inhibitors on the dead oils. The live oil depressurization test was used for the live oil sample. The tests were performed at varying inhibitor concentration.

3 RESULTS AND DISCUSSION

3.1 Stability Analysis

3.1.1 Analysis of Dead Oil. Table 1, shows the SARA data for the crude oil. The saturate content approximately equals that of aromatics. The asphaltene content is only slightly lower than that of the resins - the peptizers. This is the first indication of potential instability.

Table 1   *SARA Data for Crude Oil*

Pseudocomponent	Weight Percent
Saturates	39.3
Aromatics	38.1
Resins	13.6
Asphaltenes	9.0

Table 2 shows the stability data for the crude oil. It is apparent from the data that the oil is unstable. The oil has a CII of 0.92 and an asphaltene-resin ratio of 0.7. This asphaltene-resin ratio is well above the threshold value of instability of 0.35. The Oliensis Spot Test Number is 1 indicating a very unstable oil.

Figure 1, shows a plot of the normalised-intensity of the NIR laser as a function of the APDT number. The APDT number is defined as the ratio of heptane titrated in ml, divided by the initial mass of oil (16.5 grams). A very unstable oil usually has an APDT value between 0 and 1. A stable oil generally has an APDT number of 2 or higher at inflexion. Between 1 and 2, the oil is considered moderately unstable. The crude oil under consideration has an APDT number of 1.25, which puts it under the moderately unstable category. From the four stability tests conducted, it was concluded that the oil is unstable and could cause potential asphaltene deposition problems.

Table 2   *Stability Data for Crude Oil*

Stability Criteria	Value	Range of Instability
Colloidal Instability Index	0.92	>0.9
Asphaltene-Resin Ratio	0.7	>0.35
APDT Number	1.25	<2.0
Spot Test Number	1.0	<12

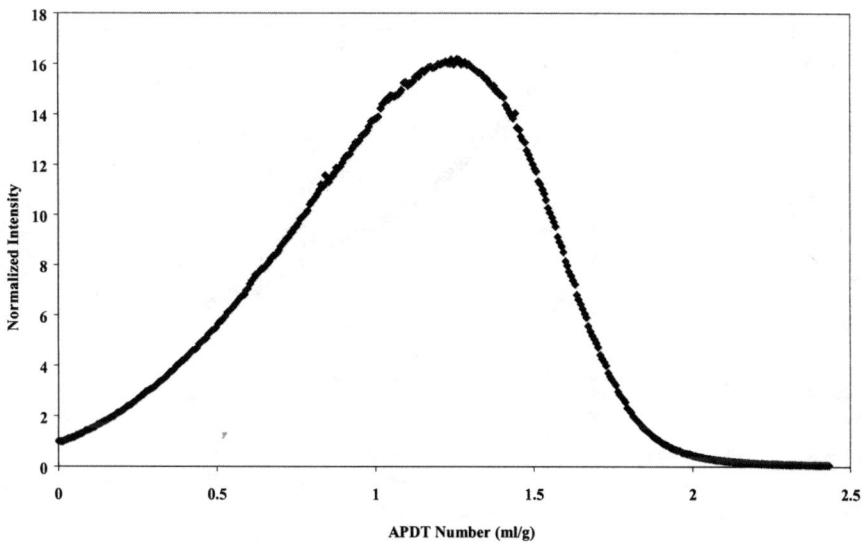

Figure 1 *Normalised Intensity of NIR laser versus APDT Number*

3.1.2 Analysis of Live Oil Figure 2, a plot of the power of transmitted light versus pressure for the live oil, shows that the asphaltenes in the oil flocculate when the pressure of the oil is depleted. Most importantly, the steadily decreasing behaviour of the portion of the curve after the onset point shows that the asphaltenes aggregate into bigger flocs. This is a sign that the oil is unstable and could potentially have asphaltene deposition problems. Table 3 shows the depressurization and deposition parameters for the oil. It has an onset point of asphaltene flocculation of 4500 psia and a saturation pressure of 4000 psia. The onset pressure is very near the saturation pressure and about 2000 psia below the bottomhole pressure. Given the depth of the well (17,958'), it is reasonable to conclude from these data that asphaltene deposition could begin in the wellbore, which could cause substantial operational problems.

From Table 3, out of 3.44g of asphaltenes in the oil, about 0.286g or 8.3% came out of solution and about 50% of this deposited with the remaining staying in solution as precipitates. The asphaltene deposits and precipitates were about 4% each of the total asphaltenes in the oil. This is not very much but given the large flow rate of the well (1950 BOPD), the amount of deposits could lead to very serious problems over time.

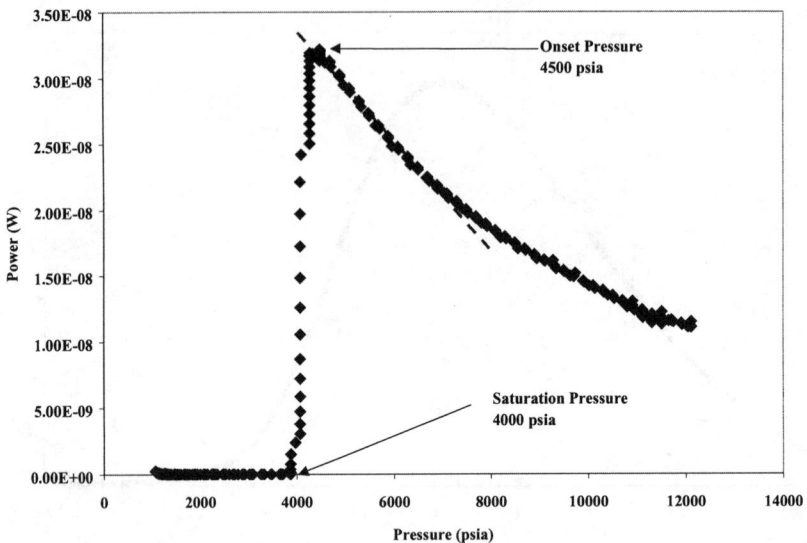

Figure 2 *Live Oil Depressurization of Latin American Crude*

Table 3 *Deposition and Depressurization Parameters for Oil*

Parameter	Value
Bottomhole pressure	6539 psia
Initial depressurization pressure	12000 psia
Asphaltene onset pressure	4500 psia
Saturation pressure	4000 psia
Total amount of asphaltenes in oil*	3.44g
Amount of asphaltenes precipitating*	0.138 g
Amount of asphaltenes depositing*	0.148 g

* - explained in section 2.1.2

3.2 Asphaltene Inhibitor Selection with Dead Oil

Results from the heptane precipitation test are shown in Table 4. The Relative Performance is calculated, using the transmittances of the solutions on a spectrophotometer, as:

$$\text{Relative Performance} = 100*\{1 - (T_{sample}/T_{blank})\} \qquad (2)$$

All the asphaltene inhibitors tested inhibit, to some extent, the asphaltenes in the crude oil with inhibitors A and B being better than C.

Table 4 *Asphaltene Inhibitor Selection with Heptane Precipitation Test*

Asphaltene Inhibitor	Relative Performance (%)
A	51.7
B	48.1
C	29.4
Blank	0

The results obtained from the APDT experiments with the crude oil treated with inhibitors A, B and C at 2000 ppm, and a blank are shown in Table 5. Comparing the blank run to the treated runs, it is clear that all inhibitors shift the APDT inflexion point from that of the blank with inhibitor A being the best. From the stability point of view, the APDT numbers in Table 5 also show that the asphaltenes become more stabilized when the oil is treated with inhibitors.

Table 5 *Asphaltene Inhibitor Selection with APDT*

Asphaltene Inhibitor	APDT Number
A	>2.5
B	1.60
C	1.45
No Inhibitor	1.25

3.3 Asphaltene Inhibitor Selection with Live Oil

Inhibitor A, the most effective chemical selected from the heptane precipitation and APDT tests, was tested on the high pressure cell at varying concentrations to determine the rate at which the inhibitor will be injected into the wellbore. Figure 3 shows a typical trace for the live oil tests. There was no drastic shift in the onset point. The depressurization parameters were similar to those for the crude without inhibitor. Table 6 shows the deposition data. The amount of asphaltene coming out of solution decreased as the inhibitor was added to the crude. The quantity of asphaltenes depositing decreased dramatically with the addition of inhibitor A. However, an increase in the concentration of inhibitor above 500 ppm did not yield substantial increases in deposit reduction. Above 500 ppm, the deposition inhibition tapers off at around 62%. The most cost-effective injection rate was determined as 500 ppm. Based on the live oil data, inhibitor A was recommended to be injected, at 500 ppm, downhole into the well. This treating rate was further optimized in the field to 300 ppm.

Table 6   *Live Oil Deposition Data for Asphaltene Inhibitor A*

Concentration of Inhibitor (ppm)	AD (g)	AP (g)	TS (AD+AP) (g)	Deposition Inhibited (%)
0	0.148	0.138	0.286	0
500	0.056	0.105	0.161	62.1
1000	0.055	0.099	0.154	62.7
2000	0.041	0.084	0.124	72.5

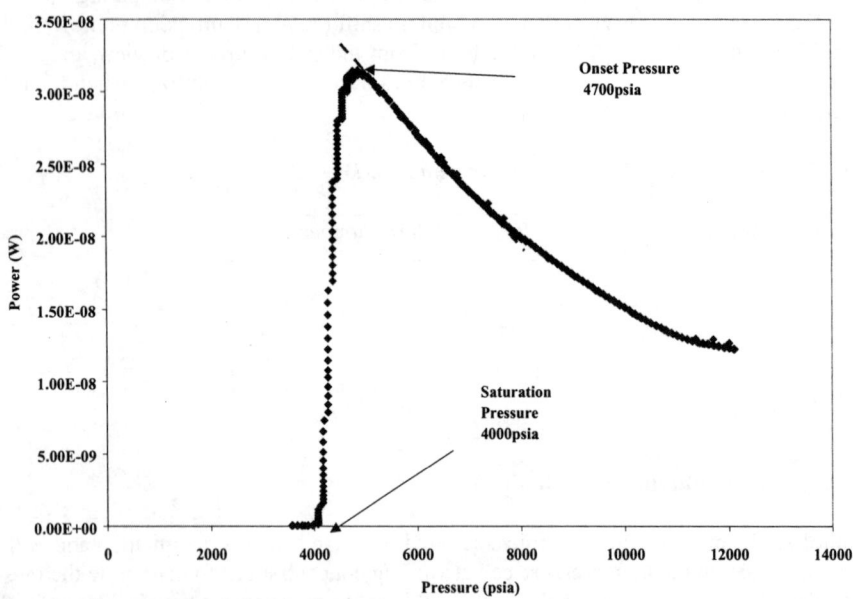

Figure 3   *Live Oil Deposition Tests for Crude treated with 500ppm of an Asphaltene Inhibitor A*

4 MONITORING OF INHIBITOR PERFORMANCE

Inhibitor A was injected continuously at the rate of 300 ppm at a depth of 14,743'. The production stabilized at 1950 BOPD, 81 BWPD and 1572 MCFD for several months and then began to decrease steadily. The well was shut in and a slickline tool with a gauge ring was sent down the well. The tool went as far as the injection point without encountering any constrictions or obstructions. Below the injection point a narrowing of the tubing diameter, believed to be a result of asphaltene deposition, was detected and the gauge ring was pulled out to avoid sticking it in the well. Large amounts of asphaltenes were circulated out of the well during cleaning of the portion of the well below the injection point.

The absence of asphaltene deposits above the injection point was testimony to the fact that chemical A successfully inhibited the asphaltene deposition and that laboratory methods could be used to pick an asphaltene inhibitor that works successfully in the field.

5 CONCLUSION

Laboratory methods consisting of both dead and live oil tests were used to pick an asphaltene inhibitor for a Latin American operator. The inhibitor was optimized in the field to a lower injection rate and continuously injected into the wellbore. The effectiveness of the inhibitor was determined through monitoring of production volumes, and the well diameter. The asphaltene inhibitor successfully inhibited asphaltene deposit formation in the tubing at all points above the inhibitor injection point. Below the injection point asphaltene deposits were encountered. The results obtained from this work show the effectiveness of the three-prong strategy and the fact that an asphaltene inhibitor selected from laboratory tests can perform successfully in the field.

Nomenclature

AD	-	Mass of asphaltene deposited
AP	-	Mass of asphaltene precipitated
APDT	-	Asphaltene precipitation detection test
BOPD	-	Barrels of oil per day
BWPD	-	Barrels of water per day
CII	-	Colloidal instability index
MCFD	-	Thousand cubic feet of gas per day
NIR	-	Near infra-red
PVT	-	Pressure-Volume-Temperature
SDS	-	Solids detection system
T_{blank}	-	Transmittance of blank sample
T_{sample}	-	Transmittance of treated sample
TS	-	Total Mass of asphaltenes out of solution

Acknowledgements

Our sincere thanks go to the management of Baker Petrolite for permission to publish this paper and to Dr. Klaus Weispfennig for helpful discussions.

References

1   C.E. Haskett and M. Tartera, *A Practical Solution to the Problem of Asphaltene Deposits – Hassi Messaoud Field, Algeria*, J. Pet. Tech., 1965, **4**, 387-391.
2   R. Thawer, D.C.A. Nicholl and G. Dick, *Asphaltene Deposition in Production Facilities*, SPE Production Engineering, 1990, 475-480.
3   G.A. Lambourn and M.Durrieu, *Fouling in Crude Oil Preheat Trans*, in *Heat Exchangers Theory and Practice*, Taborek, Hewitt and Afghan (eds.), Hemisphere Publishing Co. N.Y., (1983).

4 S. Asomaning and A.P. Watkinson, *Petroleum Stability and Heteroatom Species Effects on Fouling of Heat Exchangers by Asphaltenes*, Heat Transfer Eng., 2000, **21**, 10-16.
5 G.L. Oliensis, Proc. Assoc. Asphalt Paving Technol., 1935, **6**, 88 in *Kirk-Othmer Encyclopaedia of Chemical Technology* 3rd edn., 1992, 3, M. Howe-Grant (ed.)
6 *Baker Petrolite Level 3 Quality Manual*, Revision 0, October 1999.
7 *Methods for Analysis and Testing*, 40th Annual Edition, Institute of Petroleum, London, 1981, **1**, 143.
8 S. Asomaning and C. Gallagher, *High Pressure Asphaltene Deposition Technique for Evaluating the Deposition Tendency of Live Oil and Evaluating Inhibitor Performance*, Preprints, The Second International Conference On Petroleum and Gas Phase Behavior and Fouling, Copenhagen, Denmark, August 26-31, 2000.
9 W. K Stephenson, *Producing Asphaltenic Crude Oils: Problems and Solutions*, Petroleum Engineer, 1990, 24-31.

Subject Index